# AERONAUTICS and ASTRONAUTICS

An American Chronology
of Science and Technology in the
Exploration of Space
1915–1960

by Eugene M. Emme, NASA *Historian*

Foreword by Hugh L. Dryden, *Deputy Administrator*

NATIONAL AERONAUTICS AND SPACE ADMINISTRATION · 1961

# FOREWORD

THE NATIONAL AERONAUTICS AND SPACE ADMINISTRATION is charged by Congress with the responsibility for conducting the U.S. civilian program for the exploration and scientific investigation of space, as well as for the peaceful utilization of space for the benefit of all mankind. It also is responsible for the conduct of aeronautical research. NASA is scarcely two years old, having come into being on October 1, 1958.

SPACE EXPLORATION is an exciting and an important part of the total challenge presented to mankind in these last decades of the 20th century. With its brief history, NASA is quite aware that man's effort to navigate in space has considerable background. NASA's organizational nucleus was the National Advisory Committee for Aeronautics which spearheaded aeronautic research and development for 43 years. Many of NASA's programs, facilities, and personnel have come from activities long associated with pioneering developments in rocketry and the space sciences.

RESEARCH AND DEVELOPMENT in the conquest of space in the United States has involved scientists, engineers, civil and military activities in Government, universities, and industry. Liquid-fuel rocket propulsion, as is well known, was first demonstrated in the laboratory and in flight by Prof. Robert H. Goddard of Clark University. Rocket propulsion then came to worldwide notice in the German V-2, while subsequent ballistic missile development provided the propulsion making possible earth satellites and space probes. NASA's mission is one of joining all governmental agencies, the academic community, and industry in a national program for the peaceful conquest of space for the benefit of all mankind.

MAN'S EXPLORATION of three-dimensional space above the surface of the earth, first in and now beyond the life-giving atmosphere, has been a dramatic experience. In the scientific era in which we live, fundamental knowledge will determine our destiny more than ever before in human history. Exploration of space has provided important tools and new impetus in our scientific quest for knowledge concerning the true nature of matter, time, motion, and even life processes. We are learning hard data about many extraterrestrial realities for the first time. Unpredictable benefits for men on earth seem inevitably to result from developing the technology to explore and to learn on the newly available frontier of space. Technological progress spurts on.

SPECTACULAR and exciting events in aerospace affairs have generally been rather well publicized and widely noted, at least after the achievement of the Wright brothers became known. Behind the well-known milestones of practical flight, however, have been less publicized achievements in scientific research and engineering development making such progress possible. This volume helps to provide a fuller appreciation of events and activities already behind us. Perspective on the ever-quickening pace of events, provided by this chronology, helps provide some insight into the meaning of the events of tomorrow. The inevitability and swift pace of technical change, for example, can be more clearly appreciated.

FREE PEOPLES everywhere must retain a reliable perspective from which to discern better the future scientific, social, economic, political, and strategic consequences of dynamic advances now underway. Obviously the manner of the impact of technology upon society in the future will partly result from the broadest possible appreciation of its full significance.

HUGH L. DRYDEN, *Deputy Administrator,*
*National Aeronautics and Space Administration.*

DECEMBER 16, 1960

# CONTENTS

| | Page |
|---|---|
| FOREWORD by Dr. Hugh L. Dryden, Deputy Administrator . . | III |
| ABBREVIATIONS . . . . . . . . . . . . . . . . . . | VI |
| PREFACE . . . . . . . . . . . . . . . . . . . . . | IX |

### PART I
FROM THE FOUNDING OF NACA TO THE DAWN OF THE SPACE AGE, January 1915–October 1957 . . . . . . . . . . 1

### PART II
THE FIRST THREE YEARS OF THE SPACE AGE, October 1957–December 1960 . . . . . . . . . . . . . . . . . . . 89

APPENDICES:
- A. Chronicle of Earth Satellites and Space Probes . . . . 139
- B. Chronicle of World Airplane Records . . . . . . . . 153
- C. Chronicle of Select Balloon Flights, 1927–60 . . . . 161
- D. Select Awards and Honors in Aeronautics and Astronautics . . . . . . . . . . . . . . . . . . . 167
- E. Membership of the NACA, 1915–58 . . . . . . . . 201

SELECT BIBLIOGRAPHY . . . . . . . . . . . . . . . . . 207

SUBJECT AND NAME INDEX . . . . . . . . . . . . . . . 213

# ABBREVIATIONS

AA—Antiaircraft
AAC—Army Air Corps (USA)
AACB—Aeronautics and Astronautics Coordinating Board (NASA-DOD)
AAF—Army Air Forces (USA)
ABMA—Army Ballistic Missile Agency
AEC—Atomic Energy Commission
AF—U.S. Air Force
AFA—Air Force Association
AFB—Air Force Base
AFBMD—Ballistic Missile Division (USAF)
AFMTC—Missile Test Center (USAF)
AMAL—Aviation Medical Acceleration Laboratory (USN)
AMC—Air Materiel Command (USAF)
AMR—Atlantic Missile Range (Cape Canaveral, Florida)
ANP—Aircraft, nuclear powered
ARC—Ames Research Center (NASA)
ARDC—Air Research and Development Command (USAF)
ARPA—Advanced Research Projects Agency (DOD)
ARS—American Rocket Society

BMEWS—Ballistic Missile Early Warning System
BuAer—Bureau of Aeronautics (USN)
BuOrd—Bureau of Ordnance (USN)

CAA—Civil Aeronautics Authority (or Administration)
Cal Tech—California Institute of Technology
CNET—French Telecommunications Establishment
COSPAR—Committee on Space Research (International Council of Scientific Unions)
C/S—Chief of Staff
CSAGI—International Special Committee for the IGY

DEW—Distant Early Warning Line (DOD)
DOD—Department of Defense

FAA—Federal Aviation Agency
FAI—Fédération Aéronautique Internationale
FNRS—Fédération Nationale Researche Scientifique (Belgian)
FRC—Flight Research Center (NASA)
FY—Fiscal Year

GPO—Government Printing Office
GSFC—Goddard Space Flight Center (NASA)

IAF—International Astronautical Federation
IAS—Institute of the Aeronautical (now Aerospace) Sciences
ICAO—International Civil Aviation Organization
ICBM—Intercontinental Ballistic Missile (DOD)
IGY—International Geophysical Year
IRBM—Intermediate Range Ballistic Missile (DOD)

JANTAB—Joint Army and Navy Technical Aeronautical Board
JATO—Jet-assisted takeoff
JCS—Joint Chiefs of Staff (DOD)
JNW—Joint Committee on New Weapons and Equipment (OSRD)
JPL—Jet Propulsion Laboratory (NASA-Cal Tech)
JRDB—Joint Research and Development Board (USA-USN)

LaRC—Langley Research Center (NASA)
LMAL—Langley Memorial Aeronautical Laboratory (NACA, pre-1948)
LOD—Launch Operations Directorate (NASA)
LRC—Lewis Research Center (NASA)

MC—Medical Corps (USN, USAF)
MIT—Massachusetts Institute of Technology
MSFC—Marshall Space Flight Center (NASA)

NAA—National Aeronautic Association
NACA—National Advisory Committee for Aeronautics
NAS—Naval Air Station
NASA—National Aeronautics and Space Administration

NATO—North Atlantic Treaty Organization
NDRC—National Defense Research Committee
NERV—Nuclear Emulsion Recovery Vehicle
NIH—National Institutes of Health
NOL—Naval Ordnance Laboratory (USN)
NRL—Naval Research Laboratory (USN)
NSF—National Science Foundation

OSD—Office of the Secretary of Defense
OSRD—Office of Scientific Research and Development

PARD—Pilotless Aircraft Research Division, Langley Aeronautical Laboratory (NACA)
PMR—Pacific Missile Range (Point Mugu, Calif.)

RAF—Royal Air Force (Britain)
RCA—Radio Corp. of America
RCAF—Royal Canadian Air Force
R&D—Research and Development
RDB—Research and Development Board (DOD)
RFC—Royal Flying Corps. (Britain)

SAC—Strategic Air Command (USAF)
SAGE—Semiautomatic Ground Environment System (USAF)

SAM—School of Aviation Medicine (AAF, USAF, USN)
SER—SNAP Experimental Reactor (AEC)
SPACETRACK—National Space Surveillance Control Center (USAF)
SPASUR—Space Surveillance Detection Net (USN)
STG—Space Task Group (NASA)
STOL—Short takeoff and landing aircraft
SUI—State University of Iowa (Iowa City)

UN—United Nations Organization
USA—United States Army
USAAF—U.S. Army Air Forces
USAF—U.S. Air Force
USAS—U.S. Air Service (USA)
USIA—U.S. Information Agency
USMC—U.S. Marine Corps
USN—U.S. Navy
USNC—U.S. National Committee (IGY)

VfR—German Society for Space Travel (Verein fuer Raumschiffahrt)
VTOL—Vertical takeoff and landing (aircraft)

WADC—Wright Air Development Center (ARDC)
WDD—Western Development Division (ARDC)
WOO—Western Operations Office (NASA)
WSPG—White Sands Proving Ground

# PREFACE

FOR CENTURIES flight was demonstrated in nature by birds, winged insects, bats, and some squirrels. Men remained earthbound. The 20th century witnessed the birth of practical aviation and the profound impact which flight increasingly imposed upon the ways of mankind. The world has been shrunk in time-distance relationships by the ever-increasing speed, range, and utility of aircraft. Statesmen, scientists, artists, businessmen, and tourists acquired global wings. The art and science of war was revolutionized by airpower and military missilry. While aeronautical progress provided atmospheric stepping-stones, developments in rocket propulsion provided the means of placing manmade objects in space beyond the earth's atmosphere and effective gravitational force.

In the past three years, artificial satellites have been placed into orbit around the Earth and the Sun. Man is destined to follow his instrumented vehicles into space. By means of space vehicles, satellites, rocket-powered aircraft, and high-altitude balloons, the cataract of the earth's atmosphere has been removed for the first time from the eyes of scientists studying the universe. Revolutionary insight into physical and life science disciplines is afforded by investigation and study of the unearthly environment of space. This is the basic promise implicit in the potentialities of the scientific frontier in our time. Unforeseeable as well as predictable benefits for society on earth seem inevitable as a consequence of space exploration. Man's conquest of space presents a dramatic challenge. But we should be mindful of its historical evolution and its broadest implications.

## SCOPE OF THE CHRONOLOGY

This volume is largely a response to repeated requests for basic historical information. It is intended to serve as a ready informational reference, one from whence more detailed historical analysis should proceed.

Perspective on the dynamic pace of science and technology in aeronautics and astronautics is also served by a chronicle of events over the years. Behind the major milestones in man's conquest of space are many not-well-known events. Much more historical analysis and detail seems required to document and to assess the full history of research and development in aviation, rocketry, space flight, and their related fields. Spectacular events are but the final manifestation of the ideas, concern, and labor of many persons and institutions involved in scientific research, engineering development, finance, organization, management, and operations. With regard to space vehicles, the reader need not be reminded that the development of military missiles was a major factor in advancing the technology of rocket propulsion, a historical parallel to the first major application of aeronautics to human affairs during World War I. Yet the broad-based scientific conquest of space inherently possesses great consequences for society. Military applications of space technology in the United States are clearly the responsibility of the Department of Defense and the military services and not that of NASA. It seemed pertinent to include major developmental aspects of military technological efforts in this volume.

The scope of this chronology is defined in accordance with the following criteria:

- Emphasize scientific research and engineering development in aeronautics and astronautics as well as their related fields.
- Demonstrate historic use of aircraft, rockets, balloons, and spacecraft as tools of scientific research.
- Specialize upon U.S. efforts with inclusion of sufficient items on general historical events and foreign progress to retain an undistorted historical context.
- Illustrate practical exploitation of technological progress in aeronautics and astronautics.
- Exploit generally available source materials to insure widest and early utilization of the material compiled.

Some readers may criticize the selection of events. They may also be unappreciative of the fact that calendar-located events are not all of equal historical significance and do not necessarily appear in sequential order. Brevity and other necessary limitations also imposed certain format requirements. A detailed index has been prepared to enhance reference utility as well as topical integration; while a select bibliography serves as a launching pad for the stimulated reader.

## SOURCES AND ACKNOWLEDGMENTS

Debts were accumulated in the preparation of this volume. It is not feasible to list all indispensable contributors in NASA's Headquarters and Centers, in other governmental agencies, or in the academic and historical community. The following, however, must be named for their particularly helpful counsel and assistance: Mr. Harold Andrews (USN); Mr. David S. Akens (Marshall Space Flight Center); Mr. Joseph W. Angell (USAF); Miss Grace Bogart (NASA Technical Library); Mr. Walter T. Bonney (Aerospace Corp.); Dr. Paul Davis (AFAMC); Mr. Paul E. Garber (National Air Museum); Dr. Murray Green (USAF); Dr. Richard G. Hewlett (AEC); Mr. Milton Lehman (Goddard biographer); Prof. W. Ross Livingston (State University of Iowa); Mr. Marvin W. McFarland (Science and Technology, Library of Congress); Mr. Donald R. McVeigh (ARDC); Mr. Robert W. Mulac (LaRC); Mr. Lee Pearson (USN); Mr. Robert L. Perry (AFWADD); Dr. Nathan Reingold (Yale University); Dr. Alfred Rockefeller (AFBMD); Cdr. Malcolm D. Ross (USNR); Dr. Charles Sheldon (House Committee on Science and Astronautics); Dr. Wilfred J. Smith (Office for U.N. Conference, NASA); Mr. Raymond Snodgrass (Army Ordnance); Mr. John C. Truesdale (National Academy of Sciences); Prof. James A. Van Allen (State University of Iowa); and Dr. John F. Victory (formerly Executive Secretary of the NACA). Mrs. Anna Shade ably managed the varied and difficult chores of preparing the manuscript with able assistance of dedicated NASA typists. The responsibility for the organization and contents of this volume, however, remains solely that of the compiler.

Additional suggestions and comments will be gratefully received from any reader at any time. The depth of the historical process must inevitably be extended. Comments should be directed to the NASA Historian.

E. M. E.

JANUARY 13, 1961.

PART ONE

# From the Founding of NACA to the Dawn of the Space Age

## JANUARY 1915–OCTOBER 1957

THE FULL HISTORY of man's exploration of space might logically begin with the legend of Icarus, or with the flight of the Wright brothers in 1903, or more appropriately with their wind tunnel experiments conducted at Dayton between September and December 1901. Or, it could begin with the notebooks of Leonardo da Vinci, the laws of Johannes Kepler, not to mention the contributions of the Montgolfier brothers, Sir George Cayley, Otto Lilienthal, Octave Chanute, or Count von Zeppelin.

Regarding the early interest of governments, as such, in the promotion and exploitation of aeronautics, a chronology could start with the specifications laid down by the U.S. War Department for a military "flying machine" in 1907. Two years later, the United States became the first nation in the world to possess a military airplane, the "Wright Flyer." When the international conflict erupted in the 1914–18 war in Europe, however, the relative plight of U.S. aviation as compared to war-stimulated technical progress abroad, was clearly self-evident.

This U.S. chronology begins with the year in which the National Advisory Committee for Aeronautics was created. There was also geometric buildup of U.S. military and naval aviation as involvement in military conflict approached. Wartime progress in aeronautics was subsequently applied to the pursuits of peace. Rocket development was also to be stimulated in parallel fashion at a later date when military missile development created the propulsion necessary for the scientific exploration of space.

As is clearly self-evident, events in the history of scientific research and technical development are not isolated from organizational, political, military, or other general events, which are occasionally cited to remind the reader of the broader historical context. A chronology is, after all, but a mere recital of known calendar-located events and equal significance cannot be accorded all events listed in sequence.

# 1915

*January 15:* New official American one-man duration record of 8 hours 53 minutes set by Lt. B. Q. Jones in a Martin tractor biplane at San Diego, Calif.

——: First transcontinental telephone conversation, New York to San Francisco, by Alexander Graham Bell and Thomas A. Watson.

*January 19–20:* First German aerial bombing of Britain, by two Zeppelins, thereby opening up a new era in the exploitation of aeronautics. During World War I, a total of 56 tons of aerial bombs was dropped on London and 214 tons on the rest of Britain.

*During January:* First air-to-air combat, German airman killed by rifle fire from Allied aircraft. In February a machinegun mounted on a French aircraft, Lieutenant Garros as pilot, first shot down a German aircraft.

*February 24:* Macy automatic pilot tests were begun at San Diego, Calif.

*During February–March:* Anthony H. G. Fokker perfected synchronizing gear to allow machinegun to be fired through rotating propeller.

*March 3:* The Advisory Committee for Aeronautics (later the National Advisory Committee for Aeronautics or NACA) was established by a rider to the Naval Appropriations Act, ". . . to supervise and direct the scientific study of the problems of flight, with a view of their practical solution." The sum of $5,000 a year was appropriated for 5 years. The total appropriation for naval aeronautics was $1 million.

*March 4:* Congress passed an appropriation bill of $300,000 for Army aeronautics for fiscal year 1916.

*April 2:* President Wilson appointed the first 12 members of the National Advisory Committee for Aeronautics (NACA). Throughout the entire history of the NACA until October 1958, members served without compensation.

*April 16:* Navy AB-2 flying boat successfully catapulted from a barge, Lt. P. N. Bellinger as pilot.

*April 23:* The Secretary of War called the first meeting of the NACA in his office. Brig. Gen. George P. Scriven, Chief Signal Officer, was elected temporary Chairman, and Dr. Charles D. Walcott, Secretary of the Smithsonian Institution, was elected first Chairman of the important NACA Executive Committee. (See Appendix E.)

——: American altitude record of 10,000 feet for seaplanes was established in Burgess-Dunne AH-10 by Lt. P. N. Bellinger over Pensacola, Fla.

*May 31:* First German Zeppelin raid on London. British employed rockets in their defenses around London.

*June 1:* Navy let first contract for lighter-than-air craft in ordering one nonrigid airship from Connecticut Aircraft (later the DN-1).

*June 8:* U.S. Patent Office granted patent (No. 1142754) to Glenn H. Curtiss covering the arrangement of a step or ridge incorporated in the hull of flying boats.

*During June:* First year of formal graduate study in aeronautical engineering was completed at Massachusetts Institute of Technology, and one master of science degree was awarded.

*July 7:* Secretary of the Navy, Josephus Daniels, in a letter to Thomas A. Edison said that the Navy required "machinery and facilities for utilizing the natural inventive genius of Americans to meet the new conditions of warfare." This letter prompted creation of Naval Consulting Board of civilian advisers which functioned throughout World War I, and which included in its organization a "Committee on Aeronautics, including Aero Motors."

*July 10:* Naval Aeronautic Station, Pensacola, tested sextant equipped with a pendulum-type artificial horizon and reported that pendulum type was unsatis-

1915—Continued

factory for aircraft use, but that a sextant with a gyroscopically stabilized artificial horizon might be acceptable.

*August 11:* Naval Observatory requested Eastman Kodak to develop an aerial camera with high-speed lens suitable for photography at 1,000 or 2,000 yards' altitude.

*September 1:* Congress supplemented appropriation of Army aeronautics to $13,281,666 from $300,000 of previous fiscal year.

*October 15:* Secretary of the NACA was instructed by the committee to communicate with the Government departments, the result of investigation with regard to aeronautical activity, and to recommend or advise the Secretaries of the separate departments of the Government to continue and foster experimental development.

*November 6:* First catapult launching from a ship underway, made from the U.S.S. *North Carolina* in Pensacola Bay, by Lt. Cmdr. H. C. Mustin.

*December 3:* Lt. R. C. Saufley reached 11,975 feet over Pensacola in a Curtiss AH–14, an American altitude record for hydroaeroplanes.

*December 9:* NACA Report No. 1 was issued, a two-part "Report on Behavior of Aeroplanes in Gusts," by Jerome C. Hunsaker and E. B. Wilson of MIT.

*December 12:* An all-steel frame, fabric-covered combat plane successfully flown, one designed by Grover C. Loening and built by Sturtevant Aeroplane Co.

*During December:* All-metal fully cantilever-wing monoplane produced by Hugo Junkers in Germany, the J–1 powered by a 120-hp Mercedes, made its first successful flights.

*During 1915:* Elmer A. Sperry developed and demonstrated his drift indicator for which he received the Robert J. Collier Trophy for 1916. (See Collier Trophy, Appendix D.)

——: General Vehicle Co., Long Island City, contracted with French Government to build Gnome engines, the first radial engine produced in the United States.

——: Robert H. Goddard proved validity of rocket propulsion principles in a vacuum, at Clark University, Worcester, Mass.

# 1916

*March 15:* First U.S. tactical air unit in the field, the 1st Aero Squadron commanded by Capt. B. D. Foulois, began operations with General Pershing's expedition into Mexico.

*April 2:* American altitude record of 16,072 feet set by Lt. R. C. Saufley in a Curtiss hydroaeroplane.

*During April:* French employed first air-to-air combat rockets, four Le Prieur rockets attached to each strut of Nieuport fighter, credited with downing of German hydrogen-inflated Zeppelin LC–77. The Belgian, Willy Coppens and Briton, Albert Ball, reportedly used rockets effectively against German balloons until incendiary bullets were developed.

*May 22:* French airmen successfully destroyed five of six German balloons using Le Prieur rockets on their Nieuport fighters.

*June 8:* The NACA called the first meeting of representatives of the aircraft industry and of interested Government agencies.

*July 19:* Navy Gallauder 59A, an airplane with propeller mounted amidships in the fuselage, made preliminary flights at Norwich, Conn., Lt. (jg) G. D. Murray as pilot.

*July 22:* Navy requested Aluminum Co. of America to develop a suitable alloy for fabrication into Zeppelin-type girders.

*August 22:* President Wilson signed Navy appropriation bill, which included $3,500,000 for naval aviation.

*August 29:* The NACA requested $85,000 and received $82,515.70 for fiscal year 1917 as a part of the naval appropriation bill. $68,957.35 later went toward laboratory construction at Langley Field.

———: U.S. Army appropriations approved, which included $14,281,766 to the Signal Corps for military aeronautics.

*September 2:* Plane-to-plane radio demonstrated over North Island, Calif., at a distance of about 2 miles.

*September 2–3:* First German Zeppelin shot down by RFC aircraft over Britain; five Zeppelins were brought down over Britain during 1916.

*September 12:* Piloted hydroaeroplane equipped with automatic stabilization and direction gear developed by the Sperry Co. and P. C. Hewitt was demonstrated by Amityville, Long Island, before naval observers.

*September 21:* The National Research Council, formed at the request of President Wilson by the National Academy of Sciences, held its first meeting in New York.

*During September:* Wright-Martin Aircraft Corp. contracted with French company to manufacture the Hispano-Suiza engine in the United States.

*October 5:* The NACA first recommended inauguration of airmail service, and William F. Durand was elected Chairman of the NACA.

*October 9:* Subcommittee of the NACA appointed to consider the needs of the committee as to a site for experimental work, with authority to visit and inspect sites, and to secure the cooperation of the War and Navy Departments and the Weather Bureau.

*November 23:* The NACA recommended purchase of land north of Hampton, Va., for use as an aircraft proving ground by the Army and Navy. This site became known as Langley Field, and the location of the first NACA laboratory.

*November 28:* First airplane raid on London, by a German seaplane.

*During November:* "Design Requirements for Airplanes" (A.P. 970), a basic six-page pamphlet, was issued by the British Royal Aircraft Factory at Farnborough.

*December 20:* Army Balloon School established at Fort Omaha, Nebr.

*During 1916:* Radio-controlled pilotless monoplane, the "Aerial Target," designed by H. P. Folland with radio gear by A. M. Low, flown at the British Royal Aircraft Establishment at Farnborough.

———: Development work on air cooling of aircraft engines by means of spacing, depth and thickness of fins, and the effects of airflow, were conducted by Professor Givson at Royal Aircraft Factory at Farnborough.

———: Nine U.S. aircraft companies delivered only 64 out of 336 aircraft ordered by the Army, the performance of which compared unfavorably with European aircraft.

# 1917

*January 10:* Comptroller of the Treasury Department ruled that the NACA was an independent agency and was not an appendage of the Navy Department in spite of the fact it was originally funded under the naval appropriations bill.

*During January:* The NACA, after considering high-cost complaints of Army and Navy, recommended creation of Manufacturers Aircraft Association to effect cross-licensing of aeronautic patents. This was a milestone in preventing a virtual deadlock in aircraft construction because of patent infringement suits.

*February 2:* The NACA recommended to the President, for transmittal to Con-

## 1916—Continued

gress for approval, that the Government acquire basic aeronautical patents.

*February 13:* Aircraft Manufacturers Association formed, Frank H. Russell as president.

*March 8:* Naval Act carried appropriation of $1 million for purchase of basic aeronautical patents by the Federal Government.

*March 29:* The NACA recommended preparation of 3-year programs for aircraft production to the Secretaries of War and the Navy.

*April 6:* The United States declared war on the Central Powers. The Aviation Section of the Signal Corps consisted of 35 pilots, 1,987 enlisted men, and 55 training airplanes. Navy Aviation and Marine Corps combined had 48 officer-pilots, 239 men, 54 airplanes, 1 airship, 3 balloons, and 1 air station.

*April 10:* The NACA recommended the organization of an Aircraft Production Board, to be appointed by the Council of National Defense. Such was created on May 16.

*April 14:* Naval Consulting Board recommended to the Secretary that $50,000 be granted to carry on experimental work on aerial torpedoes in the form of automatically controlled aeroplanes or aerial machines carrying high explosives. This was origin of the Navy N-9 "flying bomb," later considered the Navy's first guided-missile effort.

*May 7:* First aerial bombing of London by German bombers at night.

*May 12:* Capt. W. A. Robertson established new American altitude record of 17,230 feet over North Island Flying School, San Diego, Calif.

*May 20:* First aircraft sinking of a submarine, the German U-36, in the North Sea by a British flying boat.

*June 2:* Aviation Section became the Airplane Division of the Army Signal Corps, and Maj. B. D. Foulois was appointed officer-in-charge on July 23.

*June 4:* Aircraft Production Board and the Joint Technical Board on Aircraft authorized the construction of five prototype models of 8- and 12-cylinder Liberty motors. Engine designs had been worked out in a Washington hotel room by J. G. Vincent of Packard Motor Car Co. and E. J. Hall of the Hall-Scott Motor Car Co. during the previous week, applying current engineering practices to mass production techniques.

*July 4:* First 8-cylinder Liberty aircraft engine arrived in Washington, D.C., for test by the National Bureau of Standards. Design, manufacture, and assembly of this motor had required less than 6 weeks.

*July 24:* Manufacturers Aircraft Association formed to handle cross-licensing patents between all manufacturers.

——: $640 million aviation bill became law, the largest U.S. appropriation for aviation to date.

*July 27:* Secretary of the Navy authorized a naval aircraft factory in Philadelphia.

——: First British DH-4 arrived in United States and became model for the first combat aircraft produced in volume in the United States.

*August 17:* Field Marshal Jan Christian Smuts, Chairman of a Sub-committee on Imperial Defence, submitted classic proposal for creation of an autonomous air force in the British military structure.

*August 21:* First airplane powered by Liberty engine successfully flown, the L. W. F. Engineering Co.'s "Model F" biplane.

*August 25:* Navy "NC" flying boat development was initiated by Chief Constructor of the Navy, D. W. Taylor, in a memo outlining general requirements of such an aircraft to combat the submarine menace and "to fly across the Atlantic to avoid difficulties of delivery, etc." Acting Secretary of the Navy, F. D. Roosevelt, authorized development of "NC" flying boats capable of flying the Atlantic.

——: 12-cylinder Liberty motor passed a 50-hour test with power output of over 300 hp prior to being ordered into mass production.

*During August:* The NACA recommended funds be given Weather Bureau to promote safety in aerial navigation.

*September 3:* Brig. Gen. W. L. Kenly appointed Chief of the Air Service, AEF, the first time control of Army air activities was placed under a single head.

*September 7:* Radio signals sent from a Navy R–6 seaplane flying from NAS Pensacola, were received by Naval Radio Station New Orleans, 140 miles distant, in tests.

*October 1:* Congress created the Aircraft Board.

*October 16:* Final tests of Army's airplane radiotelephone at Langley Field, Va., achieved 25 miles for plane-to-plane communication and 45 miles airplane-to-ground.

*October 18:* McCook Field, Dayton, Ohio, was established as an aeronautical experimental station by the Signal Corps.

——: British De Havilland DH–4 ordered into series production in the United States 6 months after U.S. entry into World War I. By the end of the war, about 4,500 had been built, and of the total of 1,216 American-built planes to reach the Western Front, all but three (two Le Peres and one experimental DH–9) were DH–4's.

——: The Aviation Medical Research Board was established by the Signal Corps.

*October 21:* First flight test of 12-cylinder Liberty engine in Curtiss HS–1 flying boat at Buffalo, N.Y.

*October 29:* First DH–4 completed, flown at Dayton, Ohio.

*November 15:* Committee on Light Alloys established within NACA to intensify efforts to develop new metals for aeronautical use, Constructor Jerome C. Hunsaker was Navy member.

*November 21:* A modified Navy N–9 "Flying Bomb" was demonstrated to Army, Navy and civilian observers at Amityville, Long Island.

*December 15:* U.S. Navy airplane design placed under LCdr. W. Starling Burgess, Bureau of Construction and Repair.

*December 26:* First test-run of altitude laboratory constructed at the Bureau of Standards for the NACA, one capable of testing engine performance up to one-third an atmosphere.

*During 1917:* U.S. Weather Bureau aerological specialist, William R. Blair, prepared NACA Report No. 13, "Meteorology and Aeronautics," which was widely circulated as a basic handbook.

——: At request of War Department, a member of NACA technical staff assigned to supervise altitude performance tests of the first Liberty engines at Detroit, Mich., and Pikes Peak, Colo.

——: Development work at the British Royal Aircraft Factory at Farnborough included a measured-injection carburetor, prototype design of 14-cylinder, double-row, static-radial, air-cooled engine (RAF–8), and design and construction of the SE–5 fighter.

——: Curtiss JN–4 "Jenny" became basic training airplane for thousands of American pilots.

# 1918

*January 19:* U.S. School of Aviation Medicine began operations under Maj. Williams H. Wilmer, Signal Corps, Hazelhurst Field, Mineola, N.Y. A low-pressure tank was constructed to simulate altitudes up to 30,000 feet, and some studies were conducted at Pikes Peak.

*January 23:* First American military balloon ascension in the AEF took place at Cuperly, Marne, France.

*During January:* The NACA established Office of Aeronautical Intelligence at the suggestion of the Aircraft Board to col-

## 1918—Continued

lect and distribute scientific and technical data on aeronautics.

*February 7:* The Joint Army and Navy Technical Aeronautical Board (JANTAB) passed resolution on Instrument Standardization in Army and Navy planes for incorporation in general specifications.

*February 16:* Plant for assembly of American-made airplanes began operations at Romorantin, France.

*March 6:* Navy unmanned "flying bomb" successfully launched by catapult and flown for 1,000 yards at Sperry Flying Field, Long Island.

*March 8:* Majs. E. C. Schneider and J. L. Whitney (USA) reached an artificial altitude of 34,000 feet in 24 minutes, at Signal Corps Laboratory, Mineola, N.Y.

*March 21:* "Dunkirk fighter" or Navy HA seaplane made its first flight at Port Washington, Long Island, with Curtiss pilot Roland Rohlfs as pilot.

*March 27:* First aircraft built at the Naval Aircraft Factory, the H-16 seaplane, was flown for the first time, and was later used for the antisubmarine patrol from United States and European stations.

*March 29:* Curtiss 18-T or "Kirkham" triplane fighter ordered by Navy from Curtiss Engineering.

*April 6:* Night aerial photographs taken with use of magnesium flares by Lt. J. C. McKinney (USA) and civilian pilot Norbert Carolin.

*April 15:* First Marine Aviation Force formed at NAS Miami, commanded by Capt. A. A. Cunningham.

*April 23:* First oversea shipment of Liberty motors arrived at assembly and repair station at Pauillac, France.

*April 25:* Loening M-3 first flown, equipped with Lawrence three-cylinder, air-cooled engine.

*April 27:* French-built airship AT-1, commanded by Lt. F. P. Culbert (USN), completed a 25-hour 23-minute flight out of Paimboeuf, France, longest flight on record for airship of this type.

*April 29:* Plans approved for construction of first wind (5-foot) tunnel at Langley Memorial Aeronautical Laboratory of NACA.

*May 11:* First American-made DH-4, with Liberty engine, received in the AEF.

*May 15:* Navy Bureau of Steam Engineering reported that Marconi SE-1100 radio transmitter designed for use on H-16 flying boat, had proven capable of reliable voice communications from plane to shore up to 50 nautical miles and code communications up to 120 nautical miles.

——: The Post Office's first regular airmail route, Washington to New York, was inaugurated by Army pilots.

*May 17:* First flight made in France of an American-built military aircraft, a DH-4, built by Dayton Wright Co. adapted from English design.

*May 20:* Army Aeronautics was divorced from the Signal Corps and two air departments were created: Bureau of Military Aeronautics and Bureau of Aircraft Production.

*May 24:* First consignment of American-built flying boats, six HS-1's, arrived at Pauillac, France.

*During May:* At instigation of Dr. W. F. Durand, Chairman of the NACA, General Electric assembled an experimental turbo supercharger on a Liberty engine at Dayton.

*June 19:* Naval Air Station Pensacola began taking upper atmosphere weather soundings to provide wind velocity and direction. Recording instruments were carried aloft by a kite balloon, a technique developed by the station meteorological officer Lt. W. F. Reed.

*During July:* Standard Aircraft Corp. requested to build Italian Caproni and English Handley-Page bombers.

*August 17:* American-designed bomber, Army Martin MB-1, made its first flight with T. E. Springer as pilot. It became the first standard bomber of the Air

Service but did not enter combat, while later modifications of it were used by the Post Office Department.

*September 18:* Altitude world record of 28,899 feet established by Maj. R. W. Schroeder (USA) at Dayton, Ohio.

*September 23:* Flywheel catapult used successfully to launch Navy "flying bomb" at Copiague, Long Island, a development undertaken by Sperry Co.

*September 28:* One JN4 aircraft maneuvered another JN4 in flight solely by means of radio at Langley Field, Va.

*October 1:* First bombing using electrical releases, Allied bombers in an attack on German infantry counterattack.

*October 2:* First successful flights of Army's Kettering pilotless aircraft with preset controls, "The Bug," at Dayton, Ohio; often called a "guided missile" in later years.

*October 3:* Flight refueling demonstrated in a seaplane by Lt. Godfrey L. Cabot (USNR), by snatching 155 pounds of weight from a moving sea sled.

*October 4:* Navy NC-1 flying boat, designed by Hunsaker, Richardson & Westervelt, was successfully test flown.

*October 19:* Pilotless Navy N-9 training plane, converted to automatic flying machine, flew prescribed course although distance gear failed to land the airplane at preset range of 14,500 yards.

*November 6–7:* Robert H. Goddard fired several rocket devices before representatives of the Signal Corps, Air Service, Army Ordnance, and others at Aberdeen Proving Ground, Md.

*November 11:* With the signing of the Armistice, the Army Air Service had a total of 195,024 personnel, of which 20,568 were officers, and the AEF had 3,538 airplanes while 4,865 were in service in the United States. Naval aviation consisted of 6,716 officers and 30,693 men, with 282 officers and 2,189 men in Marine Corps units with a total of 2,107 airplanes, of which 1,172 were flying boats.

*November 17:* NAS Hampton Roads reported that H-16 flying boat equipped with radio direction finder using British six-stage amplifier had received signals from Arlington, Va., a distance of 150 miles.

*November 25:* NC-1 flying boat established new world record by taking off from Rockaway Beach, N.Y., with 51 persons aboard.

*During November:* The NACA first recommended enactment of Federal legislation for civil aviation, enforcement to be under the Department of Commerce.

*December 4:* First Army transcontinental flight by four Curtiss JN4's began at San Diego, reaching Jacksonville, Fla., on December 22.

*December 31:* Altitude laboratory at Bureau of Standards completed a full year of detailed analysis of various engine performances up to 30,000-foot altitudes, which yielded many results of basic importance.

*During 1918:* Medical Research Laboratory of the Signal Corps published a manual on aviation medicine.

———: Ballistic Branch of the Army Ordnance Corps, in conjunction with the National Bureau of Standards, conducted wind tunnel tests to determine optimum shapes for artillery projectiles.

# 1919

*January 21–31:* Second Army transcontinental flight by Maj. T. C. Macauley in DH-4 Liberty, Fort Worth–San Diego–Miami–Fort Worth, which he repeated in April.

*February 5:* First civil airline with passenger service, Germany's *Deutsche Luftreederei* which operated between Berlin, Leipzig, and Weimar.

## 1919—Continued

*February 18:* Navy Bureau of Ordnance (BuOrd) continued wartime experimental work begun by Sperry Gyroscope in 1917 on the unmanned "Flying Bomb."

*February 19:* The NACA recommendations on regulating air commerce, the licensing of pilots, the inspection of aircraft, and the use of landing fields were transmitted to Congress through the Secretary of the Treasury.

*During February:* First flights of Thomas-Morse MB-3, first U.S.-designed fighter procured in quantity, which reached speed of 164 mph in early flights exceeding that of contemporary European aircraft.

*March 19:* The Aircraft Board was abolished by Presidential Executive Order.

*March 21:* First recorded flight test of gyrocompass, a Sperry instrument, by the Navy, which was unsuccessful.

*April 26:* World duration unofficial record attained by Navy F5L flying boat of 20 hours 19 minutes, with Lt. H. B. Grow as pilot.

*April 28:* Naval Observatory requested by LCdr. Richard E. Byrd to supply bubble levels for attachment to navigational sextants, thereby providing an artificial horizon for astronomical observations from aircraft.

*———:* Unofficial seaplane record made by Navy F5L piloted by Lt. H. B. Grow out of Hampton Roads, which completed a flight of 20 hours and 19 minutes, a distance of 1,250 nautical miles.

*During April:* Curtiss 18-T two-place fighter powered by a Curtiss-Kirkham K-12-350, made first flights, reached speed of 162 mph.

*May 8-29:* First transatlantic flight by LCdr. Albert C. Read and crew in Navy plane NC-4.

*May 26:* Date of Dr. Robert H. Goddard's progress report to the Smithsonian Institution entitled "A Method of Reaching Extreme Altitudes." It was published by the Smithsonian in January 1920.

*June 14-15:* First nonstop Atlantic flight from Newfoundland to Ireland, 1,936 miles, was accomplished by Capt. John Alcock and Lt. A. W. Brown of England in a Vickers-Vimy-2 Rolls 400, in 15 hours 57 minutes.

*June 25:* NAS Anacostia reported on measurements of temperature and humidity at altitudes made by special instruments on aircraft.

*June 28:* Signing of Treaty of Versailles disarmed Germany of a military air force but did not include rockets as potential weapons, thus leaving Germany free under international law to develop them.

*During June:* Paris office of the NACA opened with William Knight in charge to collect and disseminate aeronautical information in Britain, France, and Italy.

*July 2-6:* First airship crossing of the Atlantic, by British R-34.

*July 24-November 9:* "Around the rim" circuit flight of the United States, covering 9,823 miles, completed by Lt. Col. R. L. Hartz and Lt. E. E. Harmon in a Martin bomber.

*July 28:* First aerial observations of schools of fish made by U.S. Bureau of Fisheries with cooperation of naval aircraft, at Cape May, N.J.

*August 1-September 14:* First International Aircraft Exposition since Armistice, at Amsterdam, Holland.

*August 14:* First airmail delivered at sea, by Aeromarine flying boat to the White liner *Adriatic* (Br.).

*August 25:* First daily commercial air service, London to Paris, begun by British Airco DH-4a.

*September 6:* New unofficial world altitude two-man record of 28,250 feet was set by Maj. R. W. Schroeder and Lt. G. A. Elfrey in a Le Pere Liberty 400 at Dayton, Ohio. On October 4, Schroeder reached new record of 31,796 feet in same airplane.

*September 12:* The NACA coordinated the replies of the executive departments regarding provisions of the International Convention on Air Navigation meeting in Paris.

*September 18:* World altitude official record of 31,420 feet flown by Roland Rohlfs in Curtiss triplane-Curtiss-Kirkham K12–350.

*October 8–31:* Army transcontinental reliability and endurance flight from New York to San Francisco and return: 44 aircraft completed westbound; 15 eastbound; and 10 planes made round trip.

*October 9:* Charles D. Walcott, Secretary of the Smithsonian, elected Chairman of the NACA; Joseph S. Ames was elected Chairman of the Executive Committee, a post he held until October 7, 1939.

*October 13:* International Convention on Air Navigation signed in Paris, which reaffirmed the principle of national sovereignty in airspace and established a Commission for Aerial Navigation under the League of Nations to regulate international air commerce.

*October 30:* Reversible-pitch propeller tested at McCook Field, Dayton, Ohio.

*November 12–December 10:* Ross McPherson Smith completed 11,500-mile intercontinental flight in a Vickers-Vimy from Heston, London, to Port Darwin, Australia.

*December 8:* The Aeronautical Engineering Society was organized at MIT.

*December 29:* American Meteorological Society founded at St. Louis, Mo., for the development and dissemination of knowledge of meteorology in all its phases and applications.

*December 31:* Notable technical achievements of the year according to McCook Field were: development of leakproof tanks; reversible- and variable-pitch propellers; a siphon gasoline pump; fins and floats for emergency water landings; and the turbocompressor or supercharger developed by Sanford A. Moss of General Electric.

*During 1919:* Adolph Rohrbach of Germany developed smooth-surface, metal-surfaced wings, combined with metal box-spar internal construction, the beginning of the stressed-skin concept.

———: Weather Bureau expended $100,000 to improve meteorological observations to support increasing aviation requirements, an appropriation granted by Congress in 1917 upon the recommendation of the NACA.

———: Junkers of Germany produced J-13 low-set, cantilever-wing transport, which carried a crew of two and four passengers.

# 1920

*January 20:* Navy Bureau of Steam Engineering was allocated $100,000 to contract for the development and purchase of 200-hp radial aircooled engines from the Lawrance Aero Engine Corp.

*February 5:* Navy-sponsored project of developing radio-loop antennas for navigational purposes.

*February 27:* World altitude record of 33,113 feet set by Maj. R. W. Schroeder (USA) in a LePere-Liberty 400, at McCook Field, Dayton, Ohio.

*March 1:* The NACA proposed a national aviation policy establishing a Bureau of Aeronautics in the Commerce Department, authorizing airplane competition to stimulate new designs, increasing Army and Navy air appropriations, expanding the Air Mail Service, and expanding research at the Langley Memorial Aeronautical Laboratory.

*March 27:* Successful test of Sperry gyrostabilized automatic pilot system in an F5L was completed at NAS Hampton Roads.

*April 1:* The NACA approved the publication of Technical Report No. 91, "Nomenclature for Aeronautics," to assist use of uniform technical terms and symbols.

*April 2:* Successful altitude soundings of wind direction and velocity at night,

## 1920—Continued

using candle-lighted free balloons at Hampton Roads in flights since January, announced by the Navy.

*June 4:* Army Air Service (AAS) was created in the Army reorganization bill signed by President Wilson. AAS consisted of 1,516 officers and 16,000 enlisted men.

*June 8:* Lt. J. H. Wilson (USA) made a series of high-altitude jumps, parachuting from a record altitude of 19,861 feet over San Antonio, Tex.

*June 11:* The NACA's own program of aeronautics research, conducted by its own staff in its own facilities, was begun with the first operation of the first NACA 5-foot wind tunnel at Langley Laboratory.

*June 21:* Because development of military rigid airships by the Navy was considered proper, and one logically leading to the development of commercial types, the NACA urged adequate funding of the Navy program in spite of recent airship disasters.

———: Navy approved installation of J. V. Martin retractable landing gear on VE–7 Vought airplane, but no evidence indicates it was done. First U.S. retractable landing gear was used by J. V. Martin K–III in 1918–19 period.

*June 28:* The NACA formally encouraged the Army and Navy to detail officers to the Massachusetts Institute of Technology for aeronautical engineering study and offered use of its facilities and personnel to further research and experimental work outside of Government.

*July 1:* Wright Aeronautical produced a French Hisso "cannon engine" which fired 37-mm shells through the propeller shaft.

*July 7:* Navy F5L seaplane flown by means of radiocompass from Hampton Roads to U.S.S. *Ohio* at sea.

*July 13:* Cdr. J. C. Hunsaker (USN) elected Honorary Fellow of Royal Aeronautical Society of England, the first time this distinction was conferred on one not a British subject.

*During July–September:* Inaccessible parts of Alaska mapped from the air by Army Air Service pilots, headed by Capt. St. Clair Streett (USA).

*November 1:* First U.S. international passenger service started by Aeromarine West Indies Airways between Key West, Fla., and Havana, Cuba.

*November 25:* First Pulitzer race won by Lt. C. C. Mosely in a Verville-Packard 600 at Mitchel Field, N.Y., flying a distance of 132 miles at a speed of 156.54 mph.

*During 1920:* NACA Report No. 84, entitled "Data on the Design of Plywood for Aircraft," by Armin Elmendorf of the Forest Service, provided basic guidance for aircraft design as well as broader applications.

———: New aircraft engine laboratory, the second, was completed at the National Bureau of Standards capable of testing 800-hp engines. Work carried out under the direction of L. J. Briggs provided new data on the viscosity of air.

———: Wind tunnel at Leland Stanford Aerodynamic Laboratory devoted entirely to propeller tests under direction of W. F. Durand, while NACA's George DeBothezat carried on aerodynamic studies at McCook Field.

———: Secretaries of War and Navy appointed joint Aeronautical Board to consider military questions regarding use of aeronautics by both services. Having no connection with the NACA, the Aeronautical Board replaced the Joint Army and Navy Technical Aircraft Board established during the war to expedite military procurement and exploitation of aviation.

———: The NACA formulated and recommended reservations regarding the Convention on International Air Navigation (1919) to the State Department prior to U.S. ratification.

———: New aircraft engines of this year included the French Hisso-design 180- and 300-hp engines by Wright; the Aeromarine 120 and 180; the Packard 300- and 600-hp types; and the Lawrance 60- and 200-hp air-cooled engines.

———: Moon eclipse observed by Lts. J. H. Tilton and W. H. Cushing from height of 3 miles at NAS Rockaway, N.Y.

*During 1920-22:* Robert H. Goddard experimented with liquid oxygen and various liquid hydrocarbons, including gasoline and liquid propane as well as ether, as rocket fuel, under a grant by Clark University. He concluded that although oxygen and hydrogen possessed the greatest heat energy per unit mass, that liquid oxygen and liquid methane offered greatest heat value of combinations which could be used without considerable difficulty. But, he said, "the most practical combination appears to be liquid oxygen and gasoline."

# 1921

*January 10:* 700-hp aircraft engine having 18 cylinders arranged in three banks of six, tested at Engineering Division, McCook Field.

*January 25:* Committee on Law of Aviation, American Bar Association, filed initial report on the necessity of aerial law. On August 25, the ABA recommended Federal aerial legislation.

*January 26:* Post Office Department operated regular daily airmail routes over a distance of 3,460 miles.

*February 21:* First transcontinental flight within 24 hours, made by Lt. W. D. Coney in a DH-4B from San Diego, Calif., to Jacksonville, Fla., in 22 hours and 27 minutes.

———: School for Flight Surgeons at Mitchell Field recognized as a Special Service School in War Department General Order No. 7.

*March 16:* U.S. Public Health Service initiated aerial survey of the Mississippi Valley watershed.

*March 23:* Parachute jump from 23,700 feet made by Lt. A. G. Hamilton (USA) at Chanute Field, Ill.

*April 1:* President Harding directed NACA to organize an inter-departmental subcommittee to recommend Federal regulation of air navigation. After a series of meetings this committee's report was approved by the Executive Committee of NACA on April 9, and transmitted to the President.

*April 12:* President Harding recommended establishment of a Bureau of Aviation within the Department of Commerce, in his address to Congress.

*April 18:* John J. Ide appointed as technical assistant in charge of the Paris office of the NACA, a post he held until 1940 and resumed after the end of World War II.

*April 23:* Aerial photo survey of Dominican Republic coastline completed by First Air Squadron of the USMC; and in June, it completed aerial survey of Haitian coastline.

*June 8:* First flight of an Army Air Service pressurized cabin airplane was made, a D-9-A aircraft piloted by Lt. Harold R. Harris.

*June 9:* The NACA authorized construction of compressed-air wind tunnel (20 atmospheres) with a 5-foot test section at Langley Aeronautical Laboratory.

*July 9-11:* Aerial study of San Andreas rift, the line of earthquakes of 1857 and 1906 in California Coast Range, made by Prof. Bailey Willis of the Seismological Society of America.

*July 13-21:* In a series of Army-Navy bombing tests off the Virginia Capes, airplanes sank the captured German destroyer *G-102*, light cruiser *Frankfort*, and battleship *Ostfriesland*.

*July 29:* Brig. Gen. William Mitchell led 17 bombers in "raid" over New York.

*August 1:* World War I high-altitude bombsight mounted on a gyrostabilized base tested by Navy Torpedo Squadron at Yorktown, Va., marking completion of first phase of Carl L. Norden's development of a bombsight for BuOrd.

## 1921—Continued

*August 4:* 5,000 catalpa trees successfully sprayed from an airplane in 15 minutes, at Troy, Ohio.

*August 10:* The Navy Bureau of Aeronautics was established with Rear Adm. William A. Moffett as first chief.

*September 18:* Lt. J. A. Macready (USA) broke world altitude record in a Packard-LePere fighter plane by reaching 34,508 feet.

*September 23:* Day and night bombardment test flights by the U.S. Air Service were begun, which resulted in the sinking of the battleship *Alabama* in the Chesapeake Bay by a 2,000-pound bomb.

*September 30:* During forest fire season, 47 Air Service aircraft discovered 832 forest fires in 396 patrols from Pacific coast bases, flying 148,113 miles over national parks.

——: Pointing out the virtual U.S. monopoly of known sources of helium, the NACA passed a special resolution addressed to the President and the Secretaries of War and Navy urging the continuance of the U.S. airship development program.

*October 18:* A world speed record of 222.96 mph for 1 kilometer was set by Brig. Gen. William Mitchell in a Curtiss R6 Curtiss D12 375, at Mount Clemens, Mich.

*November 12:* First air-to-air refueling made when Wesley May stepped from wing of one aircraft to that of another with a 5-gallon can of gasoline strapped to his back.

*November 15:* Initial U.S. flight of airship *Roma* was made at Langley Field, Va.

*November 28:* NACA Report 116, "Applications of Modern Hydrodynamics to Aeronautics," by Ludwig Prandtl of Gottingen University in Germany, a major contribution to the basis of the theory governing fundamental aerodynamical applications, was published. His famous 1904 paper on boundary layers was translated and issued in NACA Technical Memorandum No. 452 in 1928.

*December 1:* Nonrigid Navy dirigible C-7, first to use nonflammable helium, made flight from Hampton Roads, Va., to Washington, D.C.

*December 7:* In its annual report, the NACA recommended establishment of a Federal airways system to include provision of extended weather service "indispensable to the success and safety of air navigation." It also recommended that Government policy be formulated "to sustain and stabilize the aeronautical industry."

*December 29:* World endurance record of 26 hours 18 minutes 35 seconds set at Roosevelt Field, N.Y., by Edward Stinson and Lloyd Bertaud in a Junkers-Larsen BMW 185 (imported German Junkers J-13).

*During December:* The NACA cooperated with private organizations in the formulation of an air safety code.

*During 1921:* The NACA's Office of Aeronautical Intelligence distributed 13,080 copies of technical reports and 7,108 copies of technical notes to governmental, industrial, and educational institutions.

# 1922

*February 7:* Completion of a 50-hour test of the Lawrance J-1, 200-hp radial air-cooled engine, by the Aeronautical Engine Laboratory, Washington Navy Yard, foreshadowed the successful use of radial engines in naval aircraft.

*March 20:* Navy's first aircraft carrier, U.S.S. *Langley*, was commissioned at Norfolk, Va., a converted collier, *Jupiter*.

*March 23:* NACA Report No. 159 on "Jet Propulsion for Airplanes," by Edgar

Buckingham of the Bureau of Standards, pointed out that jet fuel consumption would be four times that of propeller engine at 250 mph, but that efficiency of jet increased at higher speeds.

*April 25:* Stout ST-1 successfully test flown by Eddie Stinson, first all-metal airplane designed for the Navy.

*May 31:* First use of helium in a free balloon in Navy balloon flown by Lt. Comdr. J. P. Norfleet in National Elimination Balloon Race at Milwaukee, which did not place in the race.

*June 10:* Guglielmo Marconi of Italy stated that an apparatus could be designed to transmit radio waves from one ship in any desired direction and pick up reflections from another ship in a receiver, a device which would "thereby immediately reveal the presence and bearing of the other ship in fog or thick weather." Christian Huelsmeyer of Germany received a patent in 1904 on boat equipment which used reflected radio waves for navigational use on the Rhine River.

*June 12:* Capt. A. W. Stevens (USAS) made record parachute jump from 24,200 from a supercharged Martin bomber over McCook Field.

———: Smithsonian Institution scientists utilized Navy seaplanes in mollusk research in Florida waters, completing in days what would otherwise have required a year.

*June 16:* Helicopter flight made by Henry Berliner at College Park, Md.

———: Lt. C. L. Bissell (USAS) began a series of night cross-country flights between Bolling Field, D.C., and Langley Field, Va.

*June 26:* ZR-3 rigid airship ordered from the Zeppelin Co., Friedrichshafen, Germany, as part of World War I reparations under terms approved by the Allied Conference of Ambassadors on December 16, 1921.

*During June:* Wright E-2 engine operated continuously for 250 hours at wide-open throttle, demonstrating improved durability of intake and exhaust valves; Navy BuAer later increased engine suitability tests from 50 to 300 hours' endurance.

*July 1:* Eight naval medical officers were first to report for flight training, at NAS Pensacola, having previously completed flight surgeon's course at the Army Technical School of Aviation Medicine.

*July 16:* Berliner helicopter rose 12 feet and hovered before military observers at College Park, Md.

*July 17:* Aerial photos taken from naval aircraft to aid in location of reefs at Lahaina, Maui, Hawaii.

*August 2:* An unofficial three-man altitude record of 23,350 feet was set at McCook Field, Dayton, Ohio, by Lt. L. Wade, Capt. A. W. Stevens, and Sergeant Longham in a supercharged Air Service bomber.

*August 18:* AGA beacon (American Gas Accumulator) began operations at NAS Hampton Roads, with 6,000 candlepower, 18 flashes per minute, and an optical range of 20 miles horizontally.

*August 21:* Lawrence Sperry dropped landing wheels from plane in flight and landed it on a skid device at Farmingdale, Long Island.

*September 4:* First transcontinental flight within a single day, by Lt. J. H. Doolittle (USAS) in a modified DH-4B Liberty 400, from Pablo Beach, Fla., to Rockwell Field, San Diego, a distance of 2,163 miles in 21 hours 20 minutes.

*September 27:* Observations on overflying aircraft made by Navy scientists ultimately aiding development of radar, by Albert Hoyt Taylor and Leo C. Young of the Naval Aircraft Radio Laboratory, Anacostia, D.C.

*October 17:* First USN carrier takeoff by Lt. V. C. Griffin in Vought VE-7SF, from U.S.S. *Langley*.

*October 19:* Variable-density wind tunnel placed into operation at Langley Laboratory, although lack of adequate electric power prevented concurrent operation of both wind tunnels this year.

*October 23:* Reversible propeller demonstrated at Bolling Field, D.C., by American Propeller Co.

## 1922—Continued

*October 26:* First USN carrier landing made by Lt. Comdr. G. Chevalier in Aeromarine 39B on U.S.S. *Langley* off Cape Henry.

*November 1:* First engineer-in-charge appointed for NACA Langley Aeronautical Laboratory, Leigh M. Griffith.

*November 8:* Air Service Medical Research Laboratory and School for Flight Surgeons was designated School of Aviation Medicine.

*November 18:* First catapult launching from carrier U.S.S. *Langley* (CV-1) by Comdr. Kenneth Whiting flying a PT seaplane.

*December 4:* President Harding requested the recommendations of the NACA as to the most promising program for the Air Mail Service in the expenditure of its limited funds. The NACA, on December 20, recommended that $2,300,000 be appropriated to demonstrate feasibility of night flying on the mail service and to establish regular New York-San Francisco mail service in 36 hours or less.

*December 18:* DeBothezat helicopter, built by the Engineering Division of the Air Service at McCook Field, successfully testflown for 1 minute 42 seconds by Maj. F. H. Bane.

*During 1922:* Two wind tunnels (4 by 4 foot, and 8 by 8 foot) at the Washington Navy Yard, under the direction of A. F. Zahm, made tests on naval designs, the important results of which were usually published by NACA as technical reports.

———: As a result of Army-Navy conference, policy established that manufacturers and designers should be invited to compete in the design and construction of military aircraft, with engineers given a free hand. The only requirement was that the airplane have military utility and have a speed of more than 190 mph.

———: National Aeronautic Association was formed with Howard E. Coffin elected president.

———: U.S. Weather Bureau first prepared a "standard atmosphere" showing the relationship between pressure and temperature based on average conditions over the United States at 40° N. latitude.

*1920–22:* Goddard developed and unsuccessfully tested first liquid propellant engine, using liquid oxygen, and devised small high-pressure pumps to force fuel into the combustion chamber.

# 1923

*January 5:* Cloud seeding over McCook Field, Dayton, accomplished by Prof. W. D. Bancroft of Cornell University, from Air Service aircraft.

*February 6:* Aeronautical Engine Laboratory transferred from Washington Navy Yard to the Naval Aircraft Factory, establishing the Naval Aircraft Factory as the center of naval aeronautical development.

*February 21:* DeBothezat helicopter achieved sustained altitude of 15 feet for 2 minutes and 45 seconds in flight tests at McCook Field.

*March 5:* Auxiliary jettisonable belly tank fitted to bomb rack of MB3A at Selfridge Field, Mich., increased flying radius to about 400 miles.

*March 8:* Lunar radiation observations at an altitude of 19,000 feet made by Russell M. Otis in DH–4B over San Diego, Lt. F. W. Seifert as pilot.

*March 29:* Lt. R. L. Maitland attained world speed record of 239.95 mph in Curtiss R–6 at Dayton, Ohio.

*April 2:* First flight of all-metal pursuit monoplane, Wright H–3, 400-hp engine, at Curtiss Field.

*April 15:* Naval Research Laboratory (NRL) reported equipment for radio control of an F5L was satisfactory to a range of 10 miles, and that radio control of aircraft during landing and takeoff was feasible.

*April 20:* First aerial refueling with hose, at Rockwell Field, San Diego, between two DH-4B aircraft, under the direction of Henry H. Arnold (USAS).

*May 2-3:* First nonstop transcontinental flight of 2,520 miles from New York to San Diego flown by Lts. O. G. Kelly and J. A. Macready, in a Fokker T2-Liberty 375 in 26 hours 50 minutes.

*May 26:* Chief of Navy BuAer agreed with Chief of Army Air Service that identical aeronautic specifications would be advantageous to both the aviation industry and the military services. Lt. R. S. Barnaby was ordered to McCook Field as BuAer representative on interservice committee on standardization in December, the first of a series of annual meetings held until 1937.

*June 9:* Juan de la Cierva made first successful autogiro flights in a rotary wing aircraft, at Madrid, Spain.

*June 20:* Initial flight of all-metal airplane (Gallaudet) designed by Engineering Division at Wright Field.

*June 25:* First International Air Congress, London, England, 450 delegates from 17 nations attended.

*June 26:* First complete midair pipeline refueling between two airplanes, made by Lts. L. H. Smith and J. P. Richter (USA) at San Diego.

*August 22:* World's largest airplane, the six-engine Barling bomber, underwent first tests at McCook Field, Lt. H. R. Harris as pilot.

*August 27-28:* Capt. L. H. Smith and Lt. J. P. Richter flew a DH-4B-Liberty 400 to a world refueled duration record of 37 hours and 15 minutes, as well as a distance record of 3,293 miles at Rockwell Field, San Diego, Calif.

*September 4:* Navy airship *Shenandoah* (ZR-1) made its first flight at NAS Lakehurst, the first of the Zeppelin type to use helium gas.

*September 5:* Army bombers sank two obsolete battleships, the U.S.S. *Virginia* and the U.S.S. *New Jersey*, off Cape Hatteras.

*September 28:* Navy aircraft won first and second places in Schneider Cup international seaplane races at Cowes, England, and established new world record for seaplanes with a speed of 169.89 mph for 200 kilometers. Flying CR-3's powered by Curtiss D-12 engines, Lt. David Rittenhouse achieved 177.38 mph in the race, while Lt. Rutledge Irvine placed second with 173.46 mph.

*October 1:* Goodyear Tire & Rubber acquired Zeppelin rights for manufacture of rigid airships.

*October 6:* Lt. A. J. Williams (USN) set new world speed records of 243.8 mph for 100 kilometers, and 243.7 mph for 200 kilometers over a closed circuit, flying a Curtiss R2C-1 Racer in the Pulitzer Trophy Race, at St. Louis, Mo.

*November 1:* Robert H. Goddard successfully operated a liquid oxygen and gasoline rocket motor on a testing frame, both fuel components being supplied by pumps installed on the rocket.

*November 2:* Flexural fatigue machine for testing sheet duralumin stopped after 200 million alterations, on a 389-day nonstop run at the Bureau of Standards. Check calibration gave same reading as the original calibration on October 5, 1922.

*November 4:* Lt. Alford J. Williams (USN) established world speed record of 266.59 mph in Navy-Curtiss Racer over Mitchel Field, Long Island, which remained U.S. record until 1930.

*November 5:* Series of tests demonstrating feasibility of stowing, assembling, and launching a seaplane from a submarine were completed, which involved assembling a Martin MS-1 and launching it by submerging the submarine.

*November 23:* Concluding sentence of the annual report of the NACA for 1923 was: "Progress in aeronautics is being made at so rapid a rate that the only way to keep abreast of other nations *is actually to keep abreast*, year by year, *never falling behind.*" [Italic in original.]

## 1923—Continued

*November 23:* Aeromarine all-metal flying boat launched at Keyport, N. J.

*December 18:* Christmas aileron patent claim was settled when U.S. Government bought the patent rights for $100,000.

*End of 1923: Die Rakete zu den Planetenräumen* (The Rocket Into Interplanetary Space) by Hermann Oberth was published in Germany, and was the genesis for considerable discussion of rocket propulsion.

*During 1923:* Turbine-type supercharger with a gear drive under development at McCook Field.

———: Navy Bureau of Aeronautics abandoned water-cooled engines of less than 300 hp with the development of the Lawrance direct air-cooled J-1, 200-hp engine. Weight of water-cooling system was usually in excess of 25 percent of the total weight of the engine.

# 1924

*January 16:* President Coolidge canceled all preparations for Navy Arctic expedition in which it was intended to use airplanes and the dirigible *Shenandoah.*

*February 27:* Corp. C. E. Conrad (USAS) successfully parachuted from 21,500 feet, from DH-4B over Kelly Field, Tex.

*March 4:* Two Martin bombers and two DH-4's broke up an icejam on the Platte River at North Bend, Nebr., by bombing.

*March 7:* Lt. E. H. Barksdale and B. Jones (USAS) flew DH-4B Liberty 400 on instruments from McCook Field, Dayton, Ohio, to Mitchel Field, N.Y.

*During March:* Apparatus developed at Wright Field for scattering insecticide from the air, for use in checking spread of gypsy moth in New England.

*April 6–September 28:* The first round-the-world flight, the first transpacific flight, and the first westbound Atlantic crossing, from and returning to Seattle, by two Army Douglas "World Cruiser" biplanes, flying 26,345 miles in 363 hours' flying time, with an elapsed time of 175 days.

*During April:* Central Committee for the Study of Rocket Propulsion established in the Soviet Union.

*May 2:* Unofficial two-man altitude record of 31,540 feet set by Lts. John A. Macready and A. W. Stevens (USAS) on a flight during which an aerial photograph covering the greatest area of the earth's surface to date was obtained.

*May 19:* Lt. J. A. Macready (USAS) established new American altitude record of 35,239 feet at Dayton, in Le Pere Liberty 400.

*June 2:* Dr. C. L. Meisinger of the Weather Bureau and Lt. James T. Neely were killed by lightning in storm-riding balloon flight, near Monticello, Ill.

*June 23:* First "dawn-to-dusk" flight from New York to San Francisco, by Lt. R. L. Maugham in Curtiss Pursuit (PW-8), with five stops en route.

*July 1:* Dr. George W. Lewis appointed Director of Aeronautical Research of NACA, a post he held until 1947.

———: First continuous night-and-day transcontinental airmail service initiated between New York and San Francisco by Post Office Department pilots, a service which was first instituted on September 8, 1920, but had stopped.

*September 14:* French helicopter flown by its designer, Oehmichen, established world helicopter altitude record of 3.28 feet carrying 440.92-pound useful load.

*September 15:* Unmanned N-9 seaplane equipped with radio control successfully flown on 40-minute flight from Naval Proving Grounds, Dahlgren, and sank from damage sustained on landing.

*October 15:* ZR-3 (later renamed *Los Angeles*), a German dirigible constructed for the U.S. Navy under a reparations agreement, arrived at Lakehurst, N.J., after flying the Atlantic, by German crew under Dr. Hugo Eckener.

*October 24:* When all foreign entrants withdrew from Schneider Cup Race to be held at Bayshore Park, Md., the United States agreed to cancel race rather than win by a flyaway. Instead, Navy scheduled contestants and other naval aircraft placed 17 world records in the book for class C seaplanes.

*October 25:* Lt. R. A. Ofstie (USN) established new world seaplane speed record of 178.25 mph for 100 km.

*October 28:* Cloud formations at 13,000 feet were broken up over Bolling Field, D.C., by "blasting" with electrified silica in a fog-dispersal demonstration by Army aircraft.

*November 24:* NACA Committee on Aerodynamics summarized in its annual report that it had direct control of aerodynamic research conducted at Langley, the propeller research conducted at Stanford University under W. F. Durand, and some special investigation at the Bureau of Standards and at a number of Universities. Investigation undertaken at the Washington Navy Yard Aerodynamic Laboratory, the Engineering Division of the Army Air Service, the Bureau of Standards, and the Massachusetts Institute of Technology were reported to this Committee. Thus, it was "in close contact with all aerodynamical work being carried out in the United States."

——: NACA Subcommittee on metals concluded that duralumin girders which formed framework of the *Shenandoah* "will not fail by 'fatigue' in less than 40 years under service conditions" as a result of the "most extensive investigation of the properties of sheet metal which has been undertaken in this country," by the Bureau of Standards.

*December 2:* "Standard Atmosphere," after careful coordination, approved by Executive Committee of NACA, later adopted for use in aeronautical calculations by the War and Navy Departments, the Weather Bureau, and the Bureau of Standards; described by Lt. Walter S. Diehl of BuAer in NACA Technical Report No. 218. It gave pressures and densities for altitudes up to 20,000 meters and to 65,000 feet.

*December 9:* The Civil Aeronautics Act, proposing to establish a Bureau of Civil Aeronautics in the Department of Commerce, was reintroduced in Congress.

*During 1924:* High-speed wind tunnel (5 foot, 1,000 hp, 260 mph) at McCook Field used continuously, handling 150 tests of 17 airfoil, 24 model, and 15 fuselage tests.

——: High-speed photography of sprays produced by fuel injection valves successfully developed, and flight study of Roots-type supercharger with DH-4 and DT-2 aircraft conducted, at Langley Laboratory. Supercharging increased practical ceiling of DH-4 from 14,500 feet to 31,000 feet, and of the DT-2 from 18,500 feet to 28,000 feet.

——: NACA Report No. 207 by L. J. Briggs, G. F. Hull, and H. L. Dryden of the National Bureau of Standards, "Aerodynamic Characteristics of Airfoils at High Speeds," was major contribution reporting on tests of airfoils at near supersonic speeds.

# 1925

*January 24–25:* Twenty-five aircraft carried scientists and other observers above clouds in Connecticut to view total eclipse of the sun, while airship *Los Angeles* carried Naval Observatory scientists over Block Island, R.I.

*February 2:* President Coolidge signed the Kelly bill authorizing contract air transport of mail.

*February 18:* "Standard Altimeter Calibration" worked out by Bureau of Standards, and approved by all interested agencies, was approved by the NACA.

*April 13:* Henry Ford started an airfreight line between Detroit and Chicago, the first such commercial flights on a regular schedule.

## 1925—Continued

*April 15:* Daily flights to an altitude of 10,000 feet to obtain weather data and to test upper-air-sounding equipment begun at NAS Anacostia. In the following February, the schedule was extended by the Navy to include weekends and holidays, with the altitude being increased to 15,000 feet.

*April 27:* First trial flight of new Wright Cyclone 450-hp air-cooled engine in DT-6 torpedo plane, at Muchio's Field, N.J.

*During April:* Oleo landing gear tested by Navy on NB-1 at Seattle.

*May 20:* Air Service Technical School at Rantoul, Ill., carried on radio conversations from planes in the air, reaching Chicago 115 miles distant.

*June 12:* Daniel Guggenheim donated $500,000 toward establishment of a School of Aeronautics at New York University.

*June 25:* Construction of full-scale propeller research wind tunnel at Langley Aeronautical Laboratory was initiated, which was completed in 1927.

*During July:* First radiobeacon, one developed at McCook Field, installed in airmail plane for the Department of Commerce.

——: Small car moving on ground controlled by radio from an airplane at 2,000 feet, by Air Service at Wright Field, Dayton.

*August 1:* Naval Air Detail, under Lt. Comdr. R. E. Byrd, began aerial exploration of 30,000-square-mile area near Etah, North Greenland, with three Loening amphibians, as part of the MacMillan expedition.

——: Curtiss Condor, first of new series of night bombers, made first flight at Garden City, Long Island.

*September 3:* Navy dirigible *Shenandoah* crashed near Ava, Ohio, killing 14 of 43 persons aboard.

*September 12:* Morrow Board was appointed by President Coolidge to recommend U.S. air policy.

*October 7:* Post Office Department awarded first five contracts under the Kelly Air Mail Act for the flying of mail to private contractors on a bid basis.

*October 26:* Lt. James H. Doolittle, U.S. Air Service, won Schneider Cup Race flying Curtiss-R3 C-2 seaplane Racer, and also broke speed record for seaplanes attaining 245.7 mph, at Baltimore, Md.

*November 20:* Night photographs using 50-pound magnesium flares taken from Army Martin bomber by Lt. George W. Goddard, over Rochester, N.Y.

*November 30:* The President's Aircraft Board, better known for its senior member as the Morrow Board, submitted its report to President Coolidge. Recommendations of the NACA to the Morrow Board were important in decisions leading to the passage of the Air Commerce Act of 1926 and the appropriation of funds for the long-range development of Army and Navy aviation. With its recommendations inaugurated, NACA thereafter followed a policy of avoiding entanglement in matters not related to research.

*December 17:* Col. William Mitchell found guilty by Army General Court-Martial, in session since October 28.

*December 27:* Daniel Guggenheim created the $2,500,000 Daniel Guggenheim Fund for the Promotion of Aeronautics to speed development of civil aviation in the United States.

*During 1925:* School of Aviation Medicine began study on an objective aptitude test for flyers.

*During 1925:* Goeffrey de Havilland of Britain first produced two-seat biplane, the Moth, a small popular light airplane. War-surplus Curtiss JN4D airplanes had earlier been popular in the United States, while Taylor Cub monoplane appeared in 1931.

# 1926

*January 1:* Henry J. E. Reid appointed Engineer-in-Charge of NACA Langley Memorial Aeronautical Laboratory, a post he held until July 1960, when he retired as Director of NASA's Langley Research Center.

*January 16:* Daniel Guggenheim Fund for the Promotion of Aeronautics formally established.

*January 29:* An American altitude record of 38,704 feet was set by Lt. J. A. Macready (USAS) in an XCO5–A Liberty 400 at Dayton, Ohio.

*February 6:* Pratt & Whitney produced first Wasp engine, a nine-cylinder radial air-cooled engine of about 400 hp at 1,800 rpm.

*March 16:* Robert H. Goddard launched the world's first liquid-fueled rocket at Auburn, Mass., which traveled 184 feet in 2½ seconds. This event was the "Kitty Hawk" of rocketry.

*March 23:* Inventor of sodium-filled valves for internal combustion engines, S. D. Heron, granted exclusive license for manufacture to Rich Tool Co., later part of Eaton Manufacturing Co.

*April 16:* The Department of Agriculture purchased its first cotton-dusting plane.

*During April:* The NACA analysis of basic aeronautical legislation was accepted by Joint Senate-House conferees, leading to the Air Commerce Act of May 20, 1926. This freed NACA of responsibility for regulation of civil aviation and permitted it to concentrate upon the conduct of aeronautical research.

*May 5:* Robert H. Goddard communicated the results of his successful liquid-propellant rocket flight of March 16 to the Smithsonian Institution.

*May 9:* First flight over the North Pole, by Richard Byrd, navigator, and Floyd Bennett, pilot, in a Fokker Monoplane, from Spitsbergen.

*May 12:* Lincoln Ellsworth, American explorer, flew across the North Pole in the dirigible *Norge*, commanded by Roald Amundsen.

*May 20:* President Coolidge signed the Air Commerce Act, the first Federal legislation regulating civil aeronautics.

*May 24:* First annual inspection and conference for industrial and other governmental aeronautical persons held at NACA's Langley Laboratory. These annual events were of high importance in promoting aeronautical research in the United States.

*June 6:* Last elements of Navy Alaskan Aerial Survey Expedition departed Seattle for Alaska. Three Loening amphibians operating from tender U.S.S. *Gannet* made aerial mapping of Alaska throughout the summer and into September with the cooperation of the Department of the Interior.

*June 25:* Largest wind tunnel in the world (20-foot throat), the Propeller Research Tunnel, constructed at Langley.

*July 1:* Edward P. Warner, professor of aeronautics at MIT, nominated by President Coolidge to become Assistant Secretary of Navy in Charge of Aviation. Dr. Warner served on the NACA, 1929–46.

*July 2:* First known reforesting by airplane was carried out in Hawaii.

———: The Army Air Corps Act became law and the Air Service was redesignated the Air Corps. It also made provision for an Assistant Secretary of War for Air and for a 5-year Air Corps expansion program.

———: By act of Congress, the NACA was required to review aeronautical inventions and designs submitted to any branch of Government and submit reports to the Aeronautics Patents and Design Board.

*July 28:* Submarine S–1 surfaced and launched a Cox-Klemin XS–2 seaplane piloted by Lt. D. C. Allen. It later recovered airplane and submerged, thus carrying out first complete cycle in this series of feasibility experiments.

## 1926—Continued

*August 25:* JN training plane with large parachute floated deadstick down to a rough landing and some damage, at San Diego Naval Air Station.

*During August:* Air Corps School of Aviation Medicine moved from Mitchel Field to Brooks Field, Tex., and was subsequently moved to Randolph Field in October 1931.

*October 1:* Daniel Guggenheim Fund for the Promotion of Aeronautics made a grant to the University of Michigan for the completion of a wind tunnel and a Chair of Aeronautics.

*November 13:* Lt. C. F. Schilt (USMC) took second place in the Schneider Cup Race at Hampton Roads, Va., flying an R3C-2 with an average speed of 231 mph. This was last U.S. Navy participation in international racing competition.

*December 10-11:* Financed by the Daniel Guggenheim Fund for the Promotion of Aeronautics, a conference of representatives of MIT, New York University, Stanford University, California Institute of Technology, University of Michigan, and University of Washington was held at NACA to interchange ideas on educational methods, coordinating research work, and developing special courses in aeronautical education.

*During 1926:* Dr. Louis H. Bauer, former Commandant of the School of Aviation Medicine (1919–25), established a medical section in the Bureau of Air Commerce, Department of Commerce.

———: Lt. Col. D. A. Myers at the School of Aviation Medicine developed basic physiological principles necessary to the development and use of blind-flying instruments, work done in conjunction with research by Lt. Col. W. A. Ocker. This study was regarded as one of the greatest contributions of medicine to the technical advancement of aviation.

# 1927

*During February:* Army Air Corps completed aerial photographic survey of east and west coasts of Florida (1,284 square miles) for the U.S. Coast and Geodetic Survey.

*March 9:* Capt. H. C. Gray (AAC) ascended to 28,910 feet in a free balloon for an American altitude record. (World record held by Suring and Berson of Germany who ascended to 35,433 feet on June 30, 1901.)

*April 4:* Regular commercial airline passenger service initiated by Colonial Air Transport between New York and Boston.

*April 21:* Dr. Joseph S. Ames was elected Chairman of the NACA, to replace Dr. Charles Walcott, one of the original 12 members, who died in February.

*May 4:* Record balloon flight by Capt. H. C. Gray (AAC) reached 42,470 feet over Scott Field, Ill., but he was forced to bail out successfully so that record was not official.

*May 20-21:* The first solo nonstop transatlantic flight, New York to Paris, was completed by Charles A. Lindbergh. This was a major milestone in awakening the Nation to the full potentialities of aviation.

*May 25:* Lt. James H. Doolittle (AAC) flew the first successful outside loop.

*June 4:* Daniel Guggenheim School of Aeronautics officially opened at New York University. Daniel Guggenheim Fund for the Promotion of Aeronautics also made gifts to MIT, University of Michigan, Stanford University, and the California Institute of Technology in this time period.

*June 4-5:* Clarence D. Chamberlain and Charles A. Levine flew nonstop from New York to Eisleben, Germany, in Bellanca monoplane *Columbia*.

*June 5:* Society for Space Travel (*Verein fuer Raumschiffahrt*), known as "*VfR*," formed in Breslau, Germany.

*June 8:* Astronautics Committee of the *Société Astronomique Française* established in France.

*June 22:* John F. Victory, the first employee of NACA in 1915, and who had served as Assistant Secretary since 1917, was appointed Secretary of the NACA.

*June 29–30:* Cdr. Richard Byrd, Acosta, Noville, and B. Balchen flew Fokker monoplane *America* from New York to a crash landing in the sea off the French coast.

*July 25:* A world airplane altitude record of 38,484 feet was established by Lt. C. C. Champion (USN) in a Wright Apache P&W 425.

*August 1:* Fire damaged interior of variable-density wind tunnel at Langley Laboratory, which when reconstructed was used in conjunction with jet-type wind tunnel produced airflow in 12-inch chamber in excess of 800 mph.

*October 12:* Wright Field, Dayton, Ohio, was formally dedicated.

*November 4:* Capt. H. C. Gray (AAC) ascended to 42,470 feet, the identical altitude of his May 4 flight, but he did not survive the flight and thereby failed again to achieve official world record.

*November 6:* Lt. "Al" Williams (USN) flew Kirkham racing plane powered with 1,250-hp 24-cylinder Packard engine at unofficial speed of 322.6 mph.

*November 16:* U.S.S. *Saratoga* (CV–3) was placed in commission by the Navy.

*December 14:* U.S.S. *Lexington* (CV–2) was placed in commission.

*During 1927:* Air Corps sponsored development of Allison "X" type engine of 24 cylinders expected to develop 1,400 hp; while Navy flight tested radial air-cooled Wright R–1750, and used Pratt & Whitney Wasp in a number of service aircraft.

——: Coordination between NACA and British Aeronautical Research Committee included exchange of views at joint meetings and a program of comparative research for standarization of wind tunnel data.

——: Operation of Materiel Division wind tunnels at McCook Field handicapped by move to the new Wright Field. During the year, the new full-scale Propeller-Research Tunnel at Langley Laboratory became operational, while the Bureau of Standards tested 24 airfoil sections at various speeds up to 1.08 times the speed of sound.

——: Superchargers passed from experimental development stage to active service use on radial air-cooled engines, while both Roots-type and centrifugal-type superchargers were being tested on water-cooled engines.

——: Appearance of Lockheed Vega set pace for general-purpose aircraft, a high cantilever wing and wooden stressed-skin fuselage which permitted large interior structure for passengers as well as reducing weight and drag.

# 1928

*February 3:* At Wright Field, Lt. H. A. Sutton began a series of tests to study the spinning characteristics of planes, for which he was awarded the Mackay Trophy. (See Appendix D)

*February 28:* Navy issued contract for XPY–1 flying boat to Consolidated Aircraft, the first large monoplane flying boat procured and the initial configuration which evolved into the PBY Catalina.

*March 10:* $900,000 authorized for completion of the Wright Field Experimental Laboratory.

*March 28:* Assistant Secretary of Commerce for Aeronautics called conference of representatives of Army, Navy,

## 1928—Continued

Weather Bureau, Bureau of Standards, NACA, and Commerce Department to study cause and prevention of ice formation on aircraft.

*April 11:* First manned rocket automobile tested by Fritz von Opel, Max Valier, and others, at Berlin, Germany.

*April 12–13:* German pilots Köchl and Huenefeld, and J. Fitzmaurice made first westbound transatlantic airplane flight in Junkers *Bremen*.

*May 5:* Lt. C. C. Champion flew a Wright Apache equipped with P&W Wasp engine and NACA supercharger to new world altitude record for seaplanes of 33,455 feet.

*May 15:* NACA held third annual Engineering Research Conference at Langley Field, Va.

*May 22:* First patent on sodium-filled valves for combustion engines issued to S. D. Heron, engineer of the Materiel Division at Wright Field.

*During May:* Aeronautics Branch of Department of Commerce created Board to determine original causes of aircraft accidents.

*June 11:* Friedrich Stamer made first manned rocket-powered flight in a tailless glider from the Wasserkuppe in the Rhön Mountains of Germany. Takeoff was made by elastic launching rope assisted by 44-pound thrust rocket, another rocket was fired while airborne, and a flight of about 1 mile was achieved. This flight was a part of experimentation directed by A. Lippisch.

*June 16:* Successful tests were made of superchargers designed to give sea level pressure at 30,000 feet and a new liquid-oxygen system for high-altitude flying, at Wright Field. Lt. William H. Bleakley in XCO-5 made flight to 36,509 feet and remained there 18 minutes.

*During September:* The NACA undertook coordination of research programs in universities to promote the study of aeronautics and meteorology.

*September 19:* First diesel engine to power heavier-than-air aircraft, manufactured by Packard Motor Car Co., was flight tested at Utica, Mich.

*September 23:* Lt. James H. Doolittle accompanied by Capt. A. Stevens made altitude flight of 37,200 feet to obtain aerial photograph covering 33 square miles.

*October 4–5:* First Aeronautical Safety Conference held in New York under auspices of the Daniel Guggenheim Fund for the Promotion of Aeronautics.

*October 10:* Capts. St. Clair Streett and A. W. Stevens (USA) flew to 37,854 feet, less than 1,000 feet short of the official world record for single-occupant flight.

*During October:* At the request of the Air Coordination Committee, NACA prepared a report on "Aircraft Accident Analysis" for use by the War, Navy, and Commerce Departments.

——: Air Corps developed 84-foot-in-diameter parachute of sufficient strength to support weight of an airplane and its passengers.

*December 17:* International pilgrimage made to Kitty Hawk, N.C., to commemorate the 25th anniversary of the first airplane flight.

*December 19:* First autogiro flight in the United States was made by Harold F. Pitcairn, Willow Grove, Pa.

*During December:* Air Medical Association formed at International Aeronautics Conference.

*During 1928:* NACA developed cowling for radial air-cooled engines which increased speed of Curtiss AT–5A airplane from 118 to 137 mph with no increase in engine horsepower, Fred E. Weick and associates contributing to this development.

——: NACA's Langley Memorial Aeronautical Laboratory demonstrated high lift by boundary-layer control by means of pressure or suction slots in an airfoil in the atmospheric wind tunnel.

——: First refrigerated wind tunnel for research on prevention of icing of wings and propellers placed in operation at Langley Laboratory.

———: First of nine volumes of an encyclopedia on interplanetary travel by Prof. Nikolai A. Rynin published in the Soviet Union, the final volume of which appeared in 1932.

# 1929

*January 1-7:* An unofficial endurance record for refueled airplane flight was set by Maj. Carl Spaatz, Capt. Ira C. Eaker, and Lt. Elwood Quesada in the *Question Mark*, Fokker C2-3 Wright 220, over Los Angeles Airport, with flying time of 150 hours 40 minutes 15 seconds.

*January 23-27:* Modern aircraft carriers *Lexington* and *Saratoga* participated in fleet exercises for the first time.

*February 4-5:* Capt. Frank Hawks and O. E. Grubb established new nonstop transcontinental West-East record of 18 hours 22 minutes, in a single-engine *Lockheed Air Express*, the first practical application of NACA cowling for radial air-cooled engines.

*February 23:* Successful development of special goggles, heated gloves, and a device for warming oxygen before use announced by Wright Field.

*March 2:* Membership of the NACA increased from 12 to 15 members by act of Congress.

*May 8:* Lt. A. Soucek (USN) established world's altitude record of 39,140 feet, flying the Wright Apache over Anacostia, D.C.

*June 21:* NACA special subcommittee held initial meeting at Langley on aeronautical research in universities.

*June 27-29:* Capt. Frank Hawks broke transcontinental speed records from East to West and West to East flying the *Lockheed Air Express.*

*July 17:* A liquid-fueled, 11-foot rocket, fired by Robert Goddard at Auburn, Mass., carried a small camera, thermometer, and a barometer which were recovered intact after the flight. Much "moon rocket" publicity made of this flight.

*August 8-29:* Round-the-world flight of the German rigid airship *Graf Zeppelin.*

*August 23-October 31:* Russian plane, *Land of the Soviets*, flown on good-will tour of the United States from Moscow to Seattle, thence to New York, having covered 13,300 miles in 142 flying-hours.

*During August:* Use of a battery of solid-propellant rockets on Junkers-33 seaplane, the first recorded jet-assisted take-off of an airplane, made in tests near Dessau, Germany.

*September 22:* Second Alaska Aerial Survey completed by Navy, mapping 13,000 square miles in southeastern Alaska.

*September 24:* Lt. James H. Doolittle made the first public all-blind flight at Mitchel Field, Long Island, accompanied by a check pilot.

*September 30:* Opel Sander Rak. 1, a glider powered with 16 rockets of 50 pounds of thrust each, made successful flight of 75 seconds, covering almost 2 miles near Frankfort-am-Main, Germany, Von Opel as pilot.

*October 7:* Aero Medical Association of the United States founded by Louis H. Bauer, and the first issue of the *Journal of Aviation Medicine* was published in March 1930.

*October 15:* Premier of German movie film, *Frau im Mond* (The Girl in the Moon) directed by Fritz Lange, which assisted popular awareness of rocket potentialities in Germany.

*October 21:* German Dornier DO-X flying boat carried 169 passengers in hour flight over Lake Constance, Switzerland, the largest number of individuals ever carried in a single aircraft.

*November 28-29:* First flight over South Pole, by Comdr. Richard E. Byrd, in a

### 1929—Continued

Ford trimotor piloted by Bernt Balchen, from Little America.

*November 29:* First pursuit aircraft powered with high-temperature, liquid-cooling system designed by the Materiel Division, was completed by Curtiss and flown to Wright Field for flight testing.

*December 12:* Langley Medals were presented to Adm. Richard E. Byrd for his flights over both poles and posthumously to Charles M. Manly for his pioneer development of airplane engines. (See Appendix D.)

*December 31:* Daniel Guggenheim Fund for the Promotion of Aeronautics ended its activities.

*During 1929:* J. Jongbloed experimentally recognized the occurrence of a disease like the bends, or caisson disease, at pressures of less than 1 atmosphere.

———: U.S. Bureau of Standards developed the radio-echo altimeter.

———: NACA Annual Report indicated that aerodynamic efficiency may be increased by applying the principle of boundary-layer control to the wings and possibly other parts of an airplane.

# 1930

*January 3:* President Hoover made the presentation of the Collier Trophy for 1929 to Dr. Joseph S. Ames, Chairman of the NACA. (See Appendix D.)

*January 6:* Guggenheim Safe Aircraft Competition prize was awarded to Curtiss Tanager, which featured practical wing flaps and leading-edge Handley-Page slots.

*During January:* The world's first full-scale wind tunnel under construction at Langley Memorial Aeronautical Laboratory (30 feet high, 60 feet wide).

*February 15:* Naval Aircraft Factory authorized to begin construction of working models of retractable landing gears because of design progress.

*February 17–19:* First National Conference on Aeronautical Education held at St. Louis, Mo.

*March 21:* First Navy dive bomber designed to deliver 1,000-pound bomb, the Martin XT5M-1, met strength and performance requirements in diving tests.

*April 4:* The American Interplanetary Society, later the American Rocket Society (ARS), founded in New York City by David Lasser, G. Edward Pendray, Fletcher Pratt, and nine others, for the "promotion of interest in and experimentation toward interplanetary expeditions and travel."

*April 8:* Orville Wright received first Daniel Guggenheim Medal. (See Appendix D.)

*April 12:* Air Corps set world record for altitude formation flying when 19 planes reached a height of 30,000 feet (old record 17,000 feet).

*May 9:* Dr. Ludwig Prandtl of Germany received second Daniel Guggenheim Medal.

*May 13:* Fifth Annual Aircraft Engineering Research Conference held at Langley Laboratory.

*June 4:* Lt. Apollo Soucek flew Navy Wright Apache landplane equipped with P&W 450-hp engine to height of 43,166 feet over NAS Anacostia, regaining world record he held briefly in 1929.

*July 21:* Capt. A. H. Page (USMC) piloted an O2U from a sealed hooded cockpit on an instrument flight of near 1,000 miles from Omaha, Nebr., to Anacostia, via Chicago and Cleveland, with safety pilot Lt. V. M. Guymon landing the airplane.

*July 23:* Hermann Oberth and *VfR* successfully tested liquid oxygen and gasoline-fueled rocket motor for 90 seconds in Germany, a demonstration made before the Director of the Chemisch-Technische Reichsanstalt to secure financial support.

*September 8:* German sounding balloon released near Hamburg attained an altitude of 117,750 feet (22.4 miles).

*During September: Raketenflugplatze Berlin* established by *VfR* in Germany.

*December 17:* German Army Ordnance Office, after reviewing work of Goddard and others, decided to establish rocket program and to equip artillery proving ground at Kummersdorff to develop military missiles.

*December 30:* Robert H. Goddard fired 11-foot liquid fuel rocket to a height of 2,000 feet and a speed or near 500 mph near Roswell, N. Mex.

*December 31:* "Airworthiness Requirements for Aircraft Components and Accessories" of the Department of Commerce became effective.

*During December:* John J. Ide, NACA technical assistant in Europe, served as U.S. delegate to the First International Congress on Aerial Safety in Paris.

*During 1930:* NACA made confidential recommendations to industry and military services for best location of engine nacelles, with engines faired into leading edges of the wing, a report based on 1928 research of Donald H. Wood and others which influenced design of all multiengine aircraft thereafter.

———: Sound-locator acoustic system for detection of aircraft in flight was developed.

———: Sperry Gyroscope developed the "Gyro Horizon."

———: An increase of 300 percent in paid passengers on commercial airlines was recorded this year.

———: Frank Whittle, RAF officer and engineer, obtained British patents for turbojet engine.

———: Allison Division of General Motors began development of V–1710 12-cylinder liquid-cooled engine, the only liquid-cooled engine of U.S. design to be produced throughout World War II, which was increased in 17 years from 750 hp to 2,000 hp.

———: First vertical wind tunnel for study of airplane spinning was placed in operation at NACA's Langley Laboratory.

———: Robert Esnault-Pelterie of France published his classic work on *L'Astronautique;* he had begun his mathematical work on astronautics in 1907.

# 1931

*January 4:* William G. Swan stayed aloft for 30 minutes over Atlantic City, N.J., in a glider powered with 10 small rockets.

*January 22:* Navy ordered its first rotary-wing aircraft, the XOP–1, from Pitcairn Aircraft.

*March 4:* More than $100 million was appropriated by Congress for military, naval, and commercial aviation for the coming year.

*March 14:* First liquid-fuel rocket successfully fired in Europe, a methane-liquid oxygen rocket constructed by Johannes Winkler and flown from Dessau, Germany.

*April 2:* First Navy aircraft with retractable landing gear, the XFF–1 two-seat fighter, ordered from Grumman Aircraft.

*April 8:* Amelia Earhart established a woman's autogiro altitude record of 18,415 feet in a Whirlwind-powered Pitcairn at Willow Grove, Pa.

*April 10:* Airship subcloud observation car demonstrated by Lt. Wilfred J. Paul at Langley Field, Va.

*During April: Raktenflugplatz* in Germany was visited by Mr. and Mrs. G. Edward Pendray as official representatives of the American Interplanetary Society,

## 1931—Continued

who upon their return organized the experimental program of the society.

*May 27:* First full-scale wind tunnel for testing airplanes was dedicated at the Langley Memorial Aeronautical Laboratory of the NACA, engineer-in-charge of construction and operation, Smith J. De France, explained details to the annual Aircraft Engineering Research Conference.

——: The NACA tank to provide data on water performance of seaplanes was demonstrated by Starr Truscott. Its channel length was enlarged from 2,020 feet to 2,900 feet in October 1937.

——: Auguste Piccard, Swiss physicist, and Charles Knipfer made first balloon flight into stratosphere, reaching a height of 51,777 feet in a 17-hour flight from Augsburg, Germany, to a glacier near Innsbruck, Austria.

*May 28:* Lt. W. Lees and Ens. F. A. Brossy established world's endurance flight record without refueling of 84 hours 33 minutes, in diesel-powered Bellanca at Jacksonville, Fla.

*May 31:* A pilotless airplane was successfully flown by radio control from another plane at Houston, Tex.

*June 4:* Dornier DO–X, 12-engined German flying boat (which carried 169 passengers on its trial flight), arrived in New York after flying the south Atlantic.

*June 23–July 1:* Wiley Post and Harold Gatty lowered world circling record to 8 days 15 hours 51 minutes in the Lockheed *Winnie Mae*.

*July 24–31: Graf Zeppelin* carried 12 scientists on Arctic flight.

*July 28:* First nonstop flight across the Pacific, begun by Clyde Pangborn and Hugh Herndon in a single-engined Bellanca, who completed flight around the world in October.

*July 29–August 26:* Colonel and Mrs. Lindbergh made survey flight to Japan in Sirius seaplane, via Alaska and Siberia.

*September 4:* Maj. James H. Doolittle established a new transcontinental record from Burbank to Newark of 11 hours and 16 minutes elapsed time including three stops, flying Laird *Super-Solution*.

*September 9:* Start of official rocket-mail service between two Austrian towns by Friedrich Schmiedl; test flights began in February 1931, while rocket-mail service continued until March 16, 1933.

*October 30:* School of Aviation Medicine moved from Brooks Field to Randolph Field, Tex.

*During 1931:* NACA Report 385 presented results showing that maximum lift coefficient of a wing could be increased as much as 96 percent by use of boundary-layer control.

——: Robert Esnault-Pelterie of France demonstrated liquid-fuel rocket propulsion with a rocket motor operated on gasoline and liquid oxygen.

——: Bureau of Standards made a number of experiments to determine whether thrust reaction of a jet could be increased, and tested combinations of jets.

——: Alexander Lippisch of Germany first produced and demonstrated a practical delta-wing aircraft.

*During 1931–32:* Taylor Cub Model A, a two-seat, high-wing light airplane, first produced, and helped popularize sports flying in the United States.

# 1932

*March 26:* Navy Consolidated P2Y seaplane made first test flight.

*April 19:* First flight of Goddard rocket with gyroscopically controlled vanes for automatically stabilized flight, near Roswell, N. Mex.

*May 4:* Daniel Guggenheim Gold Medal for 1932 awarded to Juan de la Cierva for development of the autogiro. (See Appendix D.)

*May 9:* First blind solo flight (without a check pilot aboard) solely on instruments was made by Capt. A. F. Hegenberger (AAC) at Dayton, Ohio.

*June 30: Los Angeles* (ZR-3) decommissioned by the Navy for economy reasons after 8 years of service and over 5,000 hours in the air.

*July 28:* Navy BuAer initiated research program on physiological effects of high acceleration and deceleration encountered in dive-bombing and other violent maneuvers in allocation to Bureau of Medicine and Surgery. Pioneer research pointing to need for anti-g or anti-blackout equipment was subsequently performed at Harvard University School of Public Health under the direction of Dr. C. K. Drinker by Lt. Comdr. John R. Poppen (MC USN).

*During July–August: VfR* successfully fired *Mirak II* rocket to height of 200 feet, after which German Army Ordnance Office formalized rocket development program by placing Captain-Doctor Walter Dornberger in charge of Research Station West at Kummersdorf.

*August 18:* Auguste Piccard and Max Cosyns attained an altitude of 53,152 feet on second stratosphere balloon flight, landing on a glacier in the Alps.

*August 31:* Capt. A. W. Stevens and Lt. C. D. McAllister (AAC) flew 5 miles above earth's surface at Fryeburg, Maine, to photograph eclipse of the sun.

*During August:* Experimental transmission of weather maps by teletype initiated by Weather Bureau on a special circuit between Cleveland and Washington.

*September 3:* Maj. James H. Doolittle set a new world speed record for landplanes by averaging 294 mph over 3-km course at Cleveland, Ohio, in Granville Brothers Gee Bee monoplane with P&W Wasp engine.

*September 16:* Altitude record of 43,976 feet for landplanes established by Cyril F. Unwins in Vickers Vespa at Bristol, England.

*September 21:* Dr. Robert A. Millikan of California Institute of Technology completed series of tests on the intensity of cosmic rays at various altitudes with cooperation of 11th Bombardment Squadron, in a Condor Bomber from March Field, Calif.

*October 1:* Wernher von Braun joined the German Army Ordnance Office rocket program at Kummersdorf.

*October 15:* Institute of Aeronautical Sciences was incorporated in New York.

*November 12:* American Interplanetary Society performed static tests of rocket based on *VfR* design at Stockton, N.J.

*December 1:* Teletypewriter Weather Map Service was inaugurated by Aeronautics Branch, Department of Commerce.

*During 1932:* German engineer, Paul Schmidt, working from design of Lorin

1932—Continued

tube, developed and patented a ramjet engine later modified and used in the V-1 Flying Bomb.

*During 1932:* Robert H. Goddard developed component of modern ramjet engine with construction of a rocket fuel pump at Clark University.

——: Capt. John R. Poppen (MC USN), began experimentation with animals on physiological effects of high acceleration, proposing as a result of his studies that an inflatable abdominal corset be developed for use by fighter pilots.

——: Junkers Ju-52, German trimotor transport of great success, first produced.

——: Control mechanism for variable-pitch propellers developed under the direction of Frank Caldwell.

——: NACA published derivation and characteristics of the first systematic family of NACA airfoils.

——: JATO-type rockets first used in the Soviet Union, according to Moscow historians.

# 1933

*January 21:* Institute of Aeronautical Sciences (IAS) held its Founders Meeting at Columbia University under Jerome C. Hunsaker, president, and Lester D. Gardner.

*February 25:* Aircraft carrier U.S.S. *Ranger* (CV-4) launched at Newport News.

*During February:* Boeing 247, first "modern-type" airliner, first flew.

*March 11:* *Macon* dirigible christened at Akron, Ohio, and made first flight on April 21 with 105 persons aboard.

*March 28:* Aircraft engine manufacturers granted permission by the Aeronautics Branch, Department of Commerce, to conduct endurance tests on their own equipment.

*April 4:* Rear Adm. W. A. Moffett, Chief of Navy Bureau of Aeronautics, killed along with 72 others in crash of the dirigible *Akron* at sea off the coast of New Jersey. He was replaced by Rear Adm. E. J. King (USN).

*May 14:* American Interplanetary Society Rocket No. 2 successfully fired, attaining 250-foot altitude in 2 seconds, at Marine Park, Staten Island, N.Y.

*July 1–August 12:* Gen. Italo Balbo of Italian Air Force led flight of 25 Savoia-Marchetti S-55X seaplanes in mass flight from Rome to Chicago and return.

*July 9–December 19:* Col. and Mrs. Charles A. Lindbergh made 29,000-mile survey flight in their Cyclone-powered *Sirius* seaplane from New York to Labrador, Greenland, Iceland, Europe, Russia, the Azores, Africa, Brazil, and return.

*July 15–22:* Lockheed Vega, *Winnie Mae*, piloted in first round-the-world solo flight by Wiley Post, 15,596 miles in 7 days 18 hours 49½ minutes. Airplane contained new type of radiocompass developed by Wright Field engineers.

*During July:* Douglas DC-1 first flew, forerunner of the famed DC-3.

*August 17:* First Soviet liquid-propellant rocket successfully fired.

*September 30:* Russian stratosphere flight in Army balloon *USSR* attained a reported altitude of 60,695 feet, G. Prokofiev, K. Godunov, and E. Birnbaum as balloonists.

*November 20–21:* Lt. Comdr. T. G. W. Settle (USN) and Maj. Chester L. Fordney (USMC) set official world balloon altitude record of 61,237 feet over Akron, Ohio.

*During 1933:* Collier Trophy for 1933 awarded to Hamilton Standard Propeller Co., with particular credit to Frank W. Caldwell, chief engineer, for development of a controllable-pitch propeller now in general use. (See Appendix D.)

——: NACA assisted Army, Navy, and industry in the development of reliable retractable landing gears, controllable pitch propellers, more efficient wing sections, and wing flaps.

——: Harry W. Bull of Syracuse, N.Y., developed small liquid-propellant rocket engine.

——: Fred E. Weick and his associates at NACA's Langley Laboratory designed and constructed the Weick W-1 airplane which incorporated such novel features as tricycle landing gear, pusher propeller, and interconnected ailerons and rudder for simpler and safer flying.

——: Eugen Sänger of Germany published his classic *Rakatenflugtechnik*, which dealt with rocket motor design and high-speed flight in the atmosphere.

——: British Interplanetary Society organized.

# 1934

*January 10–11:* Six Navy Consolidated P2Y–1's flew nonstop from San Francisco to Pearl Harbor, Hawaii, 2,399 miles, in 24 hours 56 minutes.

*January 30:* Russian balloon reached 73,000 feet, but aeronauts Felosienko, Wasienko, and Vsyskin perished in free fall of gondola.

*February 19:* Under Presidential order the Army Air Corps started flying domestic airmail.

*During February:* Lockheed Electra first flew, featuring introduction of twin fins and rudders.

*April 6:* American Interplanetary Society renamed the American Rocket Society (ARS).

*April 11:* Comdr. Renato Donati established altitude record of 47,352 feet in Caproni aircraft, at Rome, Italy.

*April 18:* Baker Board, appointed by the Secretary of War to investigate the Army Air Corps, held its first meeting.

*May 1:* Lt. Frank Akers (USN) made hooded blind landing in an OJ–2 at College Park, Md., in demonstration of system intended for aircraft carrier use. In subsequent flights, he made takeoffs and landings between Anacostia and College Park under a hood without assistance.

*June 12:* Air Mail Act of 1934 signed by the President.

*During June:* Baker Board recommended purchase of War Department aircraft from private manufacturers, instead of building them in Government factories, by means of negotiated contract, by competitive bids, or by purchase after design competition.

*July 1:* Name of the Aeronautics Branch changed to the Bureau of Air Commerce in the Department of Commerce.

*July 24:* Air Corps began aerial photographic survey of Alaska under Lt. Col. H. H. Arnold.

*July 28:* A 60,613-foot altitude was reached in Air Corps-National Geographic Society balloon, *Explorer I*, by Maj. W. E. Kepner and Capts. A. W. Stevens and Orvil A. Anderson.

*August 18:* Jeanette and Jean Piccard flew *Century of Progress* balloon from Dearborn, Mich., to an altitude of 57,579 feet.

*During August:* Langley Memorial Aeronautical Laboratory expanded with completion of engine research laboratory, a vertical tunnel for testing spinning characteristics, and a 24-inch high-speed tunnel (700 mph).

*September 9:* ARS Rocket No. 4 launched to 400 feet altitude, at Marine Island, Staten Island, N.Y.

*September 15:* Aeromedical Laboratory founded at Wright Field, Dayton, Ohio.

## 1934—Continued

*November 18:* Navy issued contract to Northrop for the XBT-1, a two-seat scout and 1,000-pound bomb dive bomber, initial prototype of sequence that led to the SBD Dauntless series of dive bombers introduced to the fleet in 1938 and used throughout World War II.

*December 23:* Endowment given IAS by Sylvanus Albert Reed for annual award to be given "for notable contribution to the aeronautical sciences resulting from experimental or theoretical investigations, the beneficial influence of which on the development of practical aeronautics is apparent." (See Appendix D.)

*During December:* German Ordance group launch two A-2 rockets successfully to a height of 1.4 miles, on the Island of Borkum in the North Sea, before the C-in-C of the Army.

———: British War Office considered development of high-velocity rockets, and the Research Department at Woolwich Arsenal was requested to submit a program in April 1935. This led to antiaircraft rocket development, and some 2,500 test firings were made in Jamaica, 1938–39.

*During 1934:* Douglas began development of the twin-engined commercial transport, the famed DC-3.

———: H. G. Armstrong began studies on decompression sickness and showed that gas bubbles may form in the body from a drop of pressure below one atmosphere, at Aero Medical Laboratory.

# 1935

*January 5:* First assignment of a flight surgeon to Naval Aircraft Factory, Lt. Comdr. J. R. Poppen (USN), was directed to observe pilots, conduct physical examinations, and work on hygienic and physiological aspects of research and development projects.

*January 22:* Federal Aviation Commission, appointed by the President as provided in the Air Mail Act of June 12, 1934, submitted its report and set forth broad policy on all phases of aviation and the relation of Government thereto. It recommended strengthening of commercial and civil aviation, expansion of airport facilities, and establishment of more realistic procurement practices from industry. It recommended continued study of air organization toward more effective utilization and closer interagency relationships, to include expansion of experimental and development work and its close coordination with the NACA.

*February 12:* Navy dirigible *Macon* crashed at sea off the California coast.

*March 1:* GHQ Air Force established by the Army Air Corps.

*March 9:* Hermann Goering announced the existence of the German Air Force to Ward Price, correspondent of the *Daily Mail* (London), an event of considerable importance in international power politics for it implied unilateral breaking of the Treaty of Versailles prohibiting Germany possession of an air force.

*March 28:* Robert Goddard launched the first rocket equipped with gyroscopic controls, which attained a height of 4,800 feet, a horizontal distance of 13,000 feet, and a speed of 550 mph, near Roswell, N. Mex.

*April 2:* British Government disclosed that Adolf Hitler of Germany had declared that the German Air Force had reached parity with the Royal Air Force at a recent conference with British representatives in Germany. While untrue, Hitler's statement had a profound impact upon British aeronautical and defense efforts.

*April 16–23:* Pan American Airways' *Clipper* flew from California to Honolulu and returned in preliminary survey flight for transpacific air route to the Orient.

*May 18:* World's largest airplane, the Russian *Maxim Gorky,* crashed near Moscow, killing all aboard.

*May 31:* Goddard rocket attained altitude of 7,500 feet in New Mexico.

*July 2:* Historic report on radio direction finding (radar) was presented to the British Air Defense Research Committee.

———: First Interdepartmental Committee appointed by President Roosevelt to study international air transportation problems.

*July 26:* Russian balloon *USSR* successfully reached 52,000 feet, crew including Warigo, Christofil, and Prelucki.

*July 28:* Boeing Model 299, the XB–17 four-engine bomber prototype, made first flight.

*Summer 1935:* First static tests of Heinkel He-112 with rocket engines performed in Germany.

*August 28:* Automatic radio-navigation equipment—a Sperry automatic pilot mechanically linked to a standard radio-compass—tested by the Equipment Laboratory at Wright Field.

*October 30:* First B–17 prototype crashed on takeoff during flight testing at Wright Field.

*November 6:* Prototype Hawker *Hurricane* first flown, the later models of which destroyed more German aircraft in the Battle of Britain than all other British defenses, air and ground, combined.

*November 11:* A 72,395-foot world altitude record for manned balloons made by Capts. A. W. Stevens and Orvil A. Anderson, in the helium-inflated *Explorer II*, over Rapid City, S. Dak., in cooperation with National Geographic Society, a record which stood for 20 years.

*November 22–29:* Transpacific airmail flight by Pan American Airways Martin *China Clipper,* from San Francisco to Honolulu, Midway, Wake, Guam, and Manila, E. C. Musick as pilot.

*During December:* Douglas DC–3, one of the most successful airliners in history, first flew. By 1938, it carried the bulk of American air traffic. When production of the DC–3 and its derivatives ended in 1945, some 13,000 had been built.

*During 1935:* Russian liquid-propellant meteorological rocket, designed by M. K. Tikhonravov, successfully flown.

———: H. G. Armstrong published Air Corps Technical Report on physiologic requirements of sealed high-altitude aircraft compartments (including effects of sudden decompressions), findings which were incorporated in the XC–35 substratosphere plane, the first successful pressure-cabin aircraft.

———: Konstantin E. Ziolkovsky, Russian mathematician and pioneer space scientist, died at 78 years of age. The U.S.S.R. later acclaimed him as the "father of space travel."

# 1936

*January 20:* Acting in response to a request from BuAer, the Navy Bureau of Engineering endorsed support for the National Bureau of Standards for the development of radio meteorographs. Later renamed radiosondes, these instruments were sent aloft on free balloons to measure pressure, temperature, and humidity of the upper atmosphere, and transmitted these data to ground stations for use in weather forecasting and flight planning.

*February 23:* F. W. Kessler, W. Ley, and N. Carver launched two mail-carrying "rocket airplanes" at Greenwood Lake, N.Y., which traveled about 1,000 feet.

*During February:* Germans tested A–3 rocket with 3,300-pound thrust which served as basis for military weapon specifications.

*March 5:* Spitfire prototype with armament and Merlin engine first flown, pro-

## 1936—Continued

duction of the Spitfire Mark I beginning at Supermarine factory in early 1937. Spitfire's classic design was work of R. J. Mitchell, responsible for the Supermarine racing seaplanes which first won the Schneider Trophy for Great Britain in 1931. 18,298 Merlin-engined Spitfires of all Marks were built by 1945.

*March 16:* Robert H. Goddard's classic report on "Liquid Propellant Rocket Development," reviewing his liquid-fuel rocket research and flight testing since 1919, was published by the Smithsonian Institution.

*April 29:* Orville Wright was elected a member of the National Academy of Sciences.

*May 6:* Construction authorized for what later was named the David W. Taylor Model Basin, to provide a facility for use by the Navy Bureau of Construction and Repair in investigating and determining shapes and forms to be adopted for U.S. naval vessels, and including aircraft.

*May 9:* George W. Lewis, Director of NACA Aeronautical Research, received Daniel Guggenheim Medal for 1936 for direction of aeronautical research and for the development of original equipment and methods.

*May 12:* World's largest high-speed wind tunnel (8-foot throat) placed in operation at Langley Aeronautical Laboratory, under Russell G. Robinson.

*May 22:* Herrick *Vertiplane*, embodying characteristics of both airplanes and autogiros, underwent tests at Floyd Bennett Field.

*June 6:* Socony-Vacuum Oil Co., Inc., at Paulsboro, N.J., began production of aviation gasoline (100 octane) by the catalytic cracking method.

*June 7:* Maj. Ira C. Eaker (AAC) made first transcontinental blind flight, from New York to Los Angeles.

*June 15:* Vickers Wellington prototype RAF bomber made its first flight, while flight of first production model was made on December 23, 1937.

*July 18:* Spanish Civil War began, which was to involve German, Italian, and Russian air units as well as aircraft of France and the United States.

*July 21:* Lt. Comdr. D. S. Fahrney (USN) ordered to implement recommendation made to Chief of Naval Operations to develop radio-controlled aircraft for use as aerial targets. Reporting to BuAer and NRL, Fahrney subsequently reported on procedure to obtain drone target planes, but also recognized the feasibility of using such aircraft as guided missiles.

*July 23:* Navy awarded contract for XPB2Y-1 flying boat to Consolidated, which became the prototype for four-engined flying boats used throughout World War II.

*September 2:* Maj. Alexander P. de Seversky was refused permission by Army Air Corps to enter his pursuit plane in Bendix Trophy Race to Los Angeles "due to features considered a military secret."

*October 13:* Lt. John Sessums (AAC) visited Robert H. Goddard to officially assess military value of Goddard's work. He reported that there was little military value, but that rockets would appear useful to drive turbines.

*October 24:* First transpacific passenger service completed by Pan American Airways, with Martin four-engined *China Clipper* in a round trip to Manila.

*November 7:* Robert Goddard flew gyrocontrolled rocket to 7,500-foot altitude, near Roswell, N. Mex.

*December 19:* New world speed record for amphibians of 209.4 mph over a closed course set by Maj. Alexander P. de Seversky.

*During 1936:* Theodore von Kármán, Director of the Guggenheim Aeronautical Laboratory at the California Institute of Technology at Pasadena, founded group which began experiments in design fundamentals of high-altitude sounding rocket. The group, named the Cal Tech Rocket Research Project, consisted of Frank J. Malina, Hsue-Shen Tsien, A. M. O. Smith, John W. Parsons, Edward Forman, and Weld Arnold. This was the origin of the Jet Propulsion Laboratory.

———: First practical helicopter flight, German Focke-Achgelis, FA-61; in the following year it made first helicopter flight of over 1 hour.

# 1937

*January 1:* First physiological research laboratory completed at Wright Field by Air Corps to investigate and devise means to alleviate distressing symptoms occurring in flight.

*March 1:* First operational Boeing B-17 delivered to the GHQ Air Force at Langley Field, Va.

*During Spring:* Single-engine Heinkel (He-112) with Junker 650-pound thrust, liquid-fuel rocket motor successfully flown at Neuhardenberg, Germany, Capt. Erich Warsitz as pilot.

*April 12:* Frank Whittle's first gas turbine engine, the U-type, was static tested.

*May 6:* German dirigible *Hindenburg* destroyed at Lakehurst, N.J., an event which ordained the death of the large dirigibles.

*May 9:* H. F. Pierce launched liquid propellant rocket to 250-foot altitude at Old Ferris Point, N.Y.

*During May:* Joint German Army-Air Force rocket research station opened at Peenemünde on Baltic Sea; Army Ordnance rocket program under Capt. Walter Dornberger moved his staff from Kummersdorf.

*June 30:* Navy issued contract to Martin for XPBM-1 two-engine flying boat, the initial prototype for the PBM Mariner series used during and after World War II.

*July 1:* Weather Service of the Signal Corps was transferred to the Army Air Corps.

*July 2:* Amelia Earhart Putnam and copilot lost near Howland Island in the Pacific.

*July 4:* FA-61 helicopter flown in fully controlled, free flight by Hanna Reitsch, at Bremen, Germany.

*July 5-6:* PAA and Imperial Airways make joint survey flights across the North Atlantic prior to establishment of transatlantic service. Both flights were successful, marking the 11th and 12th successful nonstop transatlantic flights completed out of 85 attempts.

*July 15:* Three Soviet fliers established world distance nonstop record, flying across the North Pole from Moscow to San Jacinto, Calif., in 62 hours.

*July 27:* Japanese began aerial bombing of Chinese cities.

*August 5:* First experimental pressurized-cabin airplane, a Lockheed XC-35, made first flight at Wright Field.

*August 23:* The first wholly automatic landings in history were made at Wright Field by Capt. Carl J. Crane, inventor of the system; Capt. George Holloman, pilot; and Raymond K. Stout, project engineer.

*October 15:* Boeing XB-15 made first flight.

*During November:* Low turbulence wind tunnel for investigation of laminar flow airfoil constructed at NACA's Langley Memorial Aeronautical Laboratory.

———: Navy Grumman F4F made first test flight, standard carrier-based fighter in early World War II operations.

*December 23:* Successful unmanned radio-controlled flight made by Navy JH-1 drone, at Coast Guard Air Station, Cape May, N.J.

*During December:* Initial rocket thrust chamber tests by R. C. Truax at Annapolis, Md., using compressed air and gasoline as fuels.

*During 1937:* World's scheduled airlines carried 2,500,000 passengers in 1937, with average number of 5.3 passengers per aircraft, according to the International Civil Aviation Organization.

———: U.S.S.R. established rocket test centers at Kazan, Moscow, and Leningrad.

# 1938

*January 16:* Spanish rebel planes began daily bombing of Barcelona from Majorca.

*February 10:* British Hurricane fighter flown from Edinburgh to Northolt, near London, at an average speed of 408.75 mph, J. W. Gillan as pilot.

*February 26:* Secretary of Interior Ickes approved purchase by the Federal Government of helium plants at Dexter, Kans., thus giving the Government a virtual monopoly. On May 11, his refusal to sell helium to Germany was upheld by the President.

*February 27:* The good-will flight to Buenos Aires of six B-17's under Lt. Col. Robert D. Olds, which had left Miami on February 17, returned to Langley Field, Va.

*April 21:* Navy delivered XF2A-1 to Langley Memorial Aeronautical Laboratory of NACA, which marked initiation of full-scale wind tunnel tests, which resulted in increasing speed of the XF2A-1 by 31 mph and led to utilization of NACA testing of other high performance aircraft by both the Army and the Navy. Data thus obtained were also directly applicable to the design of new aircraft.

*June 1:* Routine use of radiosondes initiated at NAS Anacostia, Washington, D.C. By the end of the year the balloon-carried radio meteorographs were also used in Navy fleet operations.

*June 6:* The Daniel Guggenheim Medal for 1938 awarded to A. H. R. Fedden for "contributions to the development of aircraft engine design and for the specific design of the sleeve valve aircraft engine."

*June 9:* British Government announced intention to purchase U.S. Lockheed Hudsons and North American Harvards for aerial reconnaissance and training purposes.

*June 23:* President Roosevelt signed the Civil Air Authority Act.

*August 22:* The Civil Aeronautics Act became effective, coordinating all nonmilitary aviation under the Civil Aeronautics Authority.

*August 24:* First American use of drone target aircraft in antiaircraft exercises, the *Ranger* fired upon a radio-controlled JH-1 making simulated horizontal bombing attack on the fleet.

*August 29:* Maj. Alexander de Seversky set east-west transcontinental speed record of 10 hours 2 minutes 55.7 seconds in a 2,457-mile flight.

*September 12:* Wind tunnel capable of simulating altitudes to 37,000 feet dedicated at MIT as a memoral to the Wright brothers.

*September 14:* Radio-controlled Navy N2C-2 target drone made simulated dive-bombing attack on battleship *Utah* in test firing of anticraft battery.

*September 29:* Brig. Gen. Henry H. Arnold named Chief of the Army Air Corps to replace Maj. Gen. O. Westover, killed in crash on September 21.

*September 30:* Agreement signed at Munich, Germany, between Germany, Britain, France, and Italy, allowing Germany to occupy the Sudetenland of Czechoslovakia, an event in which the relative air strength of the major nations was a prominent factor.

*October 19:* Curtiss XP-40 Tomahawk made first flight.

*During October:* British Purchasing Commission ordered 200 Lockheed Hudsons (military version of Super Electra airliner), the first American-built aircraft to see operational service with the RAF in World War II.

———: All-wood British de Havilland Mosquito twin-engine bomber conceived, official order for 50 received on March 1, 1940.

*December 10:* First static test of James Wyld's regeneratively cooled rocket thrust chambers, which achieved 90-pound thrust.

——: ARS tested R. C. Truax's rocket thrust chamber at New Rochelle, N.Y., which achieved 20-pound thrust before burning through.

*December 16:* First successful test of NACA high-speed motion-picture camera developed by C. D. Miller, conducted at Langley Laboratory, later used extensively in photographic analysis of combustion and operated up to rates of 40,000 photographs per second.

——: Navy K-2 airship delivered to NAS Lakehurst for trials, the prototype for World War II K Class patrol airships, of which 135 were procured.

*December 17:* Dr. Hugh L. Dryden, National Bureau of Standards, delivered second Wright Brothers Lecture at Columbia University.

*December 30:* Special Committee on "Future Research Facilities of NACA" recomended creation of another laboratory; resulted in Ames Aeronautical Laboratory at Moffett Field.

*During 1938:* Jack Parsons of Cal Tech conceived value of slow-burning rocket propellant of constant thrust for JATO use, active development of which was undertaken by Cal Tech in 1940.

——: Vital importance of the factor of duration in pilot's exposure to hypoxia demonstrated in animal experiments by H. G. Armstrong and J. W. Heim.

——: Heinz von Diringshofen, German scientist, conducted research on human tolerance to multiple g-loads; exposed test subjects to a few seconds of subgravity by putting an aircraft through a vertical dive.

*1938–39:* NACA developed airfoils providing laminar flow to a degree far greater than previously obtainable (based in part upon Ludwig Prandtl's boundary layer theory in NACA Report 116 published in 1921); Eastman N. Jacobs developed low-drag wing sections worthy of special mention.

# 1939

*January 16:* Maj. Gen. Frank M. Andrews, Chief of Army General Headquarters Air Force, in an address to the annual convention of the National Aeronautic Association at St. Louis, said that the United States was a fifth- or sixth-rate air power.

*January 21:* Dr. George W. Lewis, NACA Director of Aeronautical Research, elected president of the IAS.

*January 31:* Dr. Edward P. Warner appointed economic and technical adviser of the CAA.

*February 11:* Lockheed P-38 Lightning first flown across the Nation from California, to a crack-up landing at Mitchel Field, Long Island, Lt. Ben Kelsey as pilot.

*During February:* Airflow Research Staff at Langley Laboratory initiated reevaluation of jet propulsion for aircraft at speeds higher than considered by Buckingham in NACA Report No. 159 published in 1923.

*March 26:* Capt. John H. Towers named Chief of Bureau of Aeronautics with rank of Rear Admiral.

*April 3:* President Roosevelt signed the National Defense Act of 1940, authorizing 6,000 airplanes and increasing personnel of Army Air Corps to 3,203 officers and 45,000 enlisted men, and appropriating $300 million for the Air Corps.

*April 7:* Amphibian version of PBY flying boat ordered by the Navy from Consolidated.

*April 20:* The free-flight tunnel placed into operation at Langley Aeronautical Laboratory.

*April 20–21:* Experiments with four-bladed controllable propeller on Curtiss P-36 begun at Wright Field.

*April 28:* Flying a Messerschmitt BF-109R, Fritz Wendel achieved record speed of 468.9 mph in level flight, at Augsburg, Germany.

## 1939—Continued

*May 5:* Kilner-Lindbergh Board was established by Gen. H. H. Arnold to revise military characteristics of all U.S. military aircraft, including the B-29 design in the AAF 5-year program. The Board was composed of Gen. W. C. Kilner, Charles A. Lindbergh, Cols. Carl Spaatz and Naiden, and Major Lyon.

*May 15:* Navy issued contract to Curtiss Wright for the SXB2C-1 dive bomber, which despite prolonged operational development became the principal carrier dive bomber in the last year of World War II known as the *Helldiver*.

*During June:* First transatlantic passenger service, by Pan American Airways with a Boeing four-engined *Yankee Clipper*.

*July 1:* National Academy of Sciences sponsored a $10,000 research program at Cal Tech Rocket Research Project for development of rockets suitable to assist Air Corps planes in takeoffs, the first U.S. rocket program.

*During Summer:* Albert Einstein, Enrico Fermi, Leo Szilard, and Eugene Wigner interested President Roosevelt, through Alexander Sachs, in the potential military importance of uranium. The President appointed an Advisory Committee on Uranium under the chairmanship of Dr. Lyman Briggs, Director of the National Bureau of Standards.

——: Total complement of NACA was 523 persons, of which only 278 were classified as technical personnel.

*August 9:* Congress authorized construction of second NACA research station at Moffett Field, Calif., which became the Ames Aeronautical Laboratory, named after Joseph S. Ames, president emeritus of Johns Hopkins University, member of the NACA from its beginning in 1915 to 1939, and Chairman of NACA from 1927 until 1939.

*August 24:* Assignment of Navy medical officer to BuAer was approved for the purpose of establishing an Aviation Medical Research Unit.

*August 27:* First complete flights of jet-propelled aircraft made secretly in Germany, a Heinkel 178 powered by the He S-3B jet engine, piloted by Erich Warsitz.

*September 1:* German blitzkrieg launched on Poland. President Roosevelt appealed to the European nations not to bomb civilian populations or unfortified cities.

*September 3–4:* RAF Bomber Command carried out first night propaganda raid, dropping leaflets over Hamburg, Bremen, and the Ruhr. On September 27, British Air Ministry announced that the RAF had dropped 18 million leaflets over Germany since the beginning of the war. When leaflet bombing was suspended on April 6, 1940, Bomber Command had dropped 65 million leaflets.

*During September:* Igor I. Sikorsky made initial flights with the first successful single-main-rotor helicopter, precursor of the R-4 two-place design procured in 1942 by the AAF.

——: World's largest balloon, the *Star of Poland*, was unable to make stratospheric flight because of the German invasion. The United States had provided helium gas in August for this Polish effort and several American experts, including A. W. Stevens, provided technical assistance.

*October 14:* Naval Aircraft Factory authorized to develop radio-control equipment for use in remote-controlled flight testing of aircraft without risking the life of a test pilot.

*October 19:* Dr. Vannevar Bush was elected Chairman of the NACA to fill the post of Dr. Joseph Ames, who resigned due to ill health.

——: Second Special Committee on "Future Research Facilities of NACA," headed by Charles A. Lindbergh, recommended that a powerplant research center be established at once, a recommendation resulting in the Aircraft Engine Research Laboratory at Cleveland, Ohio, now the Lewis Research Center.

*During October:* Germans successfully fired and recovered A-5 development rockets with gyroscopic controls and parachutes, attaining altitude of 7½ miles and a range of 11 miles.

*November 20:* Navy established its own School of Aviation Medicine at Pensacola, Fla., having previously detailed officers to the Air Corps School of Aviation Medicine.

*November 30:* U.S.S.R. invaded Finland, Soviet planes bombing Helsinki and other Finnish towns.

*December 2:* Army Air Corps authorized to begin development of a four-engine bomber with a 2,000-mile radius of action, which led to the Boeing B-29 Superfortress.

*December 29:* Consolidated-Vultee B-24 Liberator made first flight at San Diego.

*During 1939:* P-41 with R-1830 engine was provided by NACA's Pinkel, Turner, and Voss with separate stacks for each cylinder, thus providing 14 jet exhausts which increased speed of aircraft from 13 to 18 mph between 10,000 and 20,000 feet. Applied to A-20 later, an increase of 45 mph was attained.

———: Curtiss P-40 fighter powered with Allison V-1710-33, with top speed of 357 mph, first ordered in quantity.

*During 1939:* Basic concepts for NACA's combined loads testing machine were proposed by E. E. Lundquist and J. N. Kotanchik of Langley Laboratory. After refinements by others, construction was started in 1940 and much testing performed before completion and operation of the fixed-component machine in 1949. The combined loads testing machine was the first capable of applying positive and negative forces along each of three axes, and positive and negative moments about these axes, in any combination of forces and moments, each applied independently. Still in use in 1960, this machine was used extensively on combined loads and moments on shell-type structures for all types of flight vehicles.

*During 1939-40:* Original design of North American B-25 Mitchell bomber required 200,000 engineering man-hours; later wartime modification of this airplane (9,800 completed by end of 1945) accounted for a total of more than 4,830,000 engineering man-hours.

# 1940

*January 19:* Maj. James H. Doolittle elected president of the IAS.

*February 1:* Capt. G. E. Price flew Bell Airacobra through flight tests.

*February 24:* BuAer issued contract for airborne television equipment capable for use in transmitting instrument readings obtained from radio-controlled flight tests, and for providing target and guidance data should radio-controlled aircraft be converted to guided missiles.

*February 27:* Based upon research of former NACA engineer, Charles H. Zimmerman, Navy initiated development of the Flying Flapjack with award of contract to Vought-Sikorsky for design of the VS-173. Design promised high speed with low takeoff speed.

*February 29:* Navy BuAer initiated steps that led to contract with H. O. Croft, State University of Iowa, to investigate the possibilities of a turbojet propulsion unit for aircraft.

*March 9:* Beechcraft AD-17 biplane flown to altitude of 21,050 feet over the Antarctic to measure cosmic rays for the U.S. Antarctic Expedition, piloted by T. Sgt. T. A. Petras (USMC).

*March 16:* First civilian casualties in Britain due to air raids, during Luftwaffe attack on Scapa Flow.

*March 22:* Naval Aircraft Factory established project for adapting radio controls to a torpedo-carrying TG-2 airplane.

*March 26:* U.S. commercial airlines completed a full year without a fatal accident or serious injury to a passenger or crew member.

*During April:* British commission gave North American Aircraft 120 days to produce fighter prototype to specifications, which resulted in the highly successful P-51 Mustang, the first aircraft to utilize the NACA low-drag wing based on prolongation of laminar flow. Low-turbulence wind tunnel tests (completed in

## 1940—Continued

1938) had led to five different families of low-drag wings by the end of 1939.

*May 14:* German Luftwaffe bombed Rotterdam.

*May 15–16:* First large-scale RAF raids on German industrial targets when 93 bombers attacked objectives in the Ruhr.

*May 16:* President Roosevelt called for U.S. production of 50,000 planes a year.

*May 28:* Robert H. Goddard offered all his research data, patents, and facilities for use by the military services at a meeting with representatives of Army Ordnance, Army Air Corps, and Navy Bureau of Aeronautics arranged by Harry Guggenheim. Nothing resulted from this except an expression of possible use of rockets in jet-assisted take-offs of aircraft.

*May 29:* Chance Vought F4U Corsair Navy fighter with inverted full wing made first test flight.

*June 8:* Paris office of the NACA was closed.

*June 26:* Congress authorized construction of the third NACA laboratory near Cleveland, Ohio, which became Aircraft Engine Research Laboratory. In 1948, it was named for George W. Lewis, NACA Director of Aeronautical Research, 1924–47.

*June 27:* National Defense Research Committee (NDRC) created by the Council of National Defense.

*July 8:* First commercial flight of the Boeing 307-B Stratoliner, Burbank, Calif., to Long Island, N.Y., the first commercial flight to use a pressurized cabin, in record time of 12 hours 18 minutes.

*During July:* National Defense Research Committee established Jet Propulsion Research Committee under Section H of Division A, at Naval Powder Factory, Indian Head, Md., to conduct fundamental research on rocket ordnance. C. N. Hickman, who had worked with Goddard during World War I, was named as head.

*August 2:* Beginning of the Battle of Britain, which raged until the end of October.

*During August:* Sir Henry Tizard, scientific adviser to the British Ministry of Aircraft Production, headed mission of leading British and Canadian scientists to brief official American representatives on devices under active development for war use and to enlist the support of American scientists. This was the beginning of very close cooperation of Anglo-American scientists in many fields, including aeronautics and rocketry, and enabled American laboratories to catch up with war-accelerated progress.

*August 20:* Smith J. DeFrance appointed Engineer-in-Charge of the NACA Ames Aeronautical Laboratory, Moffett Field, Calif.

*August 25–26:* First RAF bombing of Berlin.

*During September:* Royal Air Force used AA rockets against Luftwaffe planes in the Battle of Britain.

——: First test firing of NDRC rocket program, at Naval Proving Ground, Dahlgren, Va., a rocket-propelled bomb to pierce 14-inch armor requested by BuOrd.

*November 25:* De Havilland all-wood Mosquito bomber made first flight, large-scale production of which began in July 1941.

*During 1940:* Committee of the National Academy of Sciences reported that operation of turbine wheels at temperatures up to 1,500° F might soon be possible because of U.S. and foreign development of high-temperature alloys.

——: Dr. Heinz von Diringshofen of Berlin, Germany, "discovered" the effect of weightlessness during flight maneuvers with high performance aircraft.

——: N. W. Thorner and F. H. Lewey demonstrated destruction of certain brain cells in experimental animals by short and severe exposures to hypoxia induced by inhalation of pure nitrogen.

——: *Graf Zeppelin I* and *II* were intentionally destroyed by the Germans and their metal used for the Reich war effort.

# 1941

*January 11:* Army Air Corps announced the control of robot planes, either by radio from the ground or from another plane, had been tested successfully.

*During January:* RCA proposed to NDRC design and development of rocket-propelled, radio-controlled aerial torpedo with TV nose, which was given code name "Dragon." The National Bureau of Standards was assigned the task of developing a suitable airframe.

*February 5:* Bureau of Standards developed photoelectric detector to simplify measurement of height of clouds.

*During February:* Army Air Corps initiated development of radio-controlled aerial gliding torpedoes, gliding bombs, and aerial mines.

*March 24:* Classic NACA report prepared by Robert R. Gilruth which provided basis for subsequent aircraft development (NACA Report No. 755, "Requirements for Satisfactory Flying Qualities of Airplanes").

*During March:* NACA established Special Committee on Jet Propulsion to review early British reports on the Whittle engine, which subsequently aided development of TG-100 turboprop engine by GE and the 19-B turbojet by Westinghouse. Dr. W. F. Durand was called out of retirement to head this committee.

*April 15:* Igor Sikorsky piloted a Vought-Sikorsky in the first officially recorded single-rotor helicopter flight longer than an hour in the Western Hemisphere; flying time, 1 hour 5 minutes 14.5 seconds; at Stratford, Conn.

*April 19:* Naval Aircraft Factory initiated development of a Glomb (glider bomb), to be towed long distances by powered aircraft and released over target and guided by radio control and target-viewing television.

*May 15:* First official flight of British turbojet, Gloster E28/39 with Whittle WIX jet engine, at Cranwell, England, flown by Flight Lt. Sayer for about 17 minutes.

*——:* British De Havilland Mosquito equipped as night fighter (W4052) made its first flight with AI radar.

*May 21:* Army Corps Ferrying Command, forerunner of AAF's Air Transport Command, was created. By V-E Day it possessed 2,461 transports, of which 798 were 4 engined.

*——:* Navy Engineering Experiment Station, Annapolis, Md., directed to undertake development of liquid-fuel rocket JATO for large flying boats.

*May 29:* Naval Powder Factory, Indian Head, developed and successfully tested 4.5-inch AA rocket.

*During May:* Republic XP-47 Thunderbolt made first flight.

*May–June:* First satisfactory spark plugs (ceramic insulated) for high-performance U.S. aircraft engines such as the P&W R-2800 were ordered in mass quantities. Plugs were developed under direction of T. T. Neill, Air Corps ignition engineer at Wright Field.

*June 20:* Establishment of the U.S. Army Air Forces (AAF), comprising the Office of the Chief of Air Corps and the Air Force Combat Command (formerly GHQ Air Force), with Maj. Gen. H. H. Arnold as Chief.

*June 22:* U.S.S.R. was attacked by Germany.

*——:* Ceramic-lined rocket thrust chamber designed by Alfred Africano generated 260-pound thrust.

*June 28:* Office of Scientific Research and Development (OSRD) in the Office of Emergency Management was created by President Roosevelt in Executive Order 8807.

*June 30:* Joint Army-Navy project contract given Northrop for design of an aircraft gas turbine developing 2,500 hp at a weight of less than 3,215 pounds.

*During June:* Col. Donald J. Keirn of Wright Field sent to England to study Gloster jet aircraft and its Whittle-I

## 1941—Continued

engine. AAF decision to produce Whittle engine made in September, and the XP-59 flew a year later.

*July 1:* First commercial television broadcast over WNBT, New York (first successful demonstration by C. F. Jenkens in United States and J. L. Baird in England was made in early 1920's).

*July 16:* Full-scale wind-tunnel tests of A-1 "power-driven controllable bomb" conducted at Langley Field.

*July 24:* Dr. Jerome C. Hunsaker was elected Chairman of the NACA and Chairman of its Executive Committee.

*During July:* Navy initiated development of Mousetrap, ship-based 7.2-inch mortar-fired bomb which became first USN rocket placed into fleet action in May 1942.

———: First successful U.S. jet-assisted takeoff accomplished in an Ercoupe at March Field by Lt. Homer A. Boushey (AAF), with pressed-powder propellant JATO rockets developed by Cal Tech.

———: Project TED (EES 3401) established at Naval Engineering Experiment Station at Annapolis by BuAer.

*August 1:* President Roosevelt prohibited export of aviation fuel outside of the Western Hemisphere, except to Britain and countries resisting aggression, an act aimed at Japan which normally imported large quantities from the United States.

———: NDL was requested to develop radar guidance equipment for assault drones, both to relay target information to a control operator and to serve as an automatic homing device.

———: Three successful tests of J. Wyld liquid fuel rocket motor were made at average thrust of 125 pounds. A year later, ARS members formed Reaction Motors, Inc., to continue development of this design.

*August 12:* Ercoupe impelled by 12 powder rockets of 50 pounds thrust each, piloted by Lt. Homer A. Boushey, first flew on rocket power alone after an initial boost from a towing automobile.

*August 19:* President Roosevelt announced that Pan American Airways would establish a ferry service to fly American aircraft to the RAF in the Middle East.

*During August:* Caproni-Campini jet-propelled plane, conventional engine with ducted fan, produced and test flown in Italy.

*During September:* Messerschmitt Me-163A powered by "cold" H. Walther rocket successfully flown at Augsburg, Germany, development of which had begun in 1937, but "cold" engine proved unreliable. Flights were also made in October which reached speeds of 1,003 km/hr, or Mach 0.85.

*During September:* Dr. Robert H. Goddard began work on liquid-propellant JATO under contract to USN and AAF, delivering a device to both agencies on September 1942.

*October 27:* Post of Air Surgeon was created within the Army Air Forces.

*During October:* Harriman mission made globe-circling flight of 24,700 miles from Washington to Moscow and return in B-24 bomber.

*November 7:* First flight of the AAF GB-1 guided glide bomb, containing preset guidance.

*November 12:* First launching of an experimental GB-8 glide bomb, incorporating radio controls.

*November 30:* Italian jet-propelled Caproni-Campini airplane flown 475 kilometers in 2 hours 11 minutes from Turin to Rome, by Mario de Barnardini.

*November–December:* Russians used AA rockets against Luftwaffe aircraft in defense of Moscow and air-to-air rockets on their Stormovik Il-2 fighters.

*December 7:* Japanese naval air units attacked Pearl Harbor.

*December 30:* USAAF requested NRDC to undertake development of controlled-trajectory bombs, the beginning of the development of Azon.

*During 1941:* Navy Bureau of Aeronautics created JATO section to accelerate USN development.

*During 1941:* Aeromedical Laboratory, in collaboration with Dr. E. A. Hooten of Harvard University, initiated anthropometric surveys of AAF flyers to facilitate design of weapons and flying gear.

*During 1941:* Research facilities at NACA's Langley and Ames Laboratories increased 100 percent over previous years by the construction of new facilities for defense application.

# 1942

*January 13:* Sikorsky XR-4, single-rotary wing, two-man helicopter, made its first successful flight.

*During January:* P-38 first placed under study of NACA Langley Laboratory to assess flow changes due to compressibility, later transferred to Ames Laboratory. Dive-recovery flap developed later applied to P-47, XP-59, F-80, and FR-1.

———: "Frigitorium" for cold testing aircraft equipment for arctic operations became operational at Wright Field.

*During February:* Douglas DC-4 Skymaster first flew, becoming prominent in the generation of four-engined American transports which revolutionized long-haul air transportation.

*April 7-24:* Douglas A-20A completed 44 successive takeoffs using liquid-propellant JATO developed by Cal Tech's Frank S. Malina.

*April 9:* Radio-controlled TG-2 Navy drone made torpedo attack on destroyer *Aaron Ward* in which television camera mounted in the drone was utilized, directed by control pilot Lt. M. B. Taylor of Project Fox.

*April 18:* First American raid on Tokyo, by 16 North American B-25 AAF medium bombers flown off carrier *Hornet*, led by Lt. Col. James H. Doolittle.

*April 19:* Two feasibility tests using drone aircraft conducted by Navy in Chesapeake Bay, the most successful being Project Fox BG-2 drone equipped with target-viewing TV camera, which was crash dived into a moving raft while under an airborne control pilot 11 miles away.

*May 8:* Research begun at the NACA Aircraft Engine Research Laboratory at Cleveland.

*May 26:* Jet-assisted takeoff of a Brewster F2A-3 using five British antiaircraft solid-propellant rockets demonstrated at NAS Anacostia, Comdr. C. Fink Fischer as pilot.

*May 30-31:* First 1,000-plane raid by RAF Bomber Command on Cologne, Germany.

*June 13:* First test of the German A-4 (V-2) rocket unsuccessful at Peenemünde, Germany.

*June 17:* National Defense Research Committee initiated development of an antisubmarine guided missile, the Pelican, under Navy BuOrd, which was a glide bomb with radar homing guidance.

*June 27:* Naval Aircraft Factory was directed to participate in development of high-altitude pressure flying suits, thus joining Army which had sponsored earlier work.

*June 30:* Brig. Gen. James H. Doolittle awarded the 1942 Guggenheim Medal "for notable achievement in the advancement of aeronautics."

*During June:* Joint Committee on New Weapons and Equipment (JNW) appointed subcommittee to review all guided-missile programs, out of which came the placement of responsibility for all controlled missiles in Division 5, Missiles, in the National Defense Research Council. Division 5 of NDRC served as principal agency outside the military services involved in U.S. missile development for the remainder of World War II.

*July 3:* First airborne test firing of a retrorocket at Goldstone Lake, Calif., from a PBY-5A piloted by Lt. Comdr. J. H. Hean (USN).

*July 6:* 4.5-inch rocket (M8-type) fired for first time in flight from a P-40.

## 1942—Continued

*July 18:* German Me-262 turbojet fighter flown on spectacular flight test, concluding a series begun in May.

*During July:* First U.S.-designed jet engine successfully demonstrated at Langley Laboratory, the NACA Jeep, which was never flown but proved invaluable for continued NACA research on gas-turbine jet propulsion.

——: 9-inch supersonic tunnel providing airspeeds up to mach 2.5 put into operation at Langley Memorial Aeronautical Laboratory.

*September 21:* Boeing XB-29 Superfortress made its first flight, an indispensible aircraft in the Pacific campaign of World War II.

*During September:* After completion of liquid-fuel JATO device for AAF and Navy, Robert H. Goddard worked on liquid-fuel engines of variable thrust while Director of Research in Jet Propulsion at Annapolis until his death in August 1945.

*October 1:* First U.S. jet-propelled aircraft flight, by an Airacomet Bell XP-59A (powered by two I-16 engines developed by General Electric from the British Whittle prototype), made at Muroc Dry Lake, Calif., with Robert Stanley as pilot.

*October 2:* Maj. J. G. Kearby reached an effective, simulated altitude of 60,200 feet in Aeromedical Laboratory altitude chamber at Wright Field, as a part of an investigation of "full pressure" suits.

*October 3:* First successful launch and flight of 5½-ton German A-4 rocket (V-2) at Peenemünde, which traveled 120 miles.

*October 22:* Westinghouse Electric authorized to construct two 19A axial-flow turbojet powerplants, thereby initiating fabrication of first practical jet engine wholly American in design.

*December 2:* First nuclear chain reaction successfully accomplished at the University of Chicago.

*December 5:* Edward R. Sharp was appointed Manager of the NACA Aircraft Engine Research Laboratory at Cleveland.

*During December:* AAF conducted first flight tests of a full-pressure altitude flight suit at Eglin Field, Fla.

*During 1942:* Aerosol bomb for disinsectation of aircraft developed at Aero Medical Laboratory by Lt. William N. Sullivan, subsequently adapted for use in foxholes, bomb shelters, barracks, and other dwellings.

# 1943

*January 8:* First aircraft takeoff in United States with permanently installed JATO rocket powerplant, an A-20A at Muroc Army Air Base, Calif.

*January 27:* First American bomber raid on Germany by USAAF against Wilhelmshaven.

*January 28:* Dr. Hugh L. Dryden, Chief of the Mechanics and Sound Division of the National Bureau of Standards, elected president of the IAS.

*During January:* Lockheed C-69 Constellation first flown, a successful postwar transport with pressurized cabin.

*February 17:* 10th German A-4 (V-2) rocket traveled 121.8 miles after launch from Peenemünde.

*During February:* Navy Engineering Experiment Station Annapolis completed development of rocket engine for Pelican radio-controlled pilotless aircraft (never used operationally).

*March 5:* Fifth prototype of Gloster Meteor, developed from the first British jet aircraft, first flew, powered by Halford H-1 turbojet engines, forerunners of the De Havilland Goblin.

*During March:* First turbojet engine developed from American design by Westinghouse, the X19A, was completed. It was the precursor of the J30, J34, J40, J46, and J54 engines.

*April 2:* Research building of the AAF School of Aviation Medicine opened officially, housing 27 officers and 35 civilian staff members, and 4 altitude decompression chambers.

*April 11:* California Rocket Society tested first hybrid rocket design in United States, using oxygen and carbon.

*April 15:* Prime Minister Winston Churchill of England was informed of reports on German experiments with long-range rockets.

*May 22:* German Messerschmitt Me-262 turbojet fighter prototype flight tested at Rechlin. Test flights continued during the year on interceptor type, while series production did not begin until spring of 1944.

*During May:* A PBY Catalina, fitted with two liquid-propellant JATO rockets developed at Annapolis, took off with 20 percent reduction in run. Liquid-propellant JATO was abandoned by Navy in 1944.

*May–June:* Germans operationally test fired over 100 V-2's from Blizna, Poland, launching 10 on one day, only a small number of which were fully successful.

*June 24:* Lt. Col. W. R. Lovelace, AAF Aeromedical Laboratory, made world record parachute jump from 40,200 feet at Ephrata, Wash.

*Mid-1943:* Navy initiated development of FR "Fireball" fighter, the only U.S. jet-and-propeller-engine fighter produced in any quantity before the end of the war. Developed by Ryan, prototype was accepted in October 1944 and production authorized in December 1944.

*July 5:* First turbojet engine completed for the Navy, the Westinghouse 19A, completed its 100-hour endurance test.

*July 7:* Adolf Hitler gave the German V-2 program highest military priority.

*July 19:* Naval Aircraft Factory authorized to develop the Gorgon, an aerial ram or air-to-air missile powered by a turbojet engine and equipped with radio control and a homing device. The Gorgon was later expanded into a broad program embracing turbojet, ramjet, pulsejet, and rocket propulsion, and a variety of structures and guidance systems.

*July 24–August 3:* German city of Hamburg subjected to a series of massive RAF attacks, totaling 3,000 planes, which exploited first use of "chaff" or "window" to saturate radar early warning and resulting in a severe "fire storm."

*During July:* Naval Air Material Center established at Johnsville, Pa., to include Naval Air Factory, Naval Aircraft Modification Unit, and Naval Air Experimental Station.

———: Serious training of units for field employment of V-2 begun at Peenemünde. In January 1944, operational command of V-2 operations given to Gen. Richard Metz, leaving Gen. Dornberger in charge of V-2 development.

———: Jet propulsion static test laboratory constructed at NACA Laboratory in Cleveland, and full studies of jet propulsion for the Army and Navy were underway by the fall.

*Summer 1943:* Messerschmitt Me-163B rocket interceptor powered by Walther "hot" engine successfully flown at Bremen, Augsburg, and near Leipzig, Germany. Over 300 Me-163B's were produced by Junkers by the end of 1944.

*August 7:* German turbojet fighter, a Messerschmitt Me-262, demonstrated before Adolf Hitler in East Prussia.

*August 17:* AAF Schweinfurt-Regensburg deep-penetration daylight raid by 376 B-17's, with heavy loss of 60 bombers.

*August 17–18:* Royal Air Force attacked Germany's Peenemünde Rocket Research Center, causing heavy damage and delaying V-weapon program by weeks or months.

*During August:* Navy initiated development of McDonnell Phantom XFD-1 fighter, first pure-jet aircraft developed for USN.

## 1943—Continued

*During August:* German aircraft launched first HS-293 radio-controlled glide bomb against British ship in Bay of Biscay, the beginning of guided-missile warfare.

*September 1:* 123,000 airplanes and 349,000 airplane motors were produced in the United States between May 1940 and this date.

*During September:* Rocket Development Branch created in Army Ordnance to direct and coordinate development on rockets.

——: U.S. Bell Airacomet first flown in England at Moreton Valenee, result of exchange for first production model of a Meteor Mk I (EE210) sent to the United States.

*Late September:* Aberdeen Ballistic Research Laboratories' Study entitled "Development of Long-Range Rocket Projectile" was submitted by the Army to the National Defense Research Committee. The project was accepted by NDRC.

*October 2:* First U.S. military rocket-powered airplane, the Rocket Ram, was tested as a glider by John Myers. It was equipped with an Aerojet XCAL-200 engine, using monoethylanline as fuel.

*October 3:* First afterburner for turbojet engines in America, built at NACA Lewis Flight Propulsion Laboratory.

*October 10:* USAAF demonstrated television control of a drone aircraft.

*October 15:* Details of gyro fluxgate compass, giving accurate readings despite violent aircraft movements, made public by Bendix Aviation.

*During October:* Division 5 of NDRC suggested to AAF the appointment of general officer to coordinate entire AAF guided-missile program.

——: Army General Staff created new Weapons Division to coordinate research and development studies and plans among the Army and other divisions.

——: Navy BuOrd established facility for testing rocket motors at Naval Gun Factory, Washington, D.C.

*November 8:* Secretary of the Navy approved Naval Ordnance Test Station to be located on west coast and to be under cognizance of BuOrd. In December, its site was selected at Inyokern, China Lake, Calif.

*November 30:* Department of Aviation Medicine and Physiological Research was authorized at NAMC Philadelphia.

*During November:* Gen. H. H. Arnold, Chief of Air Staff, directed and authorized emphasis on research, development, and procurement of guided missiles, as indicated by known German advances.

——: Theodore von Kármán submitted proposal to Army Ordnance for developing long-range surface-to-surface missiles.

——: In response to military characteristics established by the Coast Artillery Board for a radio-controlled antiaircraft projectile, Frankford Arsenal conceived a guided-missile system based on existing fire-control knowledge.

*December 24:* The first major Eighth Air Force assault on German V-weapon sites was made when 670 B-17's and B-24's bombed the Pas de Calais area.

*During December:* The rocket aircraft research program conceived by NACA's John Stack, to investigate the flight characteristics of an airplane flying beyond the speed of sound or Mach 1.

——: First turbojet light bomber flight, the German Arado Ar-234B, which was powered by two Junkers .004 engines.

*During 1943:* First jet-propelled rotor helicopter flown, the Austrian Doblhoff No. 1.

*During 1942-43:* Cal Tech studied pumping of liquid rocket propellants, particularly nitric acid, resulting in successful design in 1945, which was set aside for future use because of decision to concentrate on gas-pressurized fuel systems.

# 1944

*January 1:* At request of Army Ordnance, Cal Tech's rocket laboratory started research and development program on long-range missiles, called Project ORDCIT, which resulted in development of Private "A" and Corporal missiles.

*January 8:* First flight of Lockheed XP-80 at Muroc, which was powered by British Halford turbojet engine, the first U.S. airplane designed from the beginning for turbojet propulsion. Rushed through development in 145 days by Lockheed's Clarence L. ("Kelley") Johnson, the P-80 was not distributed to tactical units until December 1945.

*January 11:* First U.S. combat use of forward-firing rockets made by Navy TBF-1C's against a German submarine.

*February 28:* First firing of Nazi Germany's Wasserfall antiaircraft missile.

*During February:* Army Ordnance and AAF initiated development of surface-to-air high-altitude supersonic guided missile, subsequently became XSAM-A-7 Nike I.

*March 16:* At seminar at NACA Langley Laboratory, attended by AF, Navy, and NACA personnel, NACA proposed on the basis of considerable study that a jet-propelled transonic research airplane be developed. This proposal ultimately led to the X-1 research airplane project.

*During March:* First operation of a turbojet engine in an altitude facility was conducted at NACA Lewis Laboratory during tests of P-59 propulsion system, ensuing program making major contributions to U.S. turbojet engine development.

*May 9:* First flight of aircraft modified to demonstrate high-lift boundary layer control made by Lt. Col. R. E. Horner, a project initiated in May 1942 by USAAF contract.

*May 10:* Bell helicopter made an indoor demonstration flight at Buffalo, N.Y., Floyd Carlson as pilot.

*May 28–June 1:* U.S. Navy airships K-123 and K-130 completed the first nonrigid transatlantic crossing from Boston to Port Lyautey, via Newfoundland and the Azores.

*May 31:* First launching of the experimental VB-7 vertical bomb, incorporating television.

*June 13:* The first German V-1's fired in anger, launched from France against England with 4 of the 11 striking London.

*During June:* Remains of V-2 which impacted in Sweden were flown to England for Allied analysis.

*July 5:* The MX-324, first U.S. military rocket-powered plane built by Northrop, was flown by test pilot Harry Crosby, at Harper Dry Lake, Calif.

*July 29:* First successful test of Pelican guided missile, two of four launched were hits against target ship 44 miles offshore from NAS New York.

*During July:* Robert R. Gilruth of the Langley Flight Research Division, prompted by the need for an experimental method of gathering aerodynamic data at transonic speeds, conceived the wing-flow method (utilizing the transonic-airflow field over the top surface of the wing of a high-speed subsonic airplane, usually a P-51 fighter, as a "flying wind tunnel" for testing small semispan wing and airplane models).

———: First positive identification of German turbojet interceptors used against Allied bombers.

———: RAF formed first Meteor jet squadrons for use against V-1's.

*Summer 1944:* German "Reichenberg" program began for use of manned V-1's air launched from He-111's for suicide missions; test flights were made at Peenemünde.

*August 4:* The first Aphrodite mission (radio-controlled aircraft carrying 20,000 pounds of TNT) was flown against rocket sites in the Pas de Calais area.

———: Meteor EE 216 became first British jet fighter to destroy an enemy aircraft, the destruction of a German V-1 Flying Bomb by tipping it with a wingtip.

## 1944—Continued

*August 13:* Two GB-4 glide bombs, incorporating television and radio control, launched against E-boat pens at Le Havre, France. Four additional GB-4's were sent against targets in France and Germany between 17 August and 13 September 1944.

*During August:* German Me-163B Komet rocket-powered fighters first attacked American bomber formations over Europe. The Me-163 had sweptback wings, Walther liquid-fuel rocket motor, speed of 590 mph, and powered flight duration of 8-10 minutes.

*September 3:* Torpex-laden Liberator drone flown from airfield at Feresfield, England, by Lt. Ralph Spaulding (USN), who set radio control and bailed out, after which drone was guided from parent aircraft to German airfield on Helgoland Island.

*September 6:* Navy awarded contract to McDonnell Aircraft for development of the Gargoyle or LBD-1, a radio-controlled low-wing gliding bomb fitted with a rocket booster and designed for use with carrier-based aircraft.

*September 8:* First German V-2 fired in combat exploded in suburb of Paris; the second struck London a few hours later.

*September 14:* Successful flight into hurricane for scientific data was made by Col. Floyd B. Wood, Maj. Harry Wexler, and Lt. Frank Reckord in a Douglas A-20.

*September 18:* Navy Pelican guided missile production terminated and project returned to developmental status because of tactical and logistic problems.

*During September:* Brig. Gen. W. A. Borden, Chief, New Developments Division of the War Department, made known that Ordnance would develop wingless ballistic-type missiles and the AAF would develop winged pilotless-aircraft-type missiles with mutual cooperation in the development of warheads and other equipment.

———: USAAF accelerated development of JB-2 robot bomb based on design of German V-1.

*During October:* Dr. H. J. E. Reid, Engineer-in-Charge of the Langley Laboratory, became scientific chief of the War Department's Alsos Mission charged with picking up as much information as possible on the enemy's scientific research and development.

*During Fall:* Preliminary studies were made of velocity gradients above wings of high-speed subsonic airplanes to determine feasibility of utilizing the wing-flow method in transonic model tests, at NACA Langley Laboratory. This led in the following winter to tests of a series of small airfoil models by this method, and later to use of rockets in flying aircraft models.

*November 1:* Nation's first center devoted to the research and development of rocket propulsion systems, founded at Cal Tech in 1936, reorganized and renamed the Jet Propulsion Laboratory (JPL).

*November 1-December 7:* Representatives of 52 nations (excluding Axis nations and U.S.S.R.) met in the International Civil Aviation Conference in Chicago; they turned down "blue skies" legal concept and reaffirmed doctrine of national sovereignty in air space, and established the Provisional International Civil Aviation Organization (PICAO) to regulate international air commerce.

*November 7:* Gen. H. H. Arnold requested Dr. Theodore von Kármán to "investigate all possibilities and desirabilities for postwar and future war's development as respects the AAF." Dr. von Kármán organized the AAF Scientific Advisory Group for this purpose.

*November 15:* Army Ordnance initiated Hermes program for research and development of ballistic missiles with a prime contract with General Electric Co.

*November 17:* Navy BuAer undertook feasibility studies of JB-2 Army version of German V-1, which subsequently became the Loon.

*During November:* First flight use of a radio telemeter for transmitting research data at transonic speeds, by the bomb-drop technique at NACA's Langley Laboratory.

*December 1-16:* At Camp Irwin, Calif., 24 Private "A" rockets were launched by

JPL, only 11 months after the start of Project ORDCIT.

*December 13-14:* In an AAF-NACA conference, Air Force representatives indicated strong preference for use of rocket engines instead of jets in X-1 research airplane project.

*During December:* Army Ordnance made plans under the Hermes program to study the German V-2 missile.

——: Glenn L. Martin granted $1,700,000 to the University of Maryland for establishment of a College of Engineering and Aeronautical Sciences.

*During 1944:* NACA established Special Committee on Self-Propelled Guided Missiles to recommend and coordinate research related to guided missiles.

——: USAAF VB-1 controlled-trajectory air-to-surface bomb (Azon) produced and used in European and Burma theaters.

——: Supersonic wind tunnel (Mach 1.7) completed at Aberdeen Proving Ground for use in ballistic research and development.

——: Initial contracts for rocket research aircraft development let by the AAF for the XS-1 with Bell Aircraft and by the Navy with Douglas Aircraft for the D-558-I, with NACA providing technical support under cooperative agreement.

*During 1944-1945:* First full-scale supersonic propulsion wind tunnel (8 by 6 feet) was conceived, designed, and directed by Abe Silverstein at NACA Lewis Laboratory. Capable of accommodating full-scale supersonic aircraft engines, it was the first of its size to have a flexible-wall test section, which allowed variations from Mach 1.4 to 2.

——: Japan launched approximately 10,000 *Fugo* balloons (30-foot diameter) carrying incendiaries and aimed at the North American continent.

# 1945

*January 20:* Robert T. Jones, NACA Langley aeronautical scientist, formulated sweptback-wing concept to overcome shockwave effects at critical Mach numbers, and verified it in wind-tunnel experiments in March 1945 prior to learning of parallel German work. It was subsequently checked by the wing-flow technique before the first NACA report was issued in June.

*January 24:* Germans successfully launched A-9, a winged prototype of the first ICBM (the A-10) designed to reach North America. A-9 reached a peak altitude of nearly 50 miles and a maximum speed of 2,700 mph.

*During January:* JNW created Guided Missiles Committee to formulate broad program of research and development in the guided missiles field, the committee to consist of two members from OSRD, one from NACA, three from the Army, and three from the Navy.

*During January:* German Luftwaffe formed special squadron of 16 Me-262 jet fighters, each armed with twenty-four 55-mm high-explosive rockets, which operated with high success against Allied bomber formations.

*February 20:* The Secretary of War approved Ordnance plans for the establishment of the White Sands Proving Ground (WSPG).

*During February:* Project Nike initiated by Army Ordnance with the Western Electric Co. to explore a new air defense system against high-speed and high-altitude bombers beyond the reach of conventional artillery.

——: AAF contracted with Bell for construction of three transonic flight research aircraft, to be powered by liquid rocket engines. Aircraft designated XS-1, and later X-1.

*March 8:* Navy rocket-powered Gorgon air-to-air missile launched from PBY-5A in first powered test flight off Cape May, N.J.

## 1945—Continued

*March 21:* Navy initiated development of the Lark surface-to-air guided missile in BuAer contract with Fairchild Aircraft.

*During March:* "Summary of Airfoil Data," by Ira H. Abbott, A. E. von Doenhoff, and Louis Stievers of NACA Langley Laboratory, was issued, which was considered a classic reference summarizing NACA data on airfoil sections.

———: Project Paperclip to recruit German missile scientists was initiated in the Pentagon.

*During Spring:* Supplemental appropriation passed by Congress authorized expanded research on guided missiles at NACA Langley Laboratory, including establishment of a rocket launch facility at Wallops Island, Va.

*April 1–13:* 17 JPL Private F rockets were fired at Hueco Range, Fort Bliss, Tex.

*During April:* Aberdeen Proving Ground wind-tunnel tests of sweptback wing at Mach 1.72 carried out on the suggestion of Theodore von Kármán.

*May 5:* Russian ground forces occupied Peenemünde, Germany.

*May 8:* World War II ended in Europe.

———: At time of German collapse, more than 20,000 V-weapons, V–1's and V–2's had been fired. Although figures vary, best estimate is that 1,115 V–2 ballistic rockets had been successfully fired against England and 1,675 against continental targets. Great disparity between production figures and operational missions due to fact that series production and development testing were performed concurrently, there being as many as 12 major modifications in basic design features.

*May 10:* Crash program to counter Japanese Baka (suicide) bomb, Naval Aircraft Modification Unit was authorized to develop Little Joe, ship-to-air missile powered with standard JATO unit.

*During May:* Boeing began development of Gapa (ground-to-air pilotless aircraft) antiaircraft missile for USAAF. Within 2 years 37 Gapa missiles had been fired and by October 1949 a total of 102 successful firings had taken place.

*June 19:* Dr. Frank L. Wattendorf, Engineering Division, Wright Field, and a member of AAF Scientific Advisory Group, recommended to Brig. Gen. F. O. Carroll, Chief, Engineering Division, that an Air Force Development Center, including facilities for development of supersonic aircraft and missiles, be built on a location away from Wright Field and near a large source of power.

*June 25:* Construction began at White Sands Proving Ground.

*During June:* Army Ground Forces Equipment Review Board concluded that increased emphasis should be placed on development of guided missiles.

———: XC–99, cargo version of B–36, made first flight.

*July 4:* Baby Wac rocket, one-fifth scale model of Wac Corporal proposal, flight tested at Camp Irwin by JPL.

———: First rocket launch at NACA's new Wallops Island facility for calibration of radar instrumentation.

*July 13:* White Sands Proving Ground (WSPG) was activated.

*July 14:* AAF A–20's from Hollandia set fire to Japanese oil fields at Boela, Ceram, in the first use of rocket bombs in the Southwest Pacific.

*July 16:* First test atomic device exploded in New Mexico.

*July 20:* Navy Little Joe antiaircraft missile made two successful flights at Applied Physics Laboratory test station at Island Beach, N.J.

*July 23: Life* published drawings of a manned space station as envisioned by the German rocket scientists of Peenemünde.

*During July:* First launching of a two-stage rocket-propelled research model, the Tiamat missile, which employed six rockets as boosters, had automatic stabilization, its maneuvers were programed, and its testing was the first research program of the NACA's Wallops Island Station.

*August 6:* First atomic bomb was dropped on Hiroshima.

*August 9:* Second atomic bomb was dropped on Nagasaki.

*August 14:* World War II ended with Japanese surrender.

———: Team of American scientists was dispatched to Europe to collect information and equipment relating to German rocket progress.

*August 24:* First successful use of a telemetry system in a rocket-propelled flight research model, the two-stage Tiamat at NACA Wallops Island, Va.

*During August:* First successful U.S. chemical gas, generator-driven, turbopump fed, regeneratively cooled rocket engine (XCALT-6000), delivered to AAF by Aerojet-General Corp.

———: Components for approximately 100 V-2 ballistic missiles were shipped from Germany to White Sands Proving Ground.

———: Joint Army-Navy Aeronautical Board established Research Committee to investigate and report on matters affecting research, development, and testing of aircraft, including liaison with NACA and industry, and to recommend action to foster aeronautical research and development.

*September 8:* William F. Durand, one of the original members of the NACA in 1915, retired.

*September 20:* First flight of airplane powered by propeller-turbine engines, made in England by experimental Gloster Meteor powered with Rolls Royce Trent-engines with five-bladed propellers.

*September 26:* The Navy publicly demonstrated the Ryan Fireball FR-1 at NAS Anacostia, the first propeller-and-jet-powered airplane designed for aircraft carriers.

———: Army Wac Corporal, first development flight, fired at White Sands, established U.S. record of 43.5 miles height, and was the first U.S. liquid-propellant rocket developed with Government funds (constructed by Douglas and Aerojet under JPL Project).

*During September:* First volume of the *Toward New Horizons* reports of the Army Air Forces Scientific Advisory Group (headed by Von Kármán), entitled *Science: The Key to Air Supremacy*, was submitted to the Commanding General of the AAF. These reports prepared by leading scientists are classic in their assessment of future developments emerging out of World War II advancements.

*October 3:* A Navy Committee for Evaluating the Feasibility of Space Rocketry (CEFSR) was established by BuAer. In November 1945, CEFSR recommended high priority for satellite development and estimated cost between $5 and $8 million.

*October 11:* First launch of full Wac Corporal (WAC-A) at WSPG attained an altitude of 235,000 feet.

*October 18:* NACA Langley's Pilotless Aircraft Research Division (PARD) launched the first successful drag research vehicle for wing and body research, forerunner of a large series of flight tests of various wings and bodies in a combination of transonic and supersonic speeds providing basic design information later applied on all later supersonic aircraft and missiles.

*October 30:* Chief of Army Ordnance invited Secretary of the Navy to utilize the White Sands Proving Ground (WSPG) as a test range for naval-guided missiles (BuOrd) and for pilotless aircraft (BuAer).

*During October:* Secretary of War Patterson approved plan to bring top German scientists to United States to aid military research and development. Small group of German rocket specialists brought to United States under Project Paperclip to work on missile development at Fort Bliss and White Sands Proving Ground.

———: Navy BuOrd established Guided Missiles, Jet Propulsion and Countermeasures Section in its Research and Development Division.

*November 6:* The first jet landing on an aircraft carrier was made by Ens. Jake C. West, USN, in an FR-1 Navy turbojet and conventional reciprocating-engine fighter.

## 1945—Continued

*November 7:* Bell Aircraft Corp. announced successful test flights of a jet-propelled P-59 by remote control; television was used to read the instruments.

*During November:* Guided Missiles Committee of the Joint Committee on New Weapons and Equipment (JNW) drafted Dewey Report on "A National Program for Guided Missiles."

*December 3:* The first USAAF jet fighter unit, the 412th Fighter Group, received its first Lockheed P-80 aircraft at March Field, Calif.

*December 9:* First Stratovision flight test made at Middle River, Md., by Westinghouse Electric Corp. and Glenn L. Martin Co. Telecasts were made from the airplane flying in the stratosphere.

*December 14:* AAF contracted with Bell for development of three supersonic flight research aircraft, powered by liquid rockets. Designated XS-2, and later X-2.

*December 17:* Rocket-Sonde Research Branch constituted in Naval Research Laboratory to conduct scientific exploration of the upper atmosphere.

*December 19:* President Truman submitted his plan to Congress for the unification of the armed services.

*During December:* Office of Deputy Chief of Air Staff for Research and Development created in Hq. USAAF, headed by Maj. Gen. C. E. LeMay.

———: More than 100 German rocket scientists and engineers, who had agreed to come to the United States under Project Paperclip, arrived at Fort Bliss, Tex.

———: Navy BuAer awarded contract to Guggenheim Aeronautical Laboratory at Cal Tech to conduct research whose findings were to be used in formulating policy for a projected high-altitude earth satellite vehicle.

*During 1945:* Abe Silverstein of Lewis Laboratory made basic application of ramjet technology to the problem of afterburner design, leading to the first full-scale afterburner tests.

———: New wind tunnels placed under construction at NACA's Ames Laboratory at Moffett Field, Langley Laboratory at Hampton, Va., and Propulsion Laboratory at Cleveland, to attain speeds of 1,400, 1,800, and 2,600 mph with various sized throats.

———: German Heinkel He-162 Salamander or "Volksjaeger" jet fighter appeared operationally, while the prototype of a heavy jet bomber appeared in the Junkers Ju-287 (four-engine) with auxiliary take-off rockets, sweptforward wings, speed over 550 mph, and bomb load of 8,800 pounds.

*End of 1945:* Increase in speed of reciprocating-engined fighter aircraft by 300 to 400 mph between World War I and World War II (speed being only one military criterion) was estimated to be 75 percent gain because of increased horsepower, 25 percent from aerodynamic improvement.

———: Dr. Jerome C. Hunsaker pointed out that U.S. aeronautical research effort during World War II was based upon short-range policy of about 90 percent for specific development problems applied to help win the war and 10 percent on basic research to gain needed knowledge. The national research effort has "concentrated on the improvement of aircraft in the production program."

# 1946

*January 2:* Special investigation of high-temperature aluminium alloys begun by J. C. McGee, Wright Field engineer, which led by June 1947 to useful alloy known as "ML," named after the Materials Laboratory.

*January 10:* An Army R-5, demonstrated by C. A. Moeller and D. D. Viner, set an unofficial world helicopter record by climbing to 21,000 feet at Stratford, Conn.

*January 16:* U.S. upper atmosphere research program initiated with captured German V-2 rockets. A V-2 panel of representatives of various interested agencies was created, and a total of more than 60 V-2's were fired before the supply ran out. The Applied Physics Laboratory of Johns Hopkins University then undertook to develop a medium-altitude rocket, the Aerobee, while the Naval Research Laboratory (NRL) directed its efforts to the development of a large high-altitude rocket, first called the Neptune, later the Viking.

*January 19:* First glide flight of AAF-NACA XS-1 rocket research airplane (No. 1 of the original three X-1's built), by Jack Woolams, Bell Aircraft test pilot, at Pinecastle Army Air Base, Fla.

*January 26:* Army announced creation by AAF of the First Experimental Guided Missiles Group to develop and test rocket missiles at Eglin Field, Fla.

———: Naval Aviation Ordnance Test Station was established at NAAS Chincoteague to develop aviation ordnance and guided missiles.

*During January:* First missile launched at Naval Air Facility, Point Mugu, Calif., a KVW-1 Loon, USN name for AAF robot bomb (JB-2) modeled on the German V-1.

*February 3:* Development of a plane with automatic devices to preset takeoff, flight, and landing, with the pilot doing nothing except monitoring the equipment, disclosed by AAF.

*February 19:* S. Paul Johnston appointed Director of the IAS to replace Lester D. Gardner, retiring after 15 years of service.

*March 7:* BuAer Committee for Evaluating the Feasibility of Space Rocketry (CEFSR) held joint meeting with AAF representatives to work out joint satellite development program based on BuAer proposal. Nothing resulted until a subsequent Project Rand report and Navy CEFSR proposal were presented to RDB, Committee on Guided Missiles, Technical Evaluation Group in March 1948.

*March 11:* First successful operation of afterburner at altitude conditions in America, in Lewis Altitude Wind Tunnel, and reported by Fleming and Dietz.

*March 12:* Chief of Naval Operations directed that Glomb, Gorgon II-C, and Little Joe guided missiles be discontinued and that Gargoyle, Gorgon II-A, and Dove be limited to test and research vehicles. He directed that Loon be continued as a possible interim weapon, the Bat be completed, and the Kingfisher, Bumblebee, and Lark be continued as high-priority missile developments.

*March 15:* First American-assembled V-2 static fired at White Sands Proving Ground.

*March 22:* First American rocket to escape earth's atmosphere, the JPL-Ordnance Wac, reached 50-mile height after launch from WSPG.

*During March:* Fleet Adm. William D. Leahy sent memorandum to the Secretaries of the War and Navy Departments on a national program for development of guided missiles.

———: AAF established Project Rand as separate department of Douglas Aircraft Co. plant at Santa Monica, Calif., to study supersonic aircraft, missiles, and earth satellites.

———: Navy successfully flight tested XSAM Talos surface-to-air guided missile.

———: USAAF established initial program on ballistic missile defense, a contract for study of interceptor weapon to cope with V-2-type missiles. In April a second contractor began study of defense against true ICBM.

*April 1:* Bell Aircraft Corp. contracted with the AAF (under Project MX-776) to produce a 100-mile guided missile (later designated the Rascal).

*April 16:* First flight test of American-assembled V-2 rocket launched by the Army at White Sands Proving Ground, N. Mex. In July firings, Missiles Nos. 5 and 9 set new altitude records of slightly over 100 miles, while Missile 17 set velocity record of 3,600 mph.

*April 17:* Army Ground Forces submitted to the Guided Missile Committee a summary of its program on antiaircraft, assault, antiship, air-launched close sup-

## 1946—Continued

port, and long-range strategic guided missiles.

*April 19:* Project MX-774 inaugurated by AAF with Consolidated-Vultee to study rocket capabilities with an ICBM as a final objective.

*April 22:* Glenn L. Martin Co. contracted with the AAF to produce (under Project MX-771) a surface-to-surface guided missile (later designated the Matador).

———: U.S. Weather Bureau in cooperation with Army, Navy, NACA, Air Transport Association, and several universities, began series of flights into thunderstorms with pilotless P-61 "Black Widows" and piloted sailplanes to obtain scientific data.

*May 8:* Chief of Naval Operations directed BuAer to make preliminary investigation of earth satellite vehicle, such an investigation to "contribute to the advancement of knowledge in the field of guided missiles, communications, meteorology, and other technical fields with military applications."

*May 16:* AAF established an Institute of Technology at Wright Field to graduate 350 officers annually.

*May 17:* Original design and development of Aerobee sounding rocket begun when contract was given to Aerojet Engineering Corp.

———: First flight of Douglas XB-43, light jet-propelled bomber.

*May 28:* AAF initiated study of use of atomic propulsion for aircraft, Project NEPA.

*May 29:* War Department Equipment Board concluded in its report that missiles would play a prominent role in future warfare. It established requirements for seven types of missiles, including a strategic ground-to-ground missile for use at ranges from 150 to several thousand miles.

*June 6:* Joint Army-Navy Research and Development Board created for purpose of coordinating all activities of joint interest in fields of aeronautics, atomic energy, electronics, geographical exploration, geophysical sciences, and guided missiles.

*June 14:* Navy established Naval Ordnance Missile Test Center at WSPG.

*June 17:* First meeting of the AAF Scientific Advisory Board met in the Pentagon, chaired by Theodore von Kármán.

*June 19:* NACA Langley's PARD launched first successful control-surface research vehicle at Wallops Island for evaluating controllability with a roll rate transmitter and Doppler radar.

———: AAF contracted with Sverdrup & Parcel, Inc., for study utility and cost requirements, and site surveys for both an AAF Air Engineering Development Center, and a NACA National Scientific Research Center.

*June 24:* Office of Naval Research approved program for high-altitude manned flight, Project Helios, based upon concept presented by Jean Piccard in February for using clustered plastic balloons.

*During June:* First U.S. airborne infrared tests by USAAF.

*July 6:* Antiaircraft and Guided Missile Center activated at Fort Bliss, Tex.

*July 9:* Subcommittee of the Guided Missiles Committee of the JCS recommended that location be sought for a long-range missile proving ground.

*July 21:* First U.S. all turbojet to operate from an aircraft carrier, a McDonnell XFD-1 "Phantom" from the U.S.S. *Franklin D. Roosevelt.*

*August 2:* National Air Museum was established under the Smithsonian Institution by act of Congress.

*August 6:* Two unmanned B-17 drones flown from Hilo, Hawaii, to Muroc, Calif.

*August 8:* First flight of the XB-36, the development of which had begun in 1941.

*August 17:* Sergeant Lambert of Wright Field, Ohio, became the first person in the United States to be ejected from an airplane by means of emergency escape equipment (ejected from a P-61 airplane traveling 302 miles per hour at an altitude of 7,800 feet).

*August 26:* Army Ground Forces informed Chief of Staff that development of certain missiles had reached a point where an assignment of operational responsibility was possible.

*September 17:* Experimental booster for Nike R&D system first tested at WSPG.

*September 30:* 13 engineers, instrument technicians, and technical observers were ordered TDY from Langley Laboratory to the Air Force test facility at Muroc, Calif., to assist in the X-1 flight research program. Named as the NACA Muroc Flight Test Unit, this group under Walter Williams was the origin of the NASA Flight Research Center at Edwards, Calif.

*October 1:* Naval Air Missile Test Center, Point Mugu, Calif., was established to conduct tests and evaluations of guided missiles and components.

——: Navy Lockheed PV-2, *Truculent Turtle*, set a record of nonstop long-distance flight, completing an 11,236-mile trip from Perth, Australia, to Columbus, Ohio, in 55 hours 15 minutes.

*October 7:* First of three XS-1 (later X-1) rocket research airplanes moved from Bell Aircraft's Niagara Falls plant to Muroc, Calif.

*October 11:* First glide flight of XS-1 (No. 2) by Chalmers Goodlin, Bell test pilot, at Muroc, Calif.

*October 24:* V-2 rocket No. 13 launched from WSPG carried camera which took motion pictures of the earth at approximately 65 miles altitude (pictures covered 40,000 square miles).

*During October:* Army Ordnance initiated Bumper Project for development leading to a two-stage rocket test vehicle, which resulted in use of JPL WAC Corporal as second stage of a V-2.

*During Fall:* Reaction Motors began design and development of rocket engine for the Navy Viking sounding rocket.

*During November:* First snow from a natural cloud produced by V. Schaefer of General Electric, the experiment carried out by means of dry-ice pellets dropped from a plane over Greylock Mountain, Mass.

*December 8:* First successful powered (RMI XLR-11 rocket engine) flight of an XS-1, flown by Chalmers Goodlin, Bell test pilot, reached a speed of 550 mph. This was first U.S. aircraft designed for supersonic speeds.

*December 17:* Space biological research program was initiated at Holloman AFB, N. Mex., by the National Institutes of Health.

——: Velocity and altitude record for single-stage rocket (3,600 mph and 116 miles altitude) made by V-2 at WSPG.

*During 1946:* Signal Corps by radio-echo transmissions between the Earth and the Moon, proved radio transmission across space was feasible with moderate power.

——: Jet Propulsion Laboratory under Army Ordnance contract developed the field of solid-propellant rocketry such as castable propellants, case bonding techniques, and radial burning techniques.

——: Daniel Guggenheim Medal for 1946 awarded to Frank Whittle for development of jet propulsion engines.

——: Program of transonic and hypersonic free-flight research on ramjet and rocket-propelled test vehicles launched from piloted aircraft inaugurated at NACA Lewis Laboratory.

——: Commandant of the School of Aviation Medicine, Col. H. G. Armstrong, and the AAF Air Surgeon, Brig. Gen. M. C. Grow, proposed establishment of aeromedical center for research and teaching.

——: Office of Naval Research contracted with General Mills for construction of a cluster of 100 plastic balloons for high altitude atmosphere research (Project Helios).

*During 1946–47:* Transonic bump technique—using floor- or wall-mounted airfoil surface in subsonic wind tunnel to get transonic flow—developed in 7- by 10-foot wind tunnel at NACA Langley Labora-

## 1946—Continued

tory. A similar development was conducted by Lockheed in the California Cooperative Tunnel during the same period. This technique was a logical step from the earlier wing-flow technique developed by the Langley Flight Research Division, and it permitted testing in the range of Mach numbers from low subsonic to Mach 1.2 until the slotted-throat transonic tunnel was developed and put into operation at Langley 2 years later.

# 1947

*January 8:* First experimental operation of model slotted-throat wind tunnel. Langley Laboratory's Ray H. Wright, working theoretically, and Vernon G. Ward, working experimentally with a parasite tunnel attached to the Langley 16-foot high-speed tunnel, collaborated in an effort that resulted in establishment of transonic flow with the use of longitudinal slots in the walls of the throat of a conventional subsonic tunnel. Known as the slotted-throat technique, first major installation was made in the Langley 8-foot subsonic high-speed tunnel in December 1949, a breakthrough in wind tunnel technique.

*January 23:* Telemetry operated successfully in a V-2 firing at WSPG, Army Ordnance's Hermes telemetry system.

*February 5:* President Truman directed that production of nuclear weapons continue, following the recommendations of the AEC and the Secretaries of War and Navy.

*February 12:* Navy Loon launched from submarine *Cusk* at Point Mugu, first launching of a guided missile from a submarine.

*February 17:* Wac Corporal (WAC-B), fired from WSPG, attained an altitude of 240,000 feet.

*February 20:* First of a series of V-2 firings (No. 20) known as Blossom Project, tested ejection of canister and its recovery by parachute, containing fruit flies and various types of seeds exposed to cosmic rays.

*March 4:* Air operations in the Antarctic known as Operation Highjump ended. From December 24, 1946, Navy PBM's and R4D's logged 650 hours in photographic mapping of 1,500,000 square miles of the interior and 5,500 miles of the coastline, the equivalent of about half the area of the United States and its entire coastline.

*March 6:* First four-engine jet bomber, the XB-45 built by North American, made first test flight at Muroc, Calif., with George Krebs as pilot. Its engines were arranged in pairs in single nacelles in each wing.

*March 7:* USN V-2 flight from WSPG took first photograph at 100-mile altitude.

*During March:* First test flights of plastic balloons conducted by General Mills at Minneapolis, Minn., for ONR Project Helios.

*During March:* AAF transferred facilities for testing guided missiles from Wendover Field in Utah and Tonopah in Nevada, to Alamogordo Field (subsequently renamed Holloman AFB) in New Mexico.

*April 15:* First flight of Douglas D-558-I research airplane successful, Gene May, Douglas test pilot, as pilot. Airplane developed was a Navy-NACA project and three were built.

*April 24:* French Government established rocket test range at Colomb Bechar, Algeria.

*April 25:* NACA Langley's PARD launched its first rocket-propelled model of a complete airplane for performance evaluation (AF XF-91), at Wallops Island. This was followed by flight tests of models of practically all Air Force and Navy supersonic airplanes.

*April 30:* Standard system of designating guided missiles and assigning popular names was adopted by the Army and Navy. Basic designation adopted was two-letter combination of the three letters A (Air), S (Surface), U (Underwater), the first letter indicating origin of missile, the second letter its objective, to be followed by the letter "M" for missile. Thus a surface-to-air missile was designated "SAM."

*During April:* First Deacon rocket launched at Wallops Island, which achieved a velocity of 4,200 feet per second.

*May 21:* NACA Langley Laboratory demonstrated practically noiseless airplane with five-bladed propellor and muffled exhaust.

*May 27:* Army Corporal E, first U.S. surface-to-surface ballistic guided missile, was fired with results exceeding expectations (a JPL project).

*May 29:* V-2 impacted 1½ miles south of Juarez, Mexico, resulting in new safety measures at WSPG.

*June 5:* First AAF research balloon launch (a cluster of rubber balloons) at Holloman, by New York University team under contract with Air Material Command.

*June 17:* Princeton University started construction of 4,000-mph wind tunnel.

*June 19:* World speed record regained by United States when P-80R flown by Col. Albert Boyd attained 623.8 mph at Muroc, Calif.

*June 30:* In meeting at Wright-Patterson, AAF and NACA representatives agreed to divide responsibilities for X-1 flight testing: AF exploit maximum performance in a few flights; NACA acquire detailed research information.

*July 1:* Contract with Convair for MX-774 "Upper Air Test Vehicle," predecessor of the Atlas ICBM, was cancelled by the AAF.

*July 3:* Start of polyethylene balloon operations at Holloman, a 10-balloon cluster launched by New York University staff with a payload of less than 50 pounds, which reached an altitude of 18,500 feet.

*July 9:* Subsonic ramjet engine successfully flown in Navy Gorgon IV (PTV-2) in 28-minute flight test at Naval Air Missile Test Center.

*July 18:* President Truman designated a five-man Air Policy Committee, with Thomas K. Finletter of New York as Chairman, to submit by 1 January 1948 a broad plan to give the United States the "greatest possible benefits from aviation."

*July 26:* President Truman signed the Armed Forces Unification Act, creating a Department of the Air Force, coequal with Army and Navy, and creating a National Military Establishment under the Secretary of Defense.

*During July:* USAF relinquished responsibility for Army's missile program and Army assigned primary responsibility for it to Ordnance.

——: Soviet MiG-15 first flew but engine performance was unsatisfactory, a problem solved with purchase of 55 British Derwent V and Nene (4,500-pound thrust) engines, first placed in series production, then improved with the RD-45 engine (5,000-pound thrust) and the VK-1 (6,000-pound thrust) engine.

*August 1-3:* Boeing B-29 set a new official world "distance in a closed-circuit record" with a flight of 8,854.308 miles, Lt. Col. O. P. Lassiter as pilot.

*August 8:* A. L. Berger of Wright Field received the Thurman H. Bane Award for 1947 for work in developing new types of high-temperature ceramic coatings for use in aircraft engines.

*August 16:* Physicist Martin Pomerantz announced at Swarthmore College that he had sent a flight of four free balloons, carrying cosmic ray equipment, to a record height of at least 127,000 feet.

*August 20:* Comdr. T. Caldwell (USN) flew the Douglas D-558-I (No. 1) Skystreak, powered by a General Electric TG-180 turbojet, to a new world's speed record of 640.7 mph. Five days later Maj. Marion Carl, USMC, added another 10 mph flying D-558-I (No. 2).

## 1947—Continued

*August 22:* Dr. Hugh L. Dryden appointed Director of Aeronautical Research of the NACA, replacing Dr. George W. Lewis.

*September 2-6:* First Joint Technical Sessions by the Royal Aeronautical Society, Great Britain, and the Institute of Aeronautical Sciences, held in London.

*September 6:* German V-2 rocket launched from U.S. aircraft carrier *Midway* in Atlantic tests, exploding prematurely after a 6-mile flight.

*September 22:* Air Force C-54 completed first transatlantic robot-controlled flight from Stephenville, Newfoundland, to Brize Norton, England, a distance of 2,400 miles.

*September 25:* First flight under ONR Project Skyhook, an unmanned plastic balloon, from St. Cloud, Minn.

———: First successful firing of Applied Physics Laboratory Aerobee research rocket at White Sands Proving Ground.

*September 26:* Maj. Gen. William E. Kepner, was named chief of the new atomic energy division of the USAF.

*September 30:* Research and Development Board (RDB) of DOD superseded Joint Research and Development Board, with Vannevar Bush named as Chairman.

*During September:* After completing studies, Project Rand reported that earth satellites were technically feasible.

*October 1:* First flight of the North American XF-86 Sabre Jet, classic swept-wing USAF fighter aircraft until the Century series.

*October 9:* General Electric engineers obtained first carefully instrumented heat-transfer data from supersonic flight when V-2 fired from WSPG attained 3,400 mph.

*October 10:* U.S. Patent Office issued patent on the Norden bombsight, which Carl L. Norden had applied for 17 years earlier.

*October 14:* The first supersonic flight in manned aircraft in level or climbing flight was made by Capt. Charles E. Yeager (USAF) at Muroc, Calif., in a rocket-powered NACA-USAF research plane, Bell XS-1, later the X-1 (M=1.06).

*October 30:* Dr. H. J. E. Reid, Engineer-in-Charge of the Langley Aeronautical Laboratory (1926-60), received the Medal of Merit from President Truman for wartime contributions to American airpower.

*During October:* Committee on Guided Missiles of RDB assigned responsibility for coordinating work on earth satellite program which had been conducted independently by each of the military services.

*November 14:* First complete Aerobee rocket was fired to a height of 190,000 feet from White Sands Proving Ground, N. Mex.

*November 15:* Air Force disclosed that the world's first ramjet helicopter, the McDonnell Flying Bike, had been successfully test flown for 6 months.

*November 26:* First successful hypersonic-flow wind tunnel (11 inch) placed into operation at March 7 at Langley Laboratory.

*November 28:* *Norton Sound* was assigned to Operational Development Force for use as an experimental rocket-firing ship, alterations initiated at Naval Shipyard at Philadelphia in March 1948, and completed October 1, 1948.

*December 10:* Lt. Col. John P. Stapp (USAF MC), made his first rocket-propelled research sled ride.

*December 17:* USAF Boeing XB-47 Stratojet made first flight from Seattle to Moses Lake, first medium turbojet bomber and the first with engines (six) mounted on pylons.

*During 1947:* USAF SAM initiated study of ecological conditions on other planets.

———: During a Politburo meeting reviewing the problem of developing an intercontinental ballistic missile, Premier Joseph Stalin reportedly stated that a transatlantic rocket capable of hitting New York City "would make it easier to talk with the gentleman-shopkeeper, Harry Truman, and keep him pinned down where we want him." This probably reflected the high priority accorded large rocket development in the U.S.S.R. at this time.

# 1948

*January 1:* President's Finletter Commission submitted its comprehensive report entitled "Survival in the Air Age."

*January 4:* University of California announced completion of pilot model for low-pressure supersonic wind tunnel, while NACA Ames Aeronautical Laboratory placed its low-density wind tunnel into operation about this time.

*January 12:* Northrop Aircraft Co. announced that rocket-powered test vehicles at Muroc Air Base, Calif., had attained a speed of 1,019 mph.

*January 15:* Gen. H. S. Vandenberg, Vice Chief of Staff, USAF, approved policy calling for development of earth satellite components and the initiation of satellite development at the proper time.

*January 30:* Orville Wright died in Dayton, Ohio, at the age of 76, thus ending his 28 years as a member of the NACA. In his lifetime, the speed of the airplane had been increased from 0 mph to almost 1,000 mph.

*February 4:* First flight of research airplane Douglas D-558-II (No. 1), John Martin of Douglas as pilot. Airplane had both jet and rocket engines and was flown from ground takeoff.

*February 6:* Successful electronic flight control exercised on V-2 launch to a 70-mile altitude at White Sands, N. Mex., by General Electric technicians for Army Ordnance.

*March 4:* NACA's Flight Research Division pilot, Herbert H. Hoover, became the first civilian to fly supersonic, in the XS-1 (No. 2) at Muroc, Calif.

*March 6:* ONR Aerobee sounding rocket attained an altitude of 78 miles.

*March 11-14:* Key West Agreement formulated by military service chiefs which delineated respective service roles and missions. It did not clearly assign military aeronautical and rocket research and development responsibilities to the services.

*March 18:* V-2 Upper Atmosphere Research Panel, representing all U.S. interested agencies, was renamed the Upper Atmosphere Rocket Research Panel.

*March 29:* Technical Evaluation Group of the RDB, Guided Missiles Committee, after reviewing Navy CEFSR and USAF Project Rand satellite proposals, stated that "neither the Navy nor the USAF has as yet established either a military or a scientific utility commensurate with the presently expected cost of a satellite vehicle. However, the question of utility deserves further study and examination."

*May 2:* The Navy announced successful testing of a submarine capable of firing guided missiles.

*May 3:* Howard C. Lilly killed in takeoff of D-588-I (No. 2) research airplane at Muroc, the first NACA test pilot killed in line of duty.

*May 13:* Two-stage Bumper-Wac fired at WSPG, the V-2 first stage reaching 70 miles and the Wac Corporal 79-mile altitude.

*May 23:* Army dedicated a continuous wind tunnel capable of 3,000 mph at Aberdeen, Md.

*May 26:* First North American NATIV missile launched at WSPG.

*June 10:* Air Force confirmed repeated attainment of supersonic speeds by X-1 (formerly XS-1) flown by Capt. C. E. Yeager.

*June 26:* Berlin airlift began, which continued until September 30, 1949, although the Russians ended their blockade of the city on May 12, 1949. 2,343,000 tons of supplies were airlifted on 277,000 flights.

*During June:* William H. Phillips of the Langley Flight Research Division published NACA report (TN-1627) which contained theoretical prediction of the then-not-recognized problem of roll coupling (sometimes referred to as "inertial coupling"). This phenomenon was to plague future high-speed aircraft with

## 1948—Continued

short wings and long fuselages, and almost 9 years passed before aerodynamicists were to use Phillips' theory to explain inertial coupling troubles.

*During June:* Bell Laboratories announced invention of the transistor of the point-contact type.

*July 13:* First Convair MX–774 (RTV–A–2) test rocket was successfully launched, first demonstrating use of gimballed engines and design features later incorporated in the Atlas ICBM. This was the first of three Convair-sponsored test flights.

*July 26:* Two separate rockets fired from White Sands, one a V–2 which reached an altitude of 60.3 miles, the other a Navy Aerobee which reached an altitude of 70 miles, carried cameras which photographed the curvature of the earth.

*During August:* Northrop F–89 Scorpion, an all-weather jet fighter with electronic intercept and fire control begun in 1946, first flew.

*September 1:* An XR–82 photographed a 2,700-mile strip of the United States from coast to coast in a single flight, using 390 individual frames and 325 feet of film.

*September 5:* Navy JRM–2 *Caroline Mars* carried a 68,282-pound cargo from Patuxent River, Md., to Cleveland, the heaviest payload ever lifted by an aircraft.

*September 15:* Committee on Guided Missiles of the Research and Development Board approved recommendation that Army Hermes project "be given the task of providing the National Military Establishment with a continuing analysis of the long-range rocket problem as an expansion of their task on an earth satellite vehicle."

——: A world speed record of 671 mph set by Maj. Richard L. Johnson, USAF, in F–86A at Muroc, Calif.

*September 27:* Second Corvair MX–774 test rocket fired.

*September 28:* An Army Signal Corps unmanned balloon, released at Belmar, N.J., set a 140,000-foot altitude record.

——: NACA Flight Propulsion Research Laboratory in Cleveland was redesignated the Lewis Flight Propulsion Laboratory, in memory of Dr. George W. Lewis who died on July 12, 1948.

*September 30:* Third Bumper-Wac launch from WSPG, the V–2 reaching 93.4 miles, the Wac-Corporal not firing.

*During September:* Delta-wing Convair XF–92 first flew, the precursor of the F–102A.

*October 13:* First launching of a rocket-propelled "flying wind tunnel" model by NACA Langley's PARD at Wallops Island, to measure roll damping of wings at transonic speeds.

*October 19:* Photographs of the earth's surface taken from altitudes between 60 and 70 miles by cameras installed in rockets, were released by the Navy.

*October 31:* The Air Force revealed the use of ramjet engines for the first time on piloted aircraft, a modified F–80.

*November 4:* USAF announced formation of the Rand Corp., successor to Project Rand, to assemble most advanced scientific, technical, industrial, and military knowledge available and bring it to bear on major Air Force decisions.

*November 10–12:* The first symposium on aeromedical problems of space travel was held at the School of Aviation Medicine, San Antonio, Tex.

*November 22:* The Wright Kitty Hawk airplane arrived at the Smithsonian Institution, Washington, D.C., after 20 years in the South Kensington Museum, London.

*November 30:* Curtiss-Wright demonstrated its new reversible-pitch propellers which enabled a C–54 to make a controlled descent from 15,000 to 1,000 feet in 1 minute 22 seconds.

*December 2:* Third Convair MX–774 test missile successfully fired.

*December 11:* Secretary of Defense established Weapons Systems Evaluation Group.

*December 13:* Secretary of Defense Louis Johnson directed a review of military missile programs, under the aegis of Air Force Secretary Stuart Symington.

*December 14:* Jet Propulsion Centers established at Princeton University and the California Institute of Technology by the Daniel and Florence Guggenheim Foundation to provide research facilities and graduate training for qualified young scientists and engineers in rocketry and astronautics. Robert H. Goddard Chairs were established at each center.

*December 16:* First flight of tailless X-4 (No. 1) research airplane completed, Northrop test pilot Charles Tucker as pilot. Two X-4's were built by Northrop and some 60 research flights were made by NACA at Muroc with the X-4 (No. 2) after about a dozen Air Force flights.

*December 29:* The first report of the Secretary of Defense, James Forrestal, reported that the United States had been engaged in research on an earth satellite. The Report of the Executive Secretary of the Research and Development Board, contained as an appendix, stated: "The Earth Satellite Vehicle Program, which was being carried out independently by each military service, was assigned to the Committee on Guided Missiles for coordination."

*During 1948:* First turboprop airliner flown, the Vickers Viscount.

———: Human Centrifuge became operational at Aero Medical Laboratory at Wright Field.

# 1949

*January 7:* X-1, flown by Capt. Charles E. Yeager, climbed 23,000 feet after launch, at record rate of 13,000 feet per minute, at Muroc.

*January 11:* First launching of a rocket model employing known but nonaerodynamic torque from canted rocket nozzles, for determining damping in roll of wings, at NACA's Wallops Island, Va.

*January 26:* First guided-missile test ship, U.S.S. *Norton Sound*, launched its first missile, a Loon, off NAMTC, Point Mugu, Calif.

*During January:* Army established formal requirement for a surface-to-air missile system to combat ballistic missiles.

*February 9:* The Department of Space Medicine was established at the School of Aviation Medicine, Randolph AFB, Tex.

*February 24:* An Army JPL Bumper-Wac two-stage rocket (a Wac Corporal mounted on a V-2 first stage) attained a record altitude of 244 miles and record speed of 5,150 miles per hour over White Sands, N. Mex., yielding information about ion densities in the F-region of the ionosphere.

*March 2:* At Carswell Air Force Base, Tex., USAF Boeing B-50, *Lucky Lady II*, with Capt. James Gallagher as pilot, completed the first nonstop, round-the-world flight in history, having covered 23,452 miles in 94 hours 1 minute, and having been refueled in the air over the Azores, Arabia, the Philippines, and Hawaii.

*March 4:* Navy flying boat, *Caroline Mars*, set new world passenger-load record by carrying 269 persons from San Diego to San Francisco.

*March 12:* Development of a multichannel telemetering system announced by the Navy.

*March 16:* First experimental track-type landing gear delivered to USAF, received by 314th Troop Carrier Wing from Fairchild Aviation Corp. for installation on C-82 aircraft.

*March 25:* New world helicopter speed record of 133.9 mph at Niagara Falls, N.Y., claimed by XH-12 of Bell Aircraft Co.

*March 26:* USAF B-36 with six reciprocating and four jet engines made first test flight at Forth Worth, Tex.

*March 30:* The President signed a bill providing for construction of a "perma-

## 1949—Continued

nent" radar defense network for the United States.

*During March:* Concept of launching of small high-performance rockets suspended from a balloon above most of the atmosphere (later called "Rockoons"), developed by Cmdr. Lee Lewis, Cmdr. G. Halvorson, S. F. Singer, and J. A. Van Allen during Aerobee firing cruise of U.S.S. *Norton Sound.*

*April 8:* First successful rocket-propelled RM-10 research missile for drag and heat transfer studies at transonic and supersonic speeds, making use of skin calorimeter techniques, at Wallops Island, Va.

*April 21:* First European flight of aircraft powered solely with ramjet engine made in France, an air-launched Leduc which flew for 12 minues. Rene Leduc had worked with ramjet design since 1935.

*May 3:* Naval Research Laboratory's Martin Viking rocket No. 1 fired at White Sands Proving Ground, N. Mex., reached an altitude of 51½ miles and a speed of 2,250 mph; its payload contained upper air pressure and temperature experiments.

———: President T r u m a n approved amendments to the basic legislation of 1915 covering "Rules and Regulations for the Conduct of the Work of the National Advisory Committee for Aeronautics," a basic statement of organizational responsibilities.

*May 11:* President Truman signed a bill providing a 5,000-mile guided-missile test range, which was subsequently established at Cape Canaveral, Fla.

*May 13:* Prototype of British Canberra medium jet bomber first flown, at Warton, England.

*May 24–26:* Second International Conference on Aeronautics, combining the Royal Aeronautical Society and the Institute of Aeronautical Sciences, held in New York.

*During May:* Single-stage Russian rocket attained an altitude of 68 miles with an instrument payload of 264 to 286 pounds, according to *Tass*, March 27, 1958.

———: Pratt & Whitney submitted specifications for XJ57-P-1 turbojet engine, basic design for which had begun in 1947 and for which production began in February 1953. The J57 ultimately powered the B-52, YB-60, F-100, F-101, YF-105A, KC-135, Boeing 707, F4D, and A3D, as well as the SNARK (SM-62) missile.

———: NACA Ames Aeronautical Laboratory completed a 10- by 14-inch supersonic wind tunnel with top Mach number of 5, later increased to 6.3.

*June 9:* First use of small pulse rockets in flight as disturbing impulse for evaluation of dynamic stability in a model of the Rascal missile, at NACA's Wallops Island.

*June 14:* Second V-2 flight carrying a live AF Aero Medical Laboratory monkey, Albert II, attained an altitude of 83 miles; the monkey survived but died on impact.

*June 27:* Naval Ordnance Laboratory (NOL) at White Oak, Md., dedicated new aeroballistic facilities, which included supersonic and hypersonic wind tunnels (up to Mach 10) and the first pressurized ballistic range.

*During June:* NACA's first hovering flights of a simplified propeller vertical takeoff landing (VTOL) airplane model conducted at Langley Laboratory.

*August 8:* First operational emergency use of T-1 partial pressure suit by Maj. F. K. Everest (USAF) in X-1 aircraft at 69,000 feet; suit's automatic operation saved pilot and aircraft.

*August 9:* First use in United States of a pilot ejection seat, by Lt. J. L. Fruin (USN), from F2H-1 Banshee while making over 500 knots near Waterboro, S.C.

*During August:* Wernher von Braun named an Honorary Fellow of the British Interplanetary Society.

*October 1:* Long-Range Proving Ground at Cape Canaveral was activated.

*October 27:* The Unitary Wind Tunnel Act (63 Stat. 936) authorized the construction of $136 million of new NACA facilities, $10 million for wind tunnels at universities, $6 million for a wind tunnel at the David W. Taylor Model Basin, and $100 million for the establishment of the Air Force Arnold Engineering Development Center, at Tullahoma, Tenn., in recognition of the fact that industry could not subsidize expensive wind tunnels for research in transonic and supersonic flight.

*November 3:* Charles B. Moore (General Mills) made first manned flight in a polyethelene balloon over Minneapolis Minn.

*November 10:* Piasecki HRP-2 passenger transport helicopter made first flights.

*November 21:* USAF Sikorsky H-19 12-place helicopter made first test flight.

*November 22:* D-558-II Skyrocket exceeded the speed of sound at Edwards AFB, Calif. It was powered by both a Westinghouse J-34 turbojet engine and a Reaction Motors, Inc. rocket motor.

*December 1:* Supersonic wind tunnel, capable of 3,000-mph speeds, was dedicated at MIT.

*December 2:* First firing of USAF Aerobee research rocket (RTV-A-1a) at Holloman AFB, the development of which was initiated earlier in the year.

*December 12:* Last monkey, Albert IV, launched in V-2 series of tests at WSPG, a successful flight indicating no ill effects on monkey until impact of V-2.

*December 22:* North American YF-86D completed first flight test at Edwards AFB.

*December 25:* Air Force revealed development of stupalith, a ceramic which contracts when heated and expands when cooled, and which can stand heat of 2,000°, used on jet and rocket engines.

*December 28:* USAF reported that 2-year investigation had found that there was no such thing as a "flying saucer" and that Project Saucer at Wright-Patterson AFB had been discontinued.

*During December:* First continuous transonic flow established in NACA's Langley 8-foot high-speed wind tunnel with use of slotted-throat technique. (See January 8, 1947.) This was a major milestone in wind-tunnel technique.

*During 1949:* USAF Advisory Committee headed by Louis N. Ridenour recommended that Air Force research and development be consolidated into a single command.

————: First "probe and drogue" method of contact aerial refueling performed in England (developed by Flight Refuelling, Ltd.). Early in year the USAF had issued requirement for development of a refueling method other than loop hose for use with single-seat jet fighter aircraft. After the nonstop round-the-world flight of the B-29 *Lucky Lady* using the Boeing loop-hose method in March, Boeing developed the "boom technique."

————: Complete fixed-component combined loads testing machine was completed and operated at NACA Langley Laboratory, remaining in use through 1960. It was first machine capable of applying forces along each of three axes and moments about those axes (positive and negative), in any combination of forces and moments, each applied independently.

# 1950

*January 10:* U.S.S. *Norton Sound* began 19-day firing cruise in Alaskan waters, launching two Aerobees, one Lark and one Loon. Eight scientists connected with Aerobee upper atmosphere research program and Army, Navy, and Air Force observers made the cruise.

*January 13:* First successful automatic homing flight of Navy Lark (XSAM-N-4) launched at NAMTC, making simulated interception at a range of 17,300 yards at an altitude of 7,400 feet.

*January 23:* USAF established the Air

## 1950—Continued

Research and Development Command (ARDC).

*January 29:* Remains of Wac Corporal which reached 250-mile altitude on February 24, 1949, found on desert near WSPG.

*January 30:* President Truman announced his decision to go ahead on the hydrogen bomb development program.

*During January:* Contractor study launched by USAF which led to Bomarc interceptor missile.

*February 9:* Navy's Martin Viking No. 3 successfully launched to 50-mile altitude from White Sands.

*February 10:* Secretary of the Air Force directed that the Air Engineering Development Center be renamed the Arnold Engineering Development Center in honor of the late General of the Air Force, Henry H. Arnold.

*February 17:* V-2 reached an altitude of 92 miles in launch from WSPG.

*March 2:* First full-thrust test of 75,000-pound liquid rocket engine for the Navaho (XLR43–NA–1) conducted by North American at Santa Susana, Calif.

*March 3:* Symposium on space medicine held by the University of Illinois at its Professional Colleges in Chicago.

*March 15:* The Joint Chiefs of Staff, in a basic decision on guided-missile roles and missions, gave the USAF formal and exclusive responsibility for strategic guided missiles.

*March 24:* First successful ramjet research model flown at Wallops Island by NACA Langley's PARD.

*During March:* Radiobiological Laboratory established at Austin, Tex., by the USAF School of Aviation Medicine and the University of Texas.

———: Hypersonic wind tunnel became operational at Wright-Patterson AFB.

*April 1:* Missile staff headed by Wernher von Braun was moved from White Sands to Army Ordnance's Redstone Arsenal, Huntsville, Ala.

*May 3:* Submarine *Cusk* launched a Loon guided missile and after submerging, tracked and controlled its flight to a range of 105 miles.

*May 10:* President Truman signed legislation creating the National Science Foundation.

*May 11:* NRL Viking No. 4 research rocket fired from the U.S.S. *Norton Sound* near Jarvis Island in the Pacific, at the intersection of the geographic and geomagnetic equators, obtaining cosmic-ray and pressure-temperature data. It set a 106.4-mile altitude record for an American single-stage rocket and was the first firing of the Viking from shipboard.

*May 12:* Last flight of X–1 (No. 1) rocket research airplane, for RKO motion picture "Test Pilot," which was turned over to the National Air Museum at the Smithsonian on August 28th.

*May 15:* Navy announced completion of test chamber at the Ordnance Aerophysics Laboratory at Daingerfield, Tex., capable of conducting tests of full-scale ramjet engines up to 48 inches in diameter at simulated altitudes up to 100,000 feet.

*May 19:* First Army Hermes A–1 test rocket fired at WSPG.

*During May:* New York University research balloon released from Holloman AFB drifted 7,000 miles and was recovered in Myrdal, Norway.

———: USAF SAM scientists, Drs. Fritz and Heinz Haber, delivered paper on "Possible Methods of Producing the Gravity-Free State for Medical Research," suggesting aerodynamic parabolas with use of aircraft to obtain up to 30 seconds of relative weightlessness.

*June 6:* Ramjet missile launched which accelerated under ramjet power to Mach 3.1 at 67,200-feet altitude, at NACA Wallops Island.

*June 13:* Department of Defense assigned range responsibilities to the armed services: Army: White Sands, N. Mex., Proving Ground and nearby Holloman Air

Force Base at Alamogordo; Navy: Point Mugu, Calif.; Air Force: Long-Range Proving Ground at Banana River, Fla. (now called Cape Canaveral).

*June 23:* First run of rocket-propelled research sled made on the 3,550-foot track at Holloman Air Force Base.

*June 25:* North Korea armed forces invaded South Korea.

*During June:* VfR, the German Rocket Society disestablished by Hitler in 1933, passed resolution calling for international conference of all astronautical societies.

——: Secretary of Defense created Guided Missiles Interdepartmental Operational Requirements Group.

*July 1:* Lacrosse guided-missile project, begun in 1947 by Naval Ordnance, transferred to the Department of the Army by the JCS.

*July 5:* James H. Doolittle named "aviator of the decade" (1940–49) by the Harmon International Aviation Awards Committee, while Jacqueline Cochran was named "aviatrix of the decade."

*July 21:* First polyethylene balloon launched at Holloman by USAF personnel.

*July 24:* Bumper No. 8, a German V-2 with a 700-pound Army-JPL Wac Corporal, was fired from Long-Range Proving Ground at Cape Canaveral; the first-stage V-2 climbed 10 miles, separated from the second-stage Corporal which traveled 15 more miles. This was the first missile launch from Cape Canaveral.

*July 29:* Bumper No. 7 was the second missile launch from Cape Canaveral, reached highest velocity (Mach 9) attained by a manmade object to date.

*August 1:* Patrick Air Force Base, administrative headquarters of the AFMTC at Cape Canaveral, officially named after Gen. Mason M. Patrick.

*During Fall:* Rand Corp. completed missile feasibility studies begun in 1949, which confirmed the military practicability of long-range rocket weapons.

*August 31:* Last of five Aeromedical Laboratory experiments (first four known as Albert series) fired by V-2 No. 51 from WSPG, which carried a non-anesthetized mouse photographed by a camera which survived impact.

*September 22:* Col. David C. Schilling and Lt. Col. William Ritchie flew two Republic F-84E jet fighters across the Atlantic nonstop, Schilling flying from London to New York with three in-flight refuelings, the first nonstop jet flight across the Atlantic, while Ritchie was forced to bail out over Newfoundland.

*September 28:* In a balloon launched at Holloman AFB, eight white mice survived an Aeromedical Laboratory flight to an altitude of 97,000 feet.

*September 29:* Record parachute jump from 42,449 feet made by Capt. R. V. Wheeler at Holloman AFB, N. Mex.

*September 30:* First International Congress on Astronautics held in Paris proposed creation of a permanent federation of astronautical societies.

*During September:* USAF School of Aviation Medicine's Department of Space Medicine headed by Hubertus Strughold, formulated research concept of "atmospheric space equivalence."

*October 24:* Kaufman T. Keller, president of the Chrysler Corp., appointed to the newly created position of Director of Guided Missiles for the U.S. Armed Forces.

*October 25:* The first Lark missile launched by Air Force from Cape Canaveral, the last of the three missiles launched in 1950 at the LRPG.

*October 26:* Army contracted with Douglas Aircraft for design, development, fabrication, and flight testing of rocket having Honest John specifications.

*During October:* Air Force announced program to replace all piston-engine aircraft with jet aircraft.

——: USAF canceled XF-85 parasite fighter project after flight test at Edwards AFB revealed that parasite fighter escort for B-36 was not feasible.

## 1950—Continued

*November 6-9:* USAF School of Aviation Medicine and Lovelace Foundation sponsored a "Symposium on the Physics and Medicine of the Upper Atmosphere," at San Antonio, Tex.

*November 8:* First jet airplane dogfight when USAF Lockheed F-80 piloted by Lt. J. R. Brown downed a Russian-built MiG-15 over Korea.

*November 21:* Navy Viking No. 5 attained 108-mile altitude.

*December 6:* Establishment of transonic flow in the Langley 16-foot high-speed wind tunnel following installation of a slotted-throat test section.

*December 11:* Navy Viking No. 6 in night firing attained only 40 miles altitude.

*During December:* Dr. Hugh L. Dryden, Director of NACA, awarded the Daniel Guggenheim Award for 1950 for outstanding leadership. (See Appendix D.)

———: Construction started at Grand Bahama Island for the first tracking station on the Florida Missile Test Range, later the Atlantic Missile Range.

*During 1950:* NACA Langley's Pilotless Aircraft Research Division demonstrated low drag of thin delta wing (which led to F-102, F-106, B-58) with rocket-powered model flights.

———: Worldwide analysis of atmospheric turbulence and gusts was made at Langley Aeronautical Laboratory based on data taken with NACA-developed VG and VGH recorders on commercial airline operations on transpacific and South American routes.

# 1951

*January 16:* Air Force established Project MX-1593 (Project Atlas), study phase for an intercontinental missile. Contract given Consolidated-Vultee Aircraft on January 23. This was the follow-on to Project MX-774 terminated in 1947.

*January 31:* F-51 set new London to New York speed record of 8 hours and 55 minutes.

*During January:* James Forrestal Center established at Princeton University as a jet-propulsion research center.

———: Westinghouse J-40 jet engine (7,500 pounds dry thrust) completed 150-hour Navy qualification test.

*February 14:* Republic F-84F with Wright J-65 Sapphire engine made first flight at Edwards AFB.

*During February:* Hiller Helicopters produced two-place helicopter powered by ramjet engines.

———: NACA Langley Research Center conducted first flights of man-carrying, jet-supported platform at Wallops Island in exploratory investigations. In these tests, a person was supported by a jet-thrust device attached to his feet.

*March 6:* Talos missile powered by ramjet engine launched at Naval Ordnance Test Station, and operated 2 minutes in longest full-scale ramjet flight yet achieved.

*March 29:* Navy Regulus (XSSM-N-8) operating under airborne command took off and landed at Edwards AFB, Calif.

*March 31:* Navy issued contract to Convair for the XFY-1, propellor-driven VTOL fighter.

*During March:* Pratt & Whitney began flight test of new 10,000-pound thrust J-57 jet engine, using converted B-50 as test bed.

*April 2:* USAF Air Research and Development Command (ARDC) became operational, to which was assigned: Air Development Force at Wright Field; AF Cambridge Research Division; AF Flight Test Center at Edwards AFB; and the Holloman AFB R&D establishment (later AFMDC). Later the Arnold Engineering Development Center (Tullahoma,

Tenn.); AF Armament Center (Eglin AFB, Fla.); and the AF Special Weapons Center (Kirtland AFB, N. Mex.).

*April 18:* The first Aerobee research rocket containing a biomedical experiment was launched at Holloman AFB, N. Mex.

*May 14:* Air Force Missile Test Center (AFMTC) established at Long-Range Proving Ground, and assigned to the Air Research and Development Command (ARDC).

*During May:* Drs. H. Strughold, H. Haber, and F. Haber initiated first research program on weightlessness at the USAF SAM, a study (Task No. 7758-20) suspended in 1952, and reactivated by Dr. S. J. Gerathewohl on July 1, 1955.

*June 11:* Navy D-558-II Douglas Skyrocket, flown by test pilot William Bridgeman, set a new unofficial airplane speed and altitude record at Edwards AFB, Muroc Dry Lake, Calif.; speed estimated at more than 1,200 mph; altitude estimated 70,000 feet.

*June 17:* Navy issued contract to Convair for development of delta-winged, hydroski-equipped research seaplane with fighter characteristics, subsequently known as XF2Y-1.

*June 20:* First launching of USAF B-61 Martin Matador pilotless aircraft at Missile Test Center.

———: Bell X-5 (No. 1) research airplane made first flight of 30 minutes at Edwards, Calif., with Jean Ziegler as pilot. This was first flight of an aircraft with variable-sweep, a USAF-NACA research project for investigation of various sweeps.

*June 22:* JPL fired first of a series of 3,544 Loki solid-propellant antiaircraft missiles at WSPG, the Army program ending after September 1955. Loki rocket was later used in ONR Rockoon upper atmosphere balloon-launched rocket research soundings.

*June 25:* USAF Arnold Engineering Development Center at Tullahoma, Tenn., dedicated by President Truman, to test and evaluate supersonic aircraft and guided missiles.

*June 30:* United States terminated its V-2 program, 67 V-2's having been flown since the first American launch of a V-2 on April 16, 1946.

*July 1:* Navy Air Turbine Test Station commissioned at Trenton, N.J., to test and to evaluate turbojet, turboprop, ramjet, and pulse-jet engines, accessories, and components.

*July 6:* Air-to-air refueling of jet aircraft (RF-80) in combat zone accomplished in Korea, believed the first such hookup.

*July 20:* First flight of Consolidated XF-92A, a USAF and later a NACA research airplane (predecessor of the F-102) at Edwards AFB.

*July 21:* United States-United Kingdom agreement signed and went into effect permitting the extension of the U.S. missile range southeastward from Florida on its first leg through the Bahamas.

*August 7:* A Navy Viking 7 rocket set an altitude record for single-stage rockets, climbing to 136 miles and reaching a speed of 4,100 mph, at White Sands, N. Mex., highest flight of original airframe design.

———: D-558-II Skyrocket reached maximum speed of 1,238 mph, with William Bridgeman as pilot.

*August 15:* William Bridgeman flew the D-558-II Skyrocket to 79,494 feet, highest altitude attained by a human being to date.

*August 29:* First of USAF Aeromedical Laboratory balloon flights at White Sands.

*August 30:* First successful launching of NACA Langley's PARD of an underslung or "piggyback" rocket booster system, at Wallops Island, Va.

*During August:* X-1D airplane destroyed by explosion.

*September 3:* The International Astronautical Federation was formed by scientists of 10 nations at the Second International Congress on Astronautics to coordinate responsibility on flights to the moon and planets. Predicted within the decade: a 50-ton earth satellite traveling 18,000 mph, orbiting earth at an altitude of 300 miles.

## 1951—Continued

*September 5:* USAF awarded contract to Consolidated Vultee to fly a B-36 with a nuclear reactor aboard, to be built by General Electric, for added boost.

*September 20:* USAF made first successful recovery of animals from a rocket flight when an instrumented monkey and 11 mice survived an Aerobee flight to an altitude of 236,000 feet from Holloman AFB.

*September 28:* Special meeting of the Air Force Council reviewed USAF R&D program and recommended to Chief of Staff Vandenberg the development of an intercontinental strategic weapons system.

*During September:* USAF directed all work in Project MX-1593 (Atlas) be for development of a rocket-powered ballistic missile.

*October 4:* M. K. Tikhonravov in *New York Times* said U.S.S.R. science made feasible space flight and creation of artificial earth satellite; reported U.S.S.R. rocket advance equaled or exceeded West.

*October 10:* JPL Corporal E-11 fired at WSPG, the basic configuration of the Army's Corporal tactical missile.

*October 29:* Firing of V-2, No. 66, at White Sands Proving Ground concluded U.S. use of these German missiles in upper atmosphere rocket research.

*October 31:* Responsibility for Hermes II transferred to Army Ordnance Guided Missile Center at Redstone Arsenal; Hermes II redesignated the RVA-A-3 test vehicle.

*During October:* International Council of Scientific Unions decided to hold Third International Polar Year, later to become International Geophysical Year in October 1952.

*November 8:* First successful launching of a research model propelled by the helium gun catapult, by Langley's PARD at Wallops Island.

*November 9:* X-1 (No. 3) rocket research airplane and its B-29 "mother" airplane were destroyed on the ground by explosion and fire.

*November 13:* First experimental investigation of transonic-type compressor was conducted at Lewis Laboratory, a breakthrough in compressor technology later utilized by virtually all advanced turbojet engines.

*December 16:* Navy Kaman K-225, modified as the first gas-turbine, shaft-powered helicopter, successfully completed flight test.

*During December:* Richard T. Whitcomb of NACA Langley Laboratory verified the "area rule" in NACA's new transonic wind tunnels which enabled significant gain in jet aircraft speeds with what became known as the "coke bottle" or "wasp waist" shape.

*During 1951:* NACA Lewis Laboratory completed first rocket combustion tests using the high-energy propellant liquid fluorine as an oxidant.

———: Production of J-65 Sapphire turbojet engine begun by Curtiss Wright, later fitted in some instances with afterburner in F-84F, B-57, FJ-3, F11F-1, and A4D-1.

———: USAF initiated development of liquid-propellant rocket engine with thrust of 150,000 pounds (XLR 43-NA-3).

———: NACA's A. Scott Crossfield first flew a series of aerodynamic parabolas to produce a short period of weightlessness, in a YF-84 at Edwards AFB, Calif. Maj. Charles E. Yeager (USAF) also flew some of these so-called "Keplerian trajectories."

———: Air passenger-miles (10,679,281,000) exceeded total passenger-miles traveled in Pullman cars (10,224,714,000), the first time in U.S. history.

# 1952

*February 25:* Army Nike I first test fired at WSPG.

*During February:* Establishment of Atmosphere and Astrophysics Division within Naval Research Laboratory, headed by Dr. John P. Hagen.

*March 18:* First successful solid-fuel ramjet research model flown at NACA's Wallops Island.

*March 23:* Two-place glider altitude record of 44,000 feet claimed by L. Edgar and H. Klieforth, Sacramento, Calif.

*During March:* Theodore von Kármán named Chairman of NATO's Advisory Group for Aeronautical Research and Development.

*April 15:* First flight of YB-52, first all-jet heavy bomber.

*April 21:* BOAC De Havilland Comet inaugurated first jet passenger service, between London and Rome.

*During April:* DOD directed Research and Development Board to determine whether Air Force with 39 different aircraft types ordered in 1953 procurement program, and the Navy with 27 different types, were operating too many different types of aircraft.

*May 3:* First successful North Pole landing, by a ski-and-wheel USAF C-47.

*May 7:* First flight of USAF X-17 ramjet test vehicle.

*May 16:* Special Committee for the International Geophysical Year established by the International Council of Scientific Unions to coordinate the international IGY programs. This committee was known as CSAGI after the initials of its French name.

———: Navy Terrier missile completed development program with successful destruction of two F6F-5K target drones.

*May 22:* Air Force Aerobee rocket placed an aeromedical payload containing two monkeys and two mice to an altitude of 36 miles, which were recovered unharmed and without apparent ill effect.

*May 26:* Navy's first and for many years the world's largest wind tunnel was decommissioned at the Naval Gun Factory, Washington, D.C. Completed in 1914, the wooden 8- by 8-foot wind tunnel was used over 30 years.

*June 17:* Aviation Medical Acceleration Laboratory dedicated at NADC at Johnsville, Pa., which featured human centrifuge capable of producing accelerations of up to 40 g's.

*June 18:* H. Julian Allen of NACA Ames Laboratory conceived the "blunt nose principle" which submitted that a blunt shape would absorb only one-half of 1 percent of the heat generated by the reentry of a body into the earth's atmosphere. This principle was later significant to ICBM nose cone and the Mercury capsule development.

*June 20:* Navy issued contract for construction of a transonic wind tunnel at the David Taylor Model Basin.

*June 27:* First glide flight of X-2 (No. 2) research airplane, by Jean "Skip" Ziegler, Bell test pilot.

*During June:* Goodyear delivered largest nonrigid airship built, the ZPN-1, to Lakehurst.

*July 2:* First AF fighter armed solely with rockets, a Lockheed F-94C jet, disclosed by USAF.

*July 14:* NACA's Executive Committee directed its laboratories to begin study of problems likely to be encountered in flight beyond the atmosphere, which in May 1954 resulted in decision in favor of manned research vehicle and NACA's proposal to the Air Force that such a vehicle be developed.

## 1952—Continued

*July 19:* First successful flights of balloons at controlled constant altitudes in the stratosphere for periods of more than 3 days announced by the USAF.

*July 22:* First production-line Nike made successful flight.

*July 26:* Aerobee fired capsule containing two monkeys and two mice to approximately 200,000 feet at Holloman AFB, all recovered unharmed.

*July 29:* First Rockoon (balloon-launched rocket) launched from icebreaker *Eastwind* off Greenland by ONR group under James A. Van Allen. Rockoon low-cost technique was conceived during Aerobee firing cruse of the *Norton Sound* in March 1949, and was later used by ONR and University of Iowa research groups in 1953–55 and 1957, from ships in sea between Boston and Thule, Greenland.

*July 31:* First transatlantic helicopter flight, by two AF MATS sikorsky H-19's.

*August 22:* European Office, ARCD, established in Brussels to handle USAF European research contracts.

*August 25–29:* 3,700 Moslems airlifted to Mecca by 14 USAF C-54 transports.

*August-September:* Series of Rockoon launchings from *Eastwind* in high-altitude research by ONR group.

*September 1:* Third International Congress on Astronautics adopted a constitution for the International Astronautical Federation (IAF), at Stuttgart, Germany.

*September 3:* First fully configured Sidewinder air-to-air missile successfully flown at Naval Ordnance Test Station, Inyokern, the beginning of an extensive period of developmental testing.

*September 18:* Construction begun of Thule AFB in northwestern Greenland, 930 miles from the North Pole.

*September 30:* First launching of Bell Rascal XGAM-63 air-to-surface strategic missile.

*During September:* H. Julian Allen of NACA Ames Laboratory personally imparted his findings on the blunt nose cone directly to the missile industry, which first was disseminated in official report early in 1953, later as NACA TN4047.

*October 20:* First flight of Douglas X-3, an USAF–NACA research airplane (Flying Stiletto) completed, William Bridgeman as pilot.

*October 23:* Hughes XH-17 Flying Crane helicopter completed first official flight.

*During October:* Scope of International Polar Year broadened and its name changed to International Geophysical Year (IGY) by the International Council of Scientific Unions.

——: James P. Henry of the Aeromedical Laboratory at Wright-Patterson AFB published research on behavior of animals under subgravity conditions, for which he and his associates later received the Tuttle Memorial Award.

*November 1:* First hydrogen device exploded at AEC Eniwetok proving ground.

*November 19:* At Santa Susana, Calif., a complete liquid-rocket engine assembly (Navaho) having a thrust in excess of 100,000 pounds was fired for the first time.

——: North American F-86D established official speed record of 698.505 mph at Salton Sea, Calif., Capt. J. Slade Nash (USAF) as pilot.

*November 26:* Northrop B-62 Snark, a turbojet subsonic missile with 5,500 nautical-mile range, first launched from a zero-length launcher.

*December 15:* NLR Viking No. 9 research rocket launched to an altitude of 135 miles at White Sands, and Navy revealed that it had launched rockets from balloons in the geomagnetic North Pole area for cosmic ray research.

*During December:* Republic XF-91 made its first supersonic rocket-powered flight (Reaction Motors 6,000-pound-thrust rocket engine) at Edwards AFB.

*During 1952:* Convair designers became interested in conical camber principle conceived by NACA Ames Laboratory scientist Charles F. Hall in 1949, verified by wind tunnel experiments 1950–57, and applied with success to the F-102 fighter and the B-58 bomber.

———: Transistors first placed in service in Bell Telephone System network as part of long-distance dialing service.

———: New optical and photographic methods for rapidly measuring rocket combustion temperatures and flow processes developed by NACA Lewis Laboratory.

———: NACA Lewis Propulsion Laboratory first identified high-frequency combustion-oscillations in jet engine afterburners, developed partial controls by 1954, and rational design solutions by 1958.

———: First actual animal experiments on weightlessness carried out in rocket launches by AF Aeromedical Laboratory at Holloman AFB, while E. R. Ballinger conducted first manned aircraft weightlessness experiments with instrumented humans at Wright-Patterson AFB.

———: NACA undertook studies of the problems of manned and unmanned flight in the upper atmosphere and at hypersonic speeds, such studies leading to the development of the rocket-propelled X-15 research airplane.

# 1953

*January 14–16:* USAF scientific advisory panel concluded that unidentified flying objects (UFO's): (1) held no direct physical threat; (2) were not foreign developments; (3) were not unknown phenomena requiring revision of current scientific concepts; and (4) a rash of sightings offered a threat from skillful hostile propagandists.

*Jaunary 22:* First flight test of a complete airplane model designed by "area rule" concepts propelled to supersonic speeds by rocket boosters, at Langley Wallops Island, Va.

*During January:* 10- by 10-foot jet-engine test facility at Lewis Flight Propulsion Laboratory began operations.

*February 1:* Chance Vought delivered last propeller-driven fighter, the Navy F4U Corsair, the 12,571st built since first one flew in 1940.

*February 13:* First full guidance flight of Navy Sparrow III missile at Naval Air Missile Test Center.

*February 19–26:* Six *Moby Dick* balloon flights to study high-altitude winds flown from Vernalles NAS, Calif., by USAF Cambridge Research Center, each capsule also containing fruit flies. (See Appendix C.)

*February 21:* First powered flight of the Bell X-1A research airplane was completed, Jean Ziegler as pilot.

*February 26:* Dorothy M. Simon, aeronautical research scientist with Lewis Flight Propulsion Laboratory, was recipient of 1952 Rockefeller Public Service Award "for the effective application of the physics and chemistry of combustion to flight research." (See Appendix D.)

*During February:* U.S. National Committee for the IGY established by the National Academy of Sciences.

———: Rocket test stand capable of testing engines to 400,000 pounds of thrust activated at AF Flight Test Center (AFFTC).

———: J-57 engine with a thrust of 10,000 pounds placed into production (1941 Whittle turbojet engine had 850 pounds thrust).

*During February:* American Medical Association authorized American Board of Preventive Medicine to establish aviation medicine as a distinct specialty and to grant certification for those physicians properly qualified.

*March 17:* Single-stage, air-launched rocket research vehicle exceeded Mach 5 in NACA Lewis Laboratory flight test.

*During March:* Research on 1-million-pound thrust plus engine begun at Rocketdyne, the feasibility of which was established in March 1955.

## 1953—Continued

*During March:* Boeing delivered last propeller-driven bomber, a RB–50H, to the USAF. More than 4,250 B–29 and B–50 Superforts were delivered to the AF in the last decade; more than 17,000 four-engined Boeing bombers since the first B–17 in 1935.

———: Lt. Col. John P. Stapp traveled at 421 mph on 3,500-foot track in rocket-powered sled.

*April 9:* Navy XF2Y–1 Sea Dart, an experimental delta-wing jet seaplane with hydroskis, made first flight at San Diego.

*May 8:* First launching of a cluster of three Deacon rockets as a booster at NACA's Wallops Island.

*May 12:* First Bell X–2 exploded during a captive flight killing Jean Ziegler, Bell test pilot, over Lake Ontario near Buffalo, N.Y.

*May 18:* Jacqueline Cochran became first woman to fly faster than the speed of sound, in a F–86.

*May 25:* USAF North American YF–100A made its first flight at Edwards AFB, the first service supersonic fighter.

*June 5:* Missile fired from the underground launching installation constructed by the Army Corps of Engineers, at WSPG.

*June 16:* Department of Defense Study Group on Guided Missiles established by the Armed Forces Policy Council under Secretary of Defense C. E. Wilson. This group made a technical evaluation of the missile programs of the military services. One of their recommendations was that a special evaluation of all Air Force strategic missiles be made. In the fall the Strategic Missiles Evaluation Committee, headed by John Von Neumann, made such an evaluation.

*June 30:* Department of Defense Reorganization Plan No. 6, transmitted by President Eisenhower to Congress under the Reorganization Act of 1949, abolished the Research and Development Board, the Munitions Board, the Defense Management Agency, and the Office of Director of Installations.

*July 1:* Fiscal year 1953, just concluded was first year that the United States spent as much as $1 million on strategic ballistic missile development.

*July 15:* First submarine launching of Regulus missile, from submarine *Tunny* off NAMTC.

*July 27:* Armistice signed in Korean war. USAF reported that 5th Air Force had shot down 984 Communist planes, including 823 MiG–15's. USAF lost 971 planes: 94 in aerial combat of which 58 were Sabrejets, 671 downed by ground fire, and 206 lost through other causes.

*August 3:* Fourth International Congress on Astronautics met at Zurich, at which S. F. Singer proposed Project Mouse (Minimum Orbital Unmanned Satellite Experiment).

*August 20:* Redstone missile No. 1 was fired by Army Redstone Arsenal personnel at AFMTC, Cape Canaveral, Fla.

———: Longest nonstop flight by single-engined jet fighters, made by 17 USAF F–84G Thunderjets from Albany, Ga., to Lakenheath, England, a distance of 4,485 miles.

———: USSR first announced H-bomb explosion, later reported by AEC to have occurred in U.S.S.R. on August 12.

———: First successful launching by NACA Langley's PARD of a hypersonic research vehicle for heat transfer studies consisting of a cluster of three Deacon first stage and HPAG rocket second stage, at Wallops Island, Va.

*August 21:* Flying Douglas D–558–II (No. 2) Skyrocket research aircraft which had been launched from a B–29 Superfortress at an altitude of 34,000 feet, Lt. Col. Marion E. Carl, USMC, attained an altitude of 83,235 feet at Edwards AFB, Calif.

*August 28:* At Santa Susana, Calif., a complete liquid-rocket engine assembly (Navaho) having a thrust in excess of 200,000 pounds was fired for the first time.

*September 1:* First aerial refueling of jet aircraft by jet tanker, a B–47 Stratojet by a KB–47B.

*September 9:* Trevor Gardner appointed to head a committee to eliminate inter-service competition in the development of guided missiles by Secretary of Defense Wilson.

*September 11:* First successful interception by Navy Sidewinder missile at NOTS, Inyokern.

*October 1:* First Pilotless Bomber Squadron (light) established by USAF at AFMTC in Florida.

*October 3:* New world speed record of 753.4 mph in Douglas XF4D–1 Navy Skyray fighter, Lt. Comdr. J. B. Verdin as pilot.

*October 14:* Prototype of North America's B–64 Navaho, a X–10 ramjet guided missile, made its initial flight.

*October 16:* Test pilot Robert O. Rahn, flying a Douglas XF–4D Skyray fighter at Edwards AFB, Muroc, Calif., established a world closed-course speed record of 728.11 mph.

*October 23:* The Daniel and Florence Guggenheim Institute for Flight Structures established at Columbia University for research and graduate training in flight structures, including structures intended for space flight.

*October 29:* Flying a F–100 Super Sabre at Edwards AFB, Calif., Lt. Col. Frank K. Everest, USAF, set a speed record of 755.149 mph.

*During October:* Prototype Convair F–102A delta-wing fighter first flew, a supersonic fighter featuring the NACA "wasp-waist."

———: American Astronautical Society (AAS) founded.

*November 19:* First launching of a Nike-Deacon two-stage rocket for heat transfer studies at NACA Wallops Island.

*November 20:* In a D–558–II (No. 2) which had been launched from a B–29, NACA test pilot A. Scott Crossfield established an unofficial speed record of 1,328 mph at Edwards AFB, Calif., the first Mach 2 flight (2.01).

*December 12:* In a Bell X–IA which had been launched from a B–29, Maj. Charles E. Yeager, USAF, attained a speed of 1,612 mph at Edwards AFB, Calif., about Mach 2.5.

*During December:* Nike-Ajax battalion deployed on site in Washington-Baltimore area, the first operational surface-to-air missile system in the United States.

*During 1953:* Jet Propulsion Laboratory completed development of the Corporal I, the first U.S. surface-to-surface ballistic missile, and continued with Corporal II development. Army Ordnance also asked JPL to study application of large-scale solid propellant rockets for use as surface-to-surface guided missiles.

———: Minneapolis-Honeywell Regulator Co. developed power transistor (20 watts).

———: USAF initiated meteorological survey over the United States with large plastic balloons to obtain data on winds, temperatures, and cloud formations over 50,000 feet. This survey, known as "Moby Dick," was expanded later to include other select launching areas of the world. (See February 19.)

———: Dr. Hubertus Strughold of SAM published *The Green and Red Planet: A Physiological Study of the Possibility of Life on Mars.*

———: Dr. E. G. Bowen of the Australian Radio and Physics Division of the Commonwealth Scientific and Research Organization first propounded the theory that meoteoric dust provides the nuclei for heavy rainfall. By 1956, he had collected worldwide statistics revealing correlation between heavy rainfall and showers of meteors through which the earth passed. In 1960, additional data were acquired with U–2 aircraft.

# 1954

*January 21:* First atomic-power submarine, U.S.S. *Nautilus*, launched at Groton, Conn.

*During January:* Pan American World Airways took over operations and maintenance of the Florida Missile Test Range, under AFMTC; changeover completed in March.

*February 10:* Air Force Strategic Missiles Evaluation (Teapot) Committee under Dr. John von Neumann reported possibility of major technological breakthrough on nuclear warhead size and that other technical problems associated with development of ICBM's could be resolved in a few years. It recommended that a special Air Force development-management group be established to accelerate the program

*February 12:* First flight test of a high-energy fuel made by NACA Lewis Laboratory in an air-launched test vehicle.

*February 17:* American Astronautical Society (AAS) incorporated in the State of New York.

*During February:* First flight of XF-104, powered with J-65 engine (later powered with J-79 engine).

———: Rand Corp. report recommended that Atlas ICBM program efforts be increased and its characteristics relaxed to obtain an operationally useful ICBM at an earlier date.

*March 1:* United States exploded its first hydrogen bomb in the Marshall Islands, and its second on March 20.

*March 17:* President Eisenhower signed Executive Order 10521 on the "Administration of Scientific Research by Federal Agencies," which gave the National Science Foundation major responsibility on pure scientific research.

*March 18:* First launching of a cluster of four Deacon rockets as a booster vehicle, at NACA Wallops Island.

*During March:* Work on AM-2 propulsion system for Atlas by Rocketdyne was begun, drawing upon the experience in developing the regeneratively cooled chamber developed for the Navaho.

*April 8:* Office of the Assistant Chief of Staff for Guided Missiles was established in Headquarters USAF.

*April 29:* First launching of a three-stage rocket vehicle consisting of two Nike boosters in tandem and a Deacon rocket as third stage, and also a first launching of a rocket booster system consisting of three "peelaway" Deacons as the first stage wrapped around a fourth Deacon as a second stage, and a HPAG rocket as the third stage, by NACA Langley's PARD at Wallops Island.

*During April:* Bell Laboratory announced invention of the silicon solar battery.

*May 4:* Third Symposium on Space Travel conducted at American Museum, Hayden Planetarium, New York. Harry Wexler of the Weather Bureau presented a proposal for a meteorological satellite program.

*May 7:* NRL Martin Viking No. 10, a single-stage research rocket, successfully fired to an altitude of 135 miles from White Sands with experiment instrumentation.

*May 11:* Start of 59-day special effort by ARDC, WADC, SAC, and Westinghouse combined forces to carry new radar set from initial design to flight-test status.

*May 17-25:* Navy nonrigid airship YZP6-2 established new world endurance record for unrefueled flight of 200 hours and 12 minutes, commanded by Comdr. M. H. Eppes (USN).

*May 18:* SUPER SKYHOOK, largest polyethylene balloon built to date, launched by General Mills for ONR and carried emulsions to 115,000 feet.

*May 24:* NRL Martin Viking No. 11 set an altitude record of 158 miles (834,240 feet) and attained a speed of 4,300 mph in a flight from White Sands Proving Ground, N. Mex.

*May 27:* President Eisenhower signed $5 million expansion bill for NACA to be used in research for ICBM fuel and high-speed seaplane fighters.

*June 2:* With test pilot J. F. Coleman at the controls, the Convair XFY-1, a vertical takeoff aircraft, made the first free vertical takeoff and landing at Moffett Naval Air Station, Mountain View, Calif.

*June 4:* Maj. Arthur Murray, USAF, piloted the X-1A research airplane, launched from a B-29 to a record altitude of slightly over 90,000 feet, highest so far attained by man.

*June 21:* USAF directed Air Research and Development Command to establish a special development-management group on west coast, with authority and control over all aspects of the program, to accelerate and reorient Project Atlas.

*June 25:* Project Orbiter outlined by informal committee of rocket specialists, to launch a satellite into a 200-mile orbit with a Redstone missile and a Loki second stage, which became a joint Army-Navy study project after meeting at Redstone Arsenal on August 3.

*July 1:* USAF Western Development Division (became Air Force Ballistic Missile Division in 1957) established at Inglewood, Calif., under Brig Gen. Bernard A. Schriever, with authority to direct the ballistic missile development program authorized by June 1954 directives.

*July 9:* NACA met with USAF and Navy BuAer representatives to propose the X-15 as an extension of the cooperative rocket research aircraft program. The NACA proposal was accepted as a joint effort and a memorandum of understanding was signed on December 23 naming NACA as technical director of the project, with advice from a joint Research Airplane Committee.

*July 15:* First jet-powered transport built in the United States, the prototype for the military Stratotanker and later the Boeing 707, flight tested near Seattle, Wash.

*July 25:* NRL transmitted the first voice earth-to-earth messages using the moon as a reflector of radio signals. This was later developed into the Communications Moon Relay (CMR) system, which was successfully used in November 1959 when solar disturbances in the ionosphere disrupted conventional high-frequency circuits between Washington and Hawaii.

*August 1:* Fifth International Congress on Astronautics began, at Innsbruck, Austria.

*August 3:* Navy F2Y-1 Sea Dart, a hydro-ski water-based fighter, exceeded the speed of sound at San Diego, Calif.

*August 5:* Bell X-2 (No. 2) flown on its first glide flight by Lt. Col. Frank K. Everest (USAF), at Edwards AFB.

*August 7:* The USAF revealed that the School of Aviation Medicine had previously received the "first piece of experimental equipment ever built specifically for the study of living conditions in space"—a sealed cabin, to simulate the interior of a spaceship.

*August 17:* First firing of Lacrosse "Group A" missile at WSPG.

*August 23:* First NACA flight of X-3 research airplane made by Joseph Walker at Edwards AFB, the first of 20 NACA research flights in program which concluded on May 23, 1956.

*August 24:* First flight test of the Army Dart missile at WSPG.

*August 26:* The Supplemental Appropriations Act, 1955, appropriated $2 million to the National Science Foundation to support the U.S. IGY program sponsored and coordinated by the National Academy of Sciences.

——: Major Arthur Murray (USAF) flew the Bell X-1A to 90,000-feet altitude, at Edwards AFB.

*September 24:* U.S.S.R. established Ziolkovsky Gold Medal for outstanding contribution to interplanetary communications, an award to be given every 3 years.

*September 26:* Moscow radio reported U.S.S.R. sent rockets to 240-miles height; claimed rocket for interplanetary travel designed and flight principles worked out.

## 1954—Continued

*September 29:* Army Ordnance awarded contract for Redstone missile to Chrysler Corp.

*October 4:* At meeting in Rome, launching of scientific earth satellites recommended by the Special Committee for the IGY (known as CSAGI).

*October 8:* First powered flight of Bell X-1B completed, Maj. Arthur Murray as pilot.

*October 9:* $500 million was added to the current year's budget for the guided-missile program. (In fiscal year 1950 through 1954, $700 million was spent.)

*October 14:* NACA's PARD launched four-stage, solid-fuel rocket for heat transfer data to Mach 10.4, at Wallops Island, Va.

*October 17:* Piloting a Sikorsky XH-39, Warrant Officer Billy I. Wester, USA, established a world helicopter altitude record of 24,500 feet at Bridgeport, Conn.

*October 18-19:* At the suggestion of Theodore von Kármán and following a request of Gen. H. B. Thatcher, an Ad Hoc Committee of the Scientific Advisory Board met in the Pentagon to consider the application of nuclear energy to missile propulsion. In its report, the Committee "noted that there was an almost complete hiatus in the study of the nuclear rocket from 1947 following a report by North American Aviation, until a 1953 report by the Oak Ridge National Laboratory. Because the technical problems appear so severe, and because another 6 years of no progress in this area would seem to be unfortunate," the Committee felt that a continuing study both analytical and experimental, at a modest level of effort, should be carried on.

*During October:* NRL Aerobee fired at White Sands took photographs at 100-mile altitude, first picture taken of complete hurricane, off the Texas gulf coast.

*During Fall:* U.S.S.R. created the Soviet Interdepartmental Commission on Interplanetary Communications, an action announced on April 15, 1955.

*November 2:* Test pilot J. F. Coleman flying the Convair XYF-1, took off in vertical flight, then shifted to horizontal, and finally changed back to vertical for landing at San Diego, Calif.

*November 18:* Inertial guidance system for Navaho X-10 missile tested in first flight at Downey, Calif.

*December 7:* First successful recovery of a Navaho X-10 using fully automatic approach and landing system, made at Edwards AFB, Calif.

*December 10:* On a rocket-propelled sled run, Col. John P. Stapp, USAF (MC), attained a speed of 632 mph and sustained the greatest g-force ever endured by man in recorded deceleration tests.

*December 16:* USAF announced Atlas ICBM under construction by Convair.

*December 21:* Department of Defense in a two-sentence comment reported that studies continued to be made in the earth satellite vehicle program.

*December 23:* NACA-USAF-USN Memorandum of Understanding signed for "Joint Project for a New High Speed Research Airplane," which covered what became the X-15 program. Design competition was opened by the USAF during this month.

*December 31:* Army Ordnance terminated the Hermes project, during which development of high-performance liquid-fuel rocket and first stabilized platform inertial guidance equipment had been accomplished.

*During December:* "Man in Space" produced by Walt Disney.

*During 1954:* Baffles successfully used to counter high-frequency oscillations in rocket thrust chambers, for the first time at NACA Lewis Laboratory.

———: Project Stratolab utilizing plastic balloons for scientific observations in the stratosphere initiated by ONR.

———: School of Aviation Medicine (SAM) initiated studies at the University of Texas on the use of plants for the regeneration of air in a space cabin. SAM also established a veterinary science division to support medical research involving the use of animals.

——: Development of the silicon transistor, announced by several firms during the year, while the first large transistorized calculator was demonstrated by IBM.

——: Aeromedical Laboratory biological specimens were reflown on two separate plastic balloon flights for a total of 74 hours and 35 hours at an altitude between 82,000 and 97,000 feet, mostly above 90,000 feet, at Holloman AFB.

# 1955

*January 10:* U.S.S.R. scientists stated that launching of an earth satellite was possible in the near future, according to Radio Moscow.

*January 17:* First launching of a test model towed by a rocket vehicle with a flexible towline, by Langley Laboratory's PARD at Wallops Island, Va.

*January 22:* Existence of ICBM program announced by DOD.

*February 4:* ONR Viking No. 12 research rocket attained altitude of 144 miles from White Sands.

*February 14:* Killian Committee (Technological Capabilities Panel) recommended concurrent development of IRBM of 1,500-mile range with ICBM effort.

*February 26:* First known survivor of supersonic ejection of a pilot, a North American test pilot ejected from an aircraft at Mach 1.05.

*March 1:* Trevor Gardner became the first Assistant Secretary of the Air Force for Research and Development.

*March 6:* USAF Chief of Staff, Nathan F. Twining, reported that ICBM's were receiving priority in the AF program because of known Soviet progress. Navaho, Snark, and Atlas programs accelerated.

*March 8:* First USAF unit of F-84 jet fighters formed which were capable of being launched and recovered by B-36 mother planes, the 91st Strategic Reconnaissance Squadron at Great Falls AFB.

*March 14:* U.S. National Committee for IGY completed feasibility study and endorsed earth satellite project in report to National Academy of Sciences and the National Science Foundation.

——: DOD officials announced that guided-missile spending would reach $518 million in fiscal year 1955 and $674 million in fiscal year 1956.

*March 25:* Chance Vought XF8U-1, Navy jet fighter, exceeded the speed of sound on its first flight, at Edwards AFB.

*During March:* Feasibility of F-1 rocket engine developing a million pounds of thrust in a single chamber established at Rocketdyne.

*April 6:* Launched from a B-36, an air-to-air guided missile with an atomic warhead was exploded 6 miles above Yucca Flats, Nev.

*April 15:* Soviet newspaper, *Vechernaya Moskva*, announced that an interdepartmental commission for interplanetary communication had been created to develop an earth satellite, which would improve weather forecasting by taking photographs. This commission had been established late in 1954.

*April 21:* First launching of USAF Aerobee-Hi sounding rocket (AF-55) attained height of 123 miles with a payload of 196 pounds.

*April 26:* Moscow Radio reported U.S.S.R. planned to explore moon with tank remotely controlled by radio, foresaw trips by man in 1 to 2 years, and reported formation of scientific team to devise satellite able to circle earth.

*May 2:* USAF approved Western Development Division proposals to inaugurate a second ICBM airframe, which became the Titan ICBM (SM-68).

## 1955—Continued

*May 6:* Detailed earth satellite program developed by the U.S. National Committee for IGY, forwarded by the National Academy of Sciences to the National Science Foundation for governmental consideration.

*May 10:* GE XJ-79 turbojet engine first flown in B-45 testbed, later powered the B-58 and F-104.

*May 19:* Under the Second Supplemental Appropriations Act, 1956, the National Science Foundation received an appropriation of "$27 million, to remain available until June 30, 1960," for the National Academy of Sciences' U.S.-IGY program.

*May 20:* Commission on the Organization of the Executive Branch of the Government (Hoover Commission) reported to Congress that the NACA "has a splendid record in its leadership of the Nation's aeronautical research. It justifies continued confidence and support."

*May 23-24:* Project Orbiter Conference was held at Redstone Arsenal and at Cape Canaveral.

*May 29:* U.S.S.R. reported that research was being conducted on hydrogen fusion as a means of propulsion for space applications.

*During May:* Basic study on interference lift completed by Antonio Ferri, Joseph H. Clarke, and Anthony Casaccio of Brooklyn Polytechnic Institute.

*June 1:* First experimental use at NACA Lewis Laboratory of a "boot-strap" rocket-exhaust powered ejector to permit rocket testing at simulated high-altitude conditions without complicated and expensive exhausting facilities.

*June 11:* Delivery and flight test of experimental all-magnesium F-80C aircraft, built to test weight and strength of magnesium alloys, at Wright-Patterson AFB, Ohio.

*June 24:* First Nike-Deacon sounding rocket launched at Wallops Island in cooperative USAF-NACA program of upper air density measurements.

*June 29:* First successful firing of Nike B, at WSPG.

*June 30:* The Independent Offices Appropriation Act 1956, appropriated "$10 million to remain available until June 30, 1960," for the U.S.-IGY program.

*July 1:* USAF research program on weightlessness in flight reactivated at SAM under direction of Dr. S. J. Gerathewohl, which conducted flight experiments until the spring of 1958.

*July 8:* First test run was held on the Supersonic Military Air Research Track (SMART), a 12,000-foot track for rocket-propelled sleds at Hurricane, Utah.

*July 14:* Martin P6M Seamaster, swept-wing powered with four J-71 engines, made first flight, initially demonstrating great promise for minelaying and reconnaissance missions.

*July 18:* First of Aeromedical Laboratory's 2-million-cubic-foot plastic balloons manufactured by Winzen Research, launched at Fleming Field, Minn., attained an altitude of over 120,000 feet; the second launched on the next day attained a record altitude of 126,000 feet.

*July 20:* NB-36H aircraft housing an atomic reactor made its first flight; the reactor was not activated.

*July 29:* President endorsed USNC-IGY earth satellite proposal and the White House announced that "The President has approved plans by this country for going ahead with the launching of small, unmanned, earth-circling satellites as part of the U.S. participation in the International Geophysical Year which takes place between July 1957 and December 1958." Scientific responsibility was assumed by the National Academy of Sciences, fiscal responsibility by the National Science Foundation, and responsibility for logistic and technical support by the Department of Defense.

*July 30:* U.S.S.R. announced that it planned to launch an earth satellite.

*During July-October:* Instrumented Loki I and Deacon rockets were successfully balloon launched (Rockoons) from shipboard off the coast of Greenland in cosmic-ray studies by State University of Iowa research group. Army Ordnance supplied JPL-developed Loki rockets and ONR sponsored the project.

*August 2:* L. I. Sedov, chairman of the U.S.S.R. Academy of Sciences Interdepartmental Commission on Interplanetary Communications, announced Soviet intention to launch artificial satellites during the IGY, at the Sixth International Congress on Astronautics at Copenhagen, the first IAF meeting attended by Soviet representatives.

*August 8:* X-1A exploded just prior to time of drop from "mother" B-29, NACA pilot Joseph A. Walker was saved and X-1A was jettisoned.

*August 16:* First successful demonstration of Rockair technique (research rocket launched from aircraft) by ONR and University of Maryland team, a 2.75-inch FFAR rocket fired from a Navy F2H-2 aircraft to an altitude of approximately 180,000 feet. Rockair technique first suggested by Herman Oberth (1929) and others.

———: Army Hawk missile first fired, at WSPG.

*August 2:* Col. Horace A. Hanes established a supersonic speed record for straightaway flight at 822.135 mph in a F-100 Super Sabre, at Edwards AFB.

*August 24:* Research and development Policy Council (DOD) unanimously recommended that the time-risk factor in the scientific satellite program be brought to the attention of the Secretary of Defense for determination as to whether a Redstone backup program was indicated.

*August 26:* First use of balloon target in missile testing at Holloman AFB.

*September 8:* President approved assignment of highest national priority to ICBM research and development program.

*September 9:* DOD Advisory Group known as Stewart Committee recommended that proposed Navy satellite program utilizing Viking and Aerobee-Hi rockets for satellite development proceed, with Chairman Homer J. Stewart submitting a dissenting minority report. The DOD Policy Council endorsed the majority recommendation. Designated Project Vanguard, this tri-service program was placed under Navy management and DOD monitorship. Objectives of Project Vanguard were: to develop and procure a satellite-launching vehicle; to place at least one satellite in orbit around the earth during IGY; to accomplish one scientific experiment; and to track flight to demonstrate the satellite actually attained orbit.

*September 30:* X-15 research airplane development contract let to North American Aviation.

*During fall:* Capt. Grover J. D. Schock of USAF Aeromedical Laboratory conducted subgravity flight program (Task 78501) with F-94C aircraft at Wright-Patterson AFB.

*October 2:* National Academy of Sciences' IGY Committee established Technical Panel for the Earth Satellite Program, with Richard W. Porter as Chairman, to plan the scientific aspects of the program, including the selection of experiments, the establishment of optical tracking stations, and the handling of international and interdisciplinary relations.

*October 7:* Prime contract for Project Vanguard awarded the Martin Co.

*October 15:* Douglas A4D Skyhawk set a new closed-course speed record of 695.163 mph.

*During October:* First solar-powered telephone call made by customer of regular Bell System service. During this year, fully transistorized radios and phonographs were first placed on the market.

*November 1:* U.S.S. *Boston*, first guided-missile cruiser, was placed in commission at Philadelphia Naval Shipyard.

*November 1-3:* NACA Conference on Aerodynamics of High-Speed Aircraft at Langley, at which Vernon J. Rossow presented paper on "Examples of Favorable Interference Effects on the Lift-Drag Characteristics of Aerodynamic Shapes at Supersonic Speeds."

*November 2:* First air-launched, multistage, solid-rocket-propelled vehicle flown to a Mach number greater than 8 by NACA Lewis Laboratory.

## 1955—Continued

*November 2:* The Atomic Energy Commission approved, on the basis of a statement of interest by the Department of Defense, the proposed plans of the Los Alamos Scientific and the Radiation Laboratories of the University of California, for the study and development of nuclear power for rocket propulsion.

*November 8:* Secretary of Defense approved Jupiter and Thor IRBM programs, the first based on experience gained by Redstone Arsenal team from V-2 and Redstone, the latter on experience gained from Atlas program.

*November 8–14:* Department of Defense and the Air Force established special streamlined administrative program and approved procedures (Gillettte Procedures) to prevent delays in ICBM and IRBM programs.

*November 17:* Navy created Special Projects Office under Vice Adm. W. F. Raborn to develop ship-launched missile weapon systems.

*November 18:* Air Force took action to insure earliest possible initial operational capabiilty with ICBM and IRBM.

——: First powered flight of Bell X-2 (No. 1) by Lt. Col. Frank Everest (USAF), powered by first throttlable rocket engine, the Curtiss Wright XLR25–CW–1, and Mach 0.99 was reached.

*November 22:* Republic F-105A exceeded the speed of sound in its initial flight at Edwards AFB.

*During November:* Naval Research Laboratory first transmitted transcontinental communication by means of reflecting teletype messages on the moon, from Washington, D.C., to San Diego, Calif., a technique repeated on August 12, 1960, using the ECHO I satellite for two-way reflected message transmission.

*December 1:* President Eisenhower assigned highest priority to ICBM and Thor and Jupiter IRBM programs.

*December 7:* First flight of XC-123D aircraft with boundary-layer control system in partial operation.

*December 8:* XJ-79-GE-3 turbojet engine first powered an aircraft, an XF4-D, the engine which became the primary powerplant of the B-58 and F-104.

*December 15:* First powered flight of the Bell X-1E, Joseph A. Walker, NACA test pilot, at Edwards AFB (after preliminary glide flight by Walker on December 12).

*December 20:* Secretary of Defense Wilson reported that fiscal year 1957 would have a record $1 billion for development and production of guided missiles, over the $750 million in fiscal year 1956. He also predicted an ICBM with a nuclear warhead within the next 5 years.

*December 27:* First prototype of Asp (atmospheric sounding projectile) sounding rocket, capable of payloads up to 80 pounds, launched successfully at NAMTC at Point Mugu, Calif.

*During December:* First flight of Ryan X-13, VTOL jet, at Edwards AFB.

*During 1955:* NACA Lewis Laboratory presented ARDC with results of air-breathing nuclear propulsion systems for manned applications, leading to AEC–AF Pluto project, and also initiated comparison of nuclear rocket with chemical systems for ICBM, a concept of use to Rover program.

——: Laboratory device for simulating reentry of satellites into the earth's atmosphere were first suggested by NACA Ames Laboratory scientist Eggers (Report No. RM-A55115).

——: Transistorized automatic pilot developed for USAF by Bendix Aviation.

——: Concept of nuclear reactor facility at Plum Brook proposed by NACA Lewis Research Center, construction of which was completed in 1961.

——: Army Ordnance ordered Jet Propulsion Laboratory to undertake research and development of Sergeant solid-propellant, surface-to-surface missile.

*During 1955–56:* NACA developed materials research for high-temperature jets and other structures at hypersonic speeds under direction of Robert R. Gilruth, which confirmed Redstone Arsenal's contention that ablation was sound heat protection method for reentry of nose cones and capsules.

*During 1955–58:* NACA laboratories completed basic aeronautical research supporting feasibility of B-70 supersonic bomber.

# 1956

*January 6:* President Eisenhower in his state-of-the-Union message noted the increasing importance of long-range missiles and nuclear-powered aircraft. $1.275 billion was scheduled for fiscal year 1957 production of guided missiles, with an additional $1.43 billion for military research and development.

*January 10:* First U.S.-built complete liquid-rocket engine having a thrust in excess of 400,000 pounds was fired for the first time at Santa Susana, Calif.

*January 13:* USAF Northrop Snark launched from Cape Canaveral on 2,000-mile flight.

*January 20:* ICBM Scientific Advisory Committee to the Air Force was transferred to the Office of the Secretary of Defense to assure common interchange of technical information on all missile programs.

*January 26–27:* Symposium on "The Scientific Uses of Earth Satellites" held at the University of Michigan under sponsorship of the Upper Atmosphere Rocket Research Panel, James A. Van Allen of the State University of Iowa, Chairman.

*February 1:* Army activated the Army Ballistic Missile Agency (ABMA) at Redstone Arsenal, Huntsville, Ala., to weaponize the Redstone and to develop the Jupiter IRBM.

*Early 1956:* Production of the J47 turbojet engine completed by GE, notable powerplant of B-47, F86D (also E, F, K, and L models), and FJ-2.

*March 5:* A. J. Eggers and C. A. Syvertson submitted concept for "interference lift," often referred to as "compression lift," which contributed important input for Mach 3 configurations ("Aircraft Configurations Developing High Lift-Drag Ratios at High Supersonic Speeds," NACA RM-A55I05).

*March 14:* The first Jupiter A launching, by ABMA at Cape Canaveral, Fla.

*March 20:* Ballistic Missile Committee, Office of the Secretary of Defense, approved Navy program for development of solid-propellant, ship-based ballistic missiles.

*March 27:* Secretary of Defense created Office of Special Assistant for Guided Missiles to establish more centralized controls and assist in the coordination of Army, Navy, and Air Force missile programs, including the development of earth satellite vehicles for the IGY. E. V. Murphree named to head this office.

*March 28:* Airman D. F. Smith remained in a sealed space cabin simulator for 24 hours at USAF's SAM.

*During March:* Army-Navy Ballistic Missile Committee authorized missile test launch ships, missile submarine development program, and precision navigation system for launch vehicles.

*April 3:* Navy program to procure guided missiles would jump from $126 million in fiscal year 1955, $238 million in fiscal year 1956, $353 million in fiscal year 1957, according to Navy Secretary Thomas.

*April 23:* Army informed the OSD that a Jupiter missile could be fired in an effort to orbit a small satellite in January 1957.

*April 26:* Naval Aircraft Factory at Philadelphia decommissioned, marking the passing of a name prominent in naval aviation since World War I. Naval Air Engineering Facility was established in its place to do research, engineering, design, development, and limited manufacturing of devices for launching and recovering aircraft and guided missiles.

## 1956—Continued

*April 30:* A House subcommittee heard that guided missiles, which accounted for 20.3 percent of the AF's fiscal year 1957 budget might climb to 35 percent by 1959.

*During April:* Dr. John von Neumann was awarded the Enrico Fermi Award for anticipating the importance of the high-speed computer in nuclear development programs and in the general advancement of science.

*May 3:* Plans were disclosed by the AF and Convair for a $41 million guided-missile facility at Sorento, Calif., for work on Atlas.

*May 8:* Aerobee-Hi sounding rocket reached an altitude of 116.5 miles from WSPG.

*May 18:* Development of a high-altitude research rocket, known as the Asp, for Navy's BuShips was announced.

*May 19:* National Science Foundation received an appropriation of $27 million to remain available until June 30, 1960, for the IGY, under the Second Supplemental Appropriations Act, 1956.

*May 21:* First known airborne H-bomb dropped from B-52 at approximately 50,000 feet, and exploded over the Bikini atoll in the Pacific.

*During May:* Air Force initiated a program to support the AEC's Project ROVER through application studies, propellant and materials research, and nonnuclear engine component development. Both programs were placed under a single staff in the AEC.

*June 20:* First Cajun research rocket successfully launched at NACA Wallops Island, Va.

*June 22:* Japanese Meteorological Observatory announced that the U.S.S.R. had exploded a missile-borne H-weapon at a 22-mile altitude.

*June 29:* An Aerobee-Hi rocket manufactured by Aerojet General Corp. attained an altitude of 163 miles in a launching from White Sands, N. Mex.

*July 6:* First Nike-Cajun research rocket successfully fired at Wallops Island, a cooperative NACA-University of Michigan project, attaining an altitude of 425,000 feet.

*July 14:* Navy Sidewinder missile first deployed with Sixth Fleet in Mediterranean and to the Seventh Fleet in August.

*July 23:* Lt. Col. Frank K. Everest (USAF) flew the Bell X-2 rocket-powered research plane at a record speed of just over 1,900 mph and to an altitude of 75,000 feet, at Edwards AFB, Calif.

*Mid-1956:* USAF X-17 flight test program started at Cape Canaveral to study reentry problems by simulating reentry velocities and conditions with three-stage solid-fuel Lockheed X-17. A total of 26 X-17 flights were conducted until March 1957.

*August 8:* Largest U.S. test stand for rocket motors was completed at Redstone Arsenal, slated for Jupiter IRBM.

*August 10:* Lt. Comdrs. Malcolm Ross (USNR) and L. Lewis (USN) made first stratospheric manned flight on polyethylene balloon, reached 40,000 feet in an open gondola. Flight was part of ONR Project Strato-Lab.

*August 21:* Speed record for U.S. combat aircraft of 1,015 mph set by F8U-1 Crusader flown by Comdr. R. W. Winslow (USN) over the Mojave Desert.

*August 23:* U.S. Army helicopter, the H-21, made the first transcontinental nonstop flight for helicopters, 2,610 miles from San Diego, Calif., to Washington, D.C., in 31 hours 40 minutes.

*August 24:* NACA Langley's PARD launched the world's first five-stage solid-fuel rocket to a speed in excess of Mach 15, from Wallops Island, Va.

*August 27:* First static firing of Thor rocket engine at AFFTC, Edwards AFB.

*September 2:* At the National Aircraft Show, Oklahoma City, an H-13, USA helicopter, set an endurance record in the air of 57 hours 40 minutes.

*September 7:* Capt. Iven C. Kincheloe (USAF) set new unofficial altitude record for manned flight at Edwards AFB, Calif., piloting a Bell X-2 rocket-powered aircraft to a height of 126,200 feet.

———: University of Minnesota launched ONR Mylar plastic balloon from Minneapolis, establishing unofficial world altitude record of 145,000 feet for an unmanned balloon.

*September 10–15:* Scientists from 40 nations, including the United States and U.S.S.R., at a meeting in Barcelona of the Special Committee for the IGY (CSAGI), approved resolutions calling for, among other things, countries having satellite programs to use tracking and telemetering radio systems compatible with those that have been announced at the current CSAGI meeting, and to release technical information on tracking equipment and scheduling and planning information essential to preparation for and execution of optical and radio observations.

*September 20:* First Jupiter C (a three-stage ABMA–JPL Redstone missile) was launched at Cape Canaveral, Fla., attained an altitude of 680 miles and traveled 3,300 miles downrange.

*September 21:* First flight test of a Terrapin sounding rocket at Wallops Island, which consisted of a Deacon and T55 rocket and carried a payload of 8 pounds to 400,000-feet altitude.

*September 26:* H. Froehlich and K. Long of General Mills flew ONR Strato-Lab balloon to new altitude record for an open-basket gondola of 42,000 feet.

*September 27:* After having been launched from a B–50 bomber over the Mojave Desert in California, Capt. Milburn G. Apt (USAF), flying an X–2 rocket-powered plane on its 13th powered flight, set a record speed of 2,094 mph, or Mach 3.196. In the course of the flight the aircraft crashed and the pilot was killed.

*During September:* Sperry Gyroscope delivered first experimental inertial navigation system to the Navy, for fleet missile submarines.

*October 2:* Full-scale test version of the Snark guided missile (XSM62) successfully recovered for the first time after a flight from Cape Canaveral.

*During October:* NACA scientists initiated examination of the need for a follow-on manned-rocket research vehicle to the X–15, following ARDC inquiries concerning a boost-glide vehicle.

*November 8:* Lt. Comdr. M. L. Lewis (USN) and Malcolm D. Ross established a world altitude record in a plastic STRATOLAB balloon by ascending to a height of 76,000 feet, taking off near Rapid City, S. Dak., and landing 175 miles away near Kennedy, Nebr., thus breaking the record of 72,394 feet set in 1935 by O. A. Anderson and A. W. Stevens.

*November 11:* Initial flight of Convair B–58 delta-winged Hustler, the first supersonic bomber, made at Fort Worth, Tex. B–58 incorporated the NACA "wasp waist" or "coke bottle" shape.

*November 13:* North American F–107 reached Mach 2 in flights at Edwards AFB, Calif.

*November 15:* NRL Aerobee-Hi sounding research rocket successfully fired at Fort Churchill, Canada, in a series of upper atmosphere research flights.

*November 16:* Department of Defense transferred northern portion of Camp Cooke, Calif. (now Vandenberg AFB), to the Air Force to be used as first ICBM base.

*November 26:* Secretary of Defense Wilson issued a memorandum to the Armed Forces Policy Council fixing the areas of jurisdiction of the three U.S. armed services in developing missiles of various ranges, and giving the USAF operational jurisdiction over long-range missiles, Army over missiles up to 200-mile range and for "point defense," and Navy for ship-based missiles.

*November 28:* Ryan X–13 Vertijet completed the world's first jet vertical takeoff transition flight, Peter F. Giraud of Ryan as pilot.

*November 30:* Martin TM–61 Matador, a jet-propelled missile, completed final test flight and became USAF's first tactical missile.

*During November:* Following Navy withdrawal from the Jupiter IRBM program, separate Army and Navy Ballistic Missile Committees were established under chairmanship of respective service secre-

## 1956—Continued

taries. Navy withdrawal based on interest in solid-propellant Polaris as ship-based IRBM.

*During November:* Rocket test stand capable of testing engines to 1 million pounds thrust activated at Edwards AFB, which became operational in March 1957.

*December 8:* First test rocket in the IGY–U.S. satellite program, a one-stage NRL Viking, attained an altitude of 126 miles and a speed of 4,000 mph. Viking No. 13 carried a "minitrack" radio transmitter which was ejected at 50 miles and tracked.

*December 11–18:* Twenty-four Wasp research and development chaff and parachute rockets, used to obtain wind soundings to 160,000 feet, were fired by Naval Ordnance Missile Test Facility at WSPG.

*December 17:* Navy Special Projects Office authorized Lockheed to proceed with Polaris development, having withdrawn from the Jupiter program earlier.

*December 21:* Maj. Arnold I. Beck (USAF) "soared" to a simulated altitude of 198,770 feet, the highest on record, in an Air Research and Development Command altitude chamber at Dayton, Ohio.

——: First launching of a Nike-Recruit research vehicle at NACA's Wallops Island, which reached speed of 7,600 feet per second at 13,000-feet altitude for a record dynamic pressure of 45,700 pounds per square foot.

——: The U.S. Atomic Energy Commission initiated a development program at the request of the Department of Defense to provide nuclear-electric power sources for use in Air Force satellites. The projects are designated the SNAP (Systems for Nuclear Auxiliary Power) program. Both reactor and radioisotope heat sources were specified as approaches.

*During 1956:* Research on tungsten nuclear rocket propulsion systems initiated by NACA Lewis Laboratory, and other feasible systems for practical nuclear rocket systems, such as 1958 concept of coaxial jet gaseous reactor, followed.

——: SAM and the Aeromedical Field Laboratory at Holloman carried out studies of subgravity conditions in swimming pool experiments.

——: NACA Lewis Research Laboratory completed research and development on new concepts of ramjet engine performance at altitude, increasing performance of Navaho engine experimentally approximately 40 percent and also contributing to Bomarc engine.

——: Field-effect transistor and the spacistor, which extended power and high-frequency capabilities of transistors were developed.

——: Pilotless Aircraft Research Division (PARD) of NACA Langley Aeronautical Laboratory completed solid-propellant, rocket-design studies leading to an improved Deacon rocket motor called the Cajun.

——: SAM's Department of Microbiology began micro-organism behavior studies in a "Mars Chamber," with a simulated Martian environment.

——: NACA Langley's Structures Research Division initiated electric arc-powered jets work, using DC and AC current and liquid nitrogen, liquid air and aqueous air jets. On December 19, first successful use of AC arc jet using gaseous air was performed. Twenty-four arc tunnels were subsequently developed and extensively used on many materials and structures research problems associated with reentry of bodies into the atmosphere.

# 1957

*January 10:* President Eisenhower in his State-of-the-Union message declared that "we are willing to enter any reliable agreement which would mutually control the outer space missile and satellite development."

———: Department of Defense assigned highest priority to ICBM/IRBM contracts and purchase orders.

*January 16–18:* Three of five B-52 jet bombers completed first nonstop jet flight around the world in 45 hours 20 minutes.

*January 25:* First attempted test flight of USAF Thor IRBM, only 13 months after first production contracts were signed, failed to launch.

*During January:* First of Boeing KC-135 Stratotankers placed in operational service in SAC, Castle AFB, Calif.

*February 7:* First of a series of two-stage test vehicles (RM-10) to make heat transfer studies at high speed in free flight, was launched from NACA's Pilotless Aircraft Research Station at Wallops Island, Va. Vehicle was developed by PARD of Langley Laboratory.

*February 14:* NACA established "Round Three" Steering Committee to study feasibility of a hypersonic boost-glide research airplane. "Round Three" was considered as the third major flight research program which started with the X-series of rocket-propelled supersonic research airplanes, and which considered the X-15 research airplane as the second major program. The boost-glide program eventually became known as Dyna-Soar.

*February 18:* Guggenheim Foundation granted $250,000 to Harvard University's Aviation Health and Safety Center.

*February 20:* U.S. National Committee for the IGY submitted report of its Technical Panel on the Earth Satellite Program to the National Science Foundation and the Department of Defense, which outlined a post-IGY space research program.

*March 4–15:* Navy nonrigid airship ZPB-2 completed nonstop round-trip Atlantic crossing, simultaneously establishing new world endurance record for unrefueled flight of 264 hours and 14 minutes, Comdr. J. R. Hunt commanding.

*March 10:* Ion engine research begun at NACA Lewis Laboratory.

*March 11:* Speed record for a transcontinental passenger flight was established when a Boeing 707 jet transport, with 42 passengers and a crew of 10, flew 2,335 miles from Seattle to Washington in 3 hours and 48 minutes.

*March 18:* As a result of guidance from the Secretary of Defense as to desired level of effort, the Atomic Energy Commission reduced its program on nuclear rocket propulsion to a single laboratory effort, phasing out work at the University of California Radiation Laboratory and concentrating AEC development efforts at Los Alamos Scientific Laboratory.

*During March:* NACA issued Research Memorandum entitled, "Preliminary Measurements of Atmospheric Turbulence at High Altitudes as Determined From Acceleration Measurements on Lockheed U-2 Airplane."

———: Feasibility research study instituted by USAF on the Midas early-warning satellite.

*April 8:* McDonnell F-101B Voodoo, powered by improved J-57 engine, made first flight.

*April 11:* U.S.–IGY scientific satellite equipment, including a radio transmitter and instruments for measuring temperature, pressure, cosmic rays, and meteoric dust encounters, was tested above earth for the first time, as a rocket containing this equipment was fired by the Navy to a 126-mile altitude.

———: The Ryan X-13, a jet research plane capable of vertical takeoffs and landings, flown successfully through the complete flight sequence at Edwards AFB, Calif.

## 1957—Continued

*April 19:* Douglas Thor IRBM (XSM-75) was launched at Cape Canaveral, Fla., destroyed by range safety officer.

*April 23:* Details of X-15 rocket research airplane were publicly revealed for the first time.

*April 24:* Lockheed X-17 research rocket reached 9,000 mph at Patrick AFB, Fla.

*April 30:* Aerobee-Hi No. 41 fired at White Sands reached speed of 4,900 mph and an altitude of 193 miles.

———: Naval Aviation Medical Center at Pensacola was commissioned, combining the clinical, training, and research functions of the Naval School of Aviation Medicine and the Pensacola Naval Hospital.

*During April:* Upper Atmosphere Rocket Research Panel was renamed the Rocket and Satellite Research Panel. Its chairman was James A. Van Allen of the State University of Iowa.

———: A. Dollfus flew from Paris, France, on a cluster of 100 weather balloons to an altitude of 42,000 feet.

*May 1:* Vanguard Test Vehicle (TV-1), a modified Martin Viking first-stage and Vanguard solid-propellant third-stage Grand Central Rocket as second-stage, launched with instrumented nose cone to an altitude of 121 miles and met all test objectives.

*May 6:* William M. Holaday was named as Special Assistant for Guided Missiles, Department of Defense.

*May 16:* Bomarc IM-99 ordered into production, a pilotless interceptor, which attained speeds near Mach 2 and was planned for long-range area defense.

*May 31:* Army Jupiter IRBM was fired 1,500 miles, limit of its designed range, and to an altitude of 250–300 miles, the first successful launching of an IRBM.

*June 2:* Capt. Joseph W. Kittinger, Jr. (USAF), remained aloft in plastic MAN HIGH I balloon over Minnesota for 6 hours 34 minutes, being above 92,000 feet for 2 hours and reaching 96,000 feet maximum altitude. This was first solo balloon flight into the stratosphere.

*June 10:* NACA made "Round Three" presentation on a boost-glide research airplane to ARDC.

*June 11:* First test flight of prototype WS-107A Atlas was detonated by command signal at 5,000 feet following a failure in the booster fuel system.

*June 15:* Astronomical Society of the Pacific and the International Mars Committee held a symposium on "Problems Common to Astronomy and Biology," at Flagstaff, Ariz.

*June 27:* The Goose (SM-73) became the first plastic airframe missile to fly, and reportedly the first missile to complete countdown, launch, and flight on the first attempt.

*June 28:* First phase of Project Far Side was completed, with the lifting by the world's largest balloon of a load of over a ton of military equipment and instruments to a height of more than 104,000 feet. (See Appendix C.)

*June 30:* Program to gather daily weather data over the Pacific, North America, and the Atlantic with use of transonde balloons was inaugurated with the release of first balloon from NAS Iwakuni, Japan. Preset to float at 30,000 feet, balloons carried instruments which reported pressure and temperature every 2 hours in a 5- to 8-day flight terminating short of the European coast.

*July 1:* Aerobee upper air research rocket developed by the Applied Physics Laboratory of Johns Hopkins, and first fired on September 25, 1947, completed 165 successful firings to date.

———: International Geophysical Year began. The scientists of 67 nations were to participate in a cooperative, worldwide scientific program which would last for 18 months and would be coordinated internationally by CSAGI of the International Council of Scientific Unions.

*July 10:* Convair B-58 Hustler publicly unveiled for the first time.

*July 11:* Navaho ramjet intercontinental missile program canceled by Air Force.

*July 16:* Chance Vought F8U-1 Crusader set Los Angeles to New York speed record with an average speed of 760 mph, Maj. John Glenn, Jr. (USMC), as pilot.

*July 19:* USAF fired first air-to-air nuclear warhead rocket, the Douglas MB-1 Genie, from an F-89J over Yucca Flat, Nev., during Operation Plumbob. Genie had been placed in weapon inventory of Air Defense Command in January 1957.

*July 24:* Distant Early Warning (DEW) Line, extending from Alaska's northwest coast eastward to Baffin Island, became operational.

———: Falcon GAR-2A, heat-seeking infrared missile, tested successfully.

*During July:* Examination of a satellite launch vehicle using solid fuel upper stages to achieve payload orbit with as simple a booster as possible initiated by NACA Langley, the beginning of the conception of Scout.

*During July–August:* NACA Ames Laboratory's Al Eggers worked out semiballistic design of manned reentry spacecraft.

*August 6:* First measurements of the terrestrial magnetic fields in the auroral zone, made by L. Cahill and J. A. Van Allen in firing of SUI Rockoon No. 59.

*August 7:* Army-JPL Jupiter-C fired a scale-model nose cone 1,200 miles down range from AMR with a summit altitude of 600 miles. Recovery the next day of aerodynamic nose cone using ablation, resolved reentry heating problem for Jupiter missile. Nose cone was shown to the Nation on TV by President Eisenhower on November 7.

*August 18:* Paul E. Bikle established world glider speed record of 55.02 mph over 300 km triangular course, in a Schweizer SGS 123E sailplane, from El Mirage, Calif.

*August 19:* STRATOSCOPE I, an unmanned balloon-telescope s y s t e m, launched by General Mills under Navy contract for Princeton University astronomers, which produced first "clear" photos of the sun from 80,000 feet using a 12-inch telescope.

*August 19–20:* Airborne for 32 hours in MAN HIGH II flight, Maj. David G. Simons, USAF, established a manned-balloon altitude record of 101,516 feet, ascending at Crosby, Minn., and landing at Elm Lake, S. Dak.

*August 26:* Soviet Union successfully launched a "super longdistance intercontinental multistage ballistic rocket . . . a few days ago," according to *Tass*, Soviet News Agency.

*August 28:* Supplemental Appropriation Act, 1958, appropriated $34,200,000 for the U.S. scientific satellite "to be derived by transfer from such annual appropriations available to the Department of Defense as may be determined by the Secretary of Defense, to remain available until expended."

*August 30:* Department of Defense announced that four to six Soviet ICBM tests took place in the spring of 1957.

———: USAF accepted first C-133A turboprop transport.

*During August:* Estimated operational capability date for Atlas changed from March 1959 to June 1959.

*September 3:* NACA "Study of the Feasibility of a Hypersonic Research Airplane" ("Round Three") was submitted to the Air Force.

———: Navy XKDT-1, solid-propellant, rocket-powered drone, made its first flight from F3H aircraft over NAMTC, Point Mugu, Calif.

*September 13:* 1st Missile Division of USAF activated under ARDC at Cooke AFB, Calif.

*September 20:* Complete USAF Thor IRBM first successfully launched from Cape Canaveral.

*September 26–November 9:* Thirty-six Rockoons (balloon-launched rockets) were launched from Navy icebreaker, U.S.S. *Glacier*, in Atlantic, Pacific, and Antarctic areas ranging from 75 N. to 72 S. latitude, as part of the U.S.–IGY scientific program headed by James A. Van Allen and Lawrence J. Cahill of the State University of Iowa (SUI). These were the first known upper atmosphere rocket soundings in the Antarctic area.

## 1957—Continued

*September 30–October 5:* Scientists from 12 countries, including the United States and U.S.S.R., attended International Rocket and Satellite Conference held at the National Academy of Sciences, Washington, D.C., under the sponsorship of CSAGI.

*During October:* Project Vanguard worldwide tracking system (minitrack) became operational.

PART TWO

# The First Three Years of the Space Age

OCTOBER 1957–DECEMBER 1960

THE SPACE AGE was born October 4, 1957. Launching of the first manmade object into orbit around the earth, SPUTNIK I, greatly prodded man's scientific conquest of space and animated a chain reaction of subsequent events which has not yet expired. One of the immediate consequences was the creation of the National Aeronautics and Space Administration by October 1958.

Scientific and technological progress cannot be either prevented or ignored. The pace of events since the dawn of the space age swiftly documented, as the following pages help demonstrate, that advances in aeronautics and astronautics will greatly benefit all of mankind in its peaceful pursuits. The feasibility of reliable global communications and improved weather forecasting with the use of satellites has already been demonstrated. It is often hard to realize that not until the flight of TIROS I in April, May, and June, 1960, had men viewed the earth's cloud cover from above on a global scale. Not many young Americans do not know about the Van Allen radiation belts, although it has only been since early 1958 that their existence was confirmed by EXPLORER I and PIONEER III.

In many ways, the world has been further shrunken in time-distance size and is now more clearly viewed in its celestial orbit in the minds of most of its human passengers. Rocket propulsion and associated developments provided new tools and techniques for the scientist in his quest for basic knowledge. The newly available environment of space offers, for the first time, an extraterrestrial laboratory of almost unpredictable potential. Scientists had no hard data on the space environment 22-million miles away from earth until the flight of PIONEER V in the spring of 1960. Increased understanding concerning the true nature of the earth's environment itself—geodesy, weather, ionospheres, radiation belts, Sun-Earth relationships including solar storms, and cosmic rays—appears of immediate importance to all of the physical and life sciences. Substantiation of theories concerning the origin of the universe or of the existence of life forms on nearby planets seems possible of early confirmation in the lunar and planetary exploration programs now underway.

This, then, seems the fundamental challenge presented by the recently accessible frontier in space facing mankind in this seventh decade of the 20th century, and one which requires a broad-based and sound response. It is also something entirely new and exciting in the history of mankind. Man can now physically project his vehicles and instruments, and himself soon, into that about which he could previously only observe from the surface of the earth.

Passage of time and studied analysis will inevitably provide clearer perspective from which to discern the basic significance of many of the recent events cited in the following pages. Therefore, the chronicle of events which follows must not be considered complete or explanatory. Full documentation of the events of the past three dynamic years alone will require much detailed research and evaluation. Behind almost every major event is generally a complex technical, organizational, and human story. The historical process has only been initiated here.

# 1957

*October 4:* SPUTNIK I, the first manmade earth satellite, launched by U.S.S.R. and remained in orbit until January 4, 1958. (For details see Appendix A.)

———: The National Rocket Club was organized in Washington, D.C.

*October 6:* Eighth IAF Congress began at Barcelona, Spain.

*October 9:* President Eisenhower in a White House press release congratulated the Soviet scientists on SPUTNIK I. He gave a brief history of the development of the U.S.–IGY satellite program and pointed to the separation of Project Vanguard from work on ballistic missiles.

*October 11:* Thor missile launched at Cape Canaveral, the second tested, achieved its designed 1,500-mile range.

*October 14:* USAF and NACA reviewed preliminary studies dating from 1954 on a boost-glide research vehicle to follow the X-15; all studies were combined into a single plan which was accepted by the Air Force and later designated as Dyna-Soar.

———: American Rocket Society presented to President Eisenhower a program for outer space development which proposed establishment of an Astronautical Research and Development Agency similar to NACA and AEC with responsibility for all space projects except those directly related to the military defense.

*October 16:* USAF successfully launched pellets at a speed faster than 33,000 mph (some 8,000 mph faster than the velocity necessary to escape from the earth) by an Aerobee rocket to a height of 35 miles; the nose section then ascended to a height of 54 miles where shaped charges blasted the pellets into space.

*October 18:* Lt. Comdrs. Malcolm Ross (USNR) and L. Lewis (USN) ascended to unofficial two-man altitude record of 85,700 feet in STRATO–LAB HIGH II balloon.

*October 18–20:* NACA "Round Three" Steering Committee met at Ames Laboratory.

*October 22:* Army Jupiter (IRBM) missile successfully fired at Cape Canaveral, Fla.

———: Four-stage rocket fired from a balloon at 100,000 feet above Eniwetok, in Operation Far Side, penetrated at least 2,700 miles into outer space.

*October 23:* IGY Vanguard prototype (TV–2) with simulated second and third stage successfully met test objectives, by reaching 109-mile altitude and 4,250 mph.

*October 24:* Thor long-range flight test successful from AMR, impacting 2,645 miles downrange.

*October 26:* SPUTNIK I ceased transmissions.

*October 31:* Snark intercontinental missile launched from Cape Canaveral first flew 5,000 miles, to a target near Ascension Island.

*During October:* Aerospace Medical Center's SAM continued experimental studies with space-cabin simulator with 20 Strategic Air Command volunteers, each man completing the full-scale run of 7 or 8 days of confinement in the cabin simulator.

*November 3:* SPUTNIK II, the world's second manmade satellite, launched by U.S.S.R. and remained in orbit until April 13, 1958, carrying a dog named "Laika." It was the first vehicle to carry a living organism into orbit. (See Appendix A.)

*November 7:* President Eisenhower in major address on science and security announced that scientists had solved the problem of ballistic missile reentry and showed the nose cone of an Army Jupiter-C missile which was intact after a flight through space. He announced the creation of the office of Special Assistant to the President for Science and Technology and the appointment of James R. Killian, president of the Massachusetts Institute of Technology, to the new post.

## 1957—Continued

*November 8:* Secretary of Defense McElroy directed the Department of the Army to launch a scientific satellite with the modified Jupiter-C test rocket. The satellite, carrying instruments selected by the National Academy of Sciences, would be a part of this country's contribution to the IGY. William M. Holaday, Assistant to the Secretary of Defense for Guided Missiles, was given authority for coordinating this ABMA-JPL project with the overall U.S.-IGY satellite program.

*November 10:* SPUTNIK II ceased transmissions.

*November 11:* KC-135 tanker flown 6,350 miles from Westover AFB, Mass., to Buenos Aires, in 13 hours 2 minutes, by Gen. Curtis LeMay, a world record for nonstop nonrefueled jet flight.

*November 13:* President Eisenhower, in a speech on future security, proposed adoption of a formula for decisions on undertaking space projects, which would include the following criteria: "If the project is designed solely for scientific purposes, its size and its cost must be tailored to the scientific job it is going to do. If the project has some ultimate defense value, its urgency for this purpose is to be judged in comparison with the probable value of competing defense projects."

———: 1,000-mile, Navy Regulus II fired in first launch with rocket boosters at Edwards AFB, and returned to base by control aircraft after a 48-minute flight.

*November 15:* William M. Holaday, special assistant to the Secretary of Defense, was named Director of Guided Missiles by Secretary of Defense McElroy. Under terms of the Defense Department directive: "The Director of Guided Missiles will direct all activities in the DOD relating to research, development, engineering, production, and procurement of guided missiles."

*November 19:* An ANP (Aircraft, Nuclear Powered) project, an integrated AEC-DOD atomic aircraft project within the AEC, was announced, with Maj. Gen. Donald Keirn (USAF) as its head.

*November 21:* The National Advisory Committee for Aeronautics authorized establishment of a special committee on space technology, headed by H. Guyford Stever. This committee would both supervise and help formulate a space research program and would be assisted by specialized subcommittees.

*November 22:* First hydrogen-fluorine rocket engine successfully operated at NACA Lewis Laboratory, demonstrating a 40-percent performance improvement over other propellant combinations.

*November 25:* USAF awarded contract for a surveillance satellite to Lockheed.

———: The Preparedness Investigating Subcommittee of the Senate Committee on Armed Services began extensive hearings on the Nation's satellite and missile programs.

*November 27:* Thor and Jupiter IRBM's ordered into production for ultimate deployment by the USAF.

*During November:* NACA 1957 Flight Propulsion Conference at Cleveland was review of analysis of space missions, nuclear propulsion systems, chemical propulsion systems, electrical propulsion systems, auxiliary power systems, and propellants.

———: First Baker-Nunn precision optical satellite tracking camera installed at White Sands, N. Mex., the first of 12 such optical tracking installations as a part of the IGY under the supervision of the Smithsonian Astrophysical Observatory.

———: Development of satellite launch vehicle focused upon all solid fuel systems at NACA Langley, a major step in the origin of Scout.

*December 4:* The American Rocket Society's proposal for an Astronautical Research and Development Agency, which was presented to President Eisenhower on October 14, 1957, was announced.

*December 6:* IGY Vanguard (TV-3), the first with three live stages, failed to launch a test satellite.

*December 9:* Secretary McElroy ordered acceleration of the Polaris program.

*December 13:* The Air Force order of December 10 creating a Directorate of Astronautics under Brig. Gen. Homer A. Boushey was suspended by Secretary William H. Douglas, as creation of such a group before establishment of the proposed Advanced Research Projects Agency was considered premature.

*December 17:* First successful test firing of USAF Atlas ICBM, the missile landing in the target area after a flight of some 500 miles, on the 54th anniversary of the Wright brothers' first flight.

*December 18:* First full-scale production of electricity for commercial use by civilian nuclear power station, at Shippingport, Pa.

*December 19:* A Thor missile, the eighth tested and the fourth successfully, completed the first fully-guided Thor IRBM flight using an all-inertial guidance system.

*December 23:* USAF awarded B-70 Mach 3 bomber development contract to North American Aviation.

*December 28:* World altitude record of 30,335 feet for helicopters set by Capt. J. E. Bowman (USA) in a Cessna YH41 Seneca at Wichita, Kans.

*During December:* Maxime Faget of NACA Langley proposed ballistic shape of Mercury capsule, while A. Eggers of Ames and E. S. Love and J. V. Becker of Langley proposed glider configurations of manned spacecraft later incorporated in Dyna-Soar and Apollo studies.

*During 1957:* NACA Technical Note, "A Comparative Analysis of Long-Range Hypervelocity Vehicles," by Ames scientists Eggers, Allen, and Neice prepared and issued. It was considered a landmark in the development of scientific thought on manned reentry.

———: Experiments at USAF School of Aviation Medicine showed that soil bacteria could not only survive but also multiply under certain simulated Martian atmospheric conditions.

———: NACA Lewis Laboratory completed major phases of pioneering research on high-energy turbojet and ramjet fuels including boron. This research included flight test in piloted aircraft and air-launched free flight models. Theoretical performance and experimental thrust chamber injector experiments were also performed at NACA Lewis, aiding in design of X-15 rocket engine.

———: Single-spool J93 turbojet engine placed under intensive development at General Electric. The J79 turbojet, the first high-compression variable-stator engine built in United States by GE, powered most Mach 2 U. S. aircraft, including the F-104, B-58, F11F-1F, F4H, and A3J, as well as the Regulus II missile.

———: First operation by the NACA Lewis Laboratory of a 20,000-pound thrust hydrogen-oxygen rocket engine completely self-cooled by the liquid hydrogen, which led to Centaur engine development.

———: The NACA proposed and led in the development of the Polaris reentry body based on the work done at Langley Laboratory, 1952-56.

———: State University of Iowa completed balloon-launched rocket (Rockoon) research at high latitudes begun in 1952. James A. Van Allen reported that principal scientific measurements attained included: first latitude survey of total cosmic-ray intensity at high altitude and high latitude; survey of latitude variation of heavy nuclei in primary cosmic radiation; discovery of X-radiation associated with aurorae; first arctic measurements of atmospheric density, pressure, and temperature at high altitudes; measurement of ultraviolet and soft X-radiation during solar flares; first measurements of terrestrial magnetic fields at high altitudes in the auroral zone.

# 1958

*January 1:* Strategic Air Command assigned responsibility for U.S. operational ICBM capability; while the 672nd Strategic Missile Squadron, first to be equipped with USAF Douglas Thor IRBM, was activated.

*January 4:* SPUTNIK I reentered the atmosphere and disintegrated.

———: American Rocket Society and the Rocket and Satellite Research Panel issued a summary of their proposals for a National Space Establishment. Preferably independent of the Department of Defense, but in any event not under one of the military services, this establishment would be responsible for the "broad cultural, scientific, and commercial objectives" of outer space development.

*January 9:* In his state-of-the-Union message, President Eisenhower reported: "In recognition of the need for single control in some of our most advanced development projects, the Secretary of Defense has already decided to concentrate into one organization all antimissile and satellite technology undertaken within the Department of Defense."

*January 11:* James H. Doolittle, Chairman of the National Advisory Committee for Aeronautics, announced that a special committee on space technology was formed on November 21, 1957.

*January 12:* President Eisenhower, in answering the December 10, 1957, letter of Soviet Premier Nikolai A. Bulganin regarding a summit conference and disarmament, proposed that the Soviet Union and the United States "agree that outer space should be used only for peaceful purposes." This proposal was compared with the 1946 offer of the United States to cease production of nuclear weapons and dedicate atomic energy to peaceful uses, an offer which was not accepted by the Soviet Union.

*January 13:* Secretary of Defense Neil H. McElroy testified before the House Armed Services Committee: "Such long-range programs as the antimissile missile and the military satellite programs are in the research and exploratory development stages. They are important and must be pursued, but they must not distract us from the speedy development of our other missile systems. To handle them, I am establishing within the Department of Defense an Advanced Research Projects Agency, which will be responsible to the Secretary of Defense for the unified direction and management of the antimissile missile program and for outer space projects."

———: In his budget message to Congress, President Eisenhower stated: "Funds are provided for an expanded research and development effort on military satellites and other outer space vehicles and on antimissile-missile systems, to be carried out directly under the Secretary of Defense." The budget for fiscal year 1959 showed that $340 million in new obligational authority was being asked for the Advanced Research Projects Agency. No new authorizations were sought for the International Geophysical Year, but estimated obligations for earth satellite exploration of the upper atmosphere under this program were $8,139,834 for fiscal year 1958 and $21 million for fiscal year 1959.

*January 14:* NACA issued a staff study entitled "A National Research Program for Space Technology."

———: Senator Lyndon B. Johnson in a CBS radio address urged the United States "to demonstrate its initiative before the United Nations by inviting all member nations to join in this adventure into outer space together."

*January 15:* 4751st Air Defense Missile Wing to develop and conduct training program for Bomarc units, and the 864th Strategic Missile Squadron to be equipped with Jupiter IRBM, were both activated.

*January 16:* The NACA adopted resolution recommending that national space program can be most effectively implemented by the cooperative effort of the Department of Defense, the NACA, the National Academy of Sciences, and the National Science Foundation, together with universities, research institutions, and industrial companies of the Nation, with military development and operation of space vehicles a responsibilty of the

Department of Defense, and research and scientific space operations the responsibility of the NACA.

———: Special Subcommittee on Outer Space Propulsion created by the Joint Congressional Committee on Atomic Energy, Senator Clinton P. Anderson as chairman.

———: Secretary of State Dulles proposed the formation of an international commission to insure the use of outer space exclusively for peaceful purposes.

*January 17:* First launch of Navy Polaris test vehicle at Cape Canaveral.

*January 27:* Dr. Hugh L. Dryden, Director of the NACA, in a speech to the Institute of the Aeronautical Sciences, stressed the importance of a well-planned and logical space program embracing both civilian and military uses. He stated that the national space program should be under the joint control of the Department of Defense, the NACA, the National Academy of Sciences, and the National Science Foundation; in addition to research flights, the NACA would "coordinate and conduct research in space technology in its own laboratories and by contract in support of both military and nonmilitary projects."

*January 28:* Thor IRBM successfully fired from Cape Canaveral, flew prescribed course, and impacted in preselected area.

*January 29:* The DOD announced plans to establish the National Pacific Missile Range (PMR) as part of the Naval Air Missile Test Center at Point Mugu, Calif., the range to be designed for long-range guided missile and ICBM testing.

*January 31:* EXPLORER I, first U.S. earth satellite, launched by modified ABMA–JPL Jupiter-C, with US–IGY scientific experiment of James A. Van Allen, which discovered the radiation belt around the earth. (See Appendix A.)

*February 3:* Soviet Premier Nikolai A. Bulganin in a letter to President Eisenhower stated that the Soviet Union "is ready to examine also the question of the intercontinental rockets if the Western powers are willing to reach agreement to ban atomic and hydrogen weapons, to end tests thereof, and to liquidate foreign military bases in other nations' territories. In that case, an agreement on the use of outer space for peaceful purposes only would unquestionably meet no difficulties."

———: Scientists at the Jet Propulsion Laboratory at the California Institute of Technology reported that initial data from EXPLORER I showed that cosmic radiation on its orbit did not exceed 12 times the amount on earth.

*February 4:* President Eisenhower directed James R. Killian, Jr., to head a committee to study and make recommendations on the governmental organization of the Nation's space and missile program.

*February 5:* Trial firing of IGY Vanguard (TV–3Bu) satellite failed at Cape Canaveral, Fla., 57 seconds after launch.

*February 6:* The Senate passed S. Res. 256, creating a Special Committee on Space and Astronautics to frame legislation for a national program of space exploration and development.

*February 7:* The Advanced Research Projects Agency (ARPA) was established by the DOD, and Roy W. Johnson, a vice president of General Electric Co., was appointed by Secretary of Defense McElroy as its Director. ARPA was placed in charge of the Nation's outer space program.

*February 10:* First successful radar returns from Venus (27,530,000 miles away) detected by MIT's Lincoln Laboratory Millstone Hill. It took 1 year to process confirmation of this event.

———: Airman 1/C Donald G. Farrell spent the week of February 10–16 in a space-cabin simulator at SAM, Randolph AFB, Tex.

*February 14:* "Basic Objectives of a Continuing Program of Scientific Research in Outer Space," a report by the Technical Panel on the Earth Satellite Program of the National Academy of Sciences IGY Committee, was published. It proposed a program of space research extending beyond the International Geophysical Year.

## 1958—Continued

*February 17:* In a letter to Soviet Premier Nikolai A. Bulganin, President Eisenhower repeated his plea for the dedication of outer space to peaceful uses. Denying that this proposal was intended "to gain strategic advantages for the United States," he stressed the urgency of dealing with outer space before its use for military purposes had, like nuclear weapons, advanced to the point where complete international control was almost impossible.

*February 18:* USAF revealed that an airflow speed of 32,400 mph had been attained for one-tenth of a second in a wind tunnel test at the Arnold Engineering Development Center, Tullahoma, Tenn., on an undisclosed date.

*February 21:* U.S.S.R. fired a single-stage rocket to 294-mile altitude with 3,340 pounds of experiments for measuring ion composition of the atmosphere, pressure, temperature, micrometeorites, etc., according to the Soviet IGY Committee.

*February 26:* James H. Doolittle, Chairman of the NACA, testified before Senate Committee on Appropriations that "four years ago, about 10 percent of our activities were associated with space; two years ago, about 25 percent; and in 1959 we will be devoting almost half of our time on missiles, antimissiles, and satellites and other space objectives."

*February 28:* Department of Defense assigned responsibility for land-based ICBM/IRBM development to the USAF, and directed it to develop Minuteman solid-propellant ICBM capable of being launched from underground sites.

*During February:* NACA Langley's PARD conceived and placed in operation the "opposed gun" technique for studying projectile impacts.

*March 5:* EXPLORER II launched by Army Jupiter-C failed to orbit due to failure of last stage to ignite, a joint JPL-ABMA project.

———: H. Res. 496, passed by the House of Representatives, established a Select Committee on Astronautics and Space Exploration to investigate the problems of outer space and to submit recommendations for the control and development of astronautical resources.

*March 15:* U.S.S.R. Foreign Ministry statement proposed that ban on use of outer space for military purposes, as suggested by President Eisenhower, be coupled with the liquidation of foreign military bases in Europe, the Middle East, and North Africa.

———: Contract awarded for inertial guidance system for the Titan ICBM to American Bosch Arma by the USAF.

*March 17:* Second U.S.-IGY satellite, VANGUARD I, launched into orbit with life expectancy of perhaps a 1,000 years, a highly successful scientific satellite which proved that the earth is slightly pear shaped. Operating on solar-powered batteries, it was still transmitting after 3 years in orbit. (See Appendix A.)

———: An experiment testing the behavior of crews under conditions of long confinement was concluded at Wright Air Development Center, as five Air Force officers ended a 5-day simulated space flight.

*March 18:* Dr. Herbert F. York was appointed as Chief Scientist for DOD's Advanced Research Projects Agency.

*March 19:* Space program for the United States proposed by the U.S.-IGY Satellite Panel.

*March 21:* Two-stage monorail rocket-propelled sled exceeded 2,700 mph at Holloman AFB.

*March 23:* Navy demonstrated first dummy test of Polaris missile from "popup" launcher off San Clemente Island, from submerged launching platform.

*March 26:* Third U.S.-IGY Satellite, EXPLORER III, a joint ABMA-JPL project, successfully launched by Army Juno II, yielded valuable data on radiation belt, micrometeorite impacts, and temperature before returning to earth on June 27.

———: President Eisenhower in a brief statement released the President's Science Advisory Committee's report, "Introduction to Outer Space: an Explanatory Statement." This report set forth the basic factors making the advancement of space technology a national necessity and explained to the nontechnical reader the principles and potentialities of space travel. The many uses of space technology for scientific and military purposes were summarized, and a timetable for carrying out these objectives was included.

———: Military telephone and telegraph system using the troposphere to bounce radio signals over long distances, called "White Alice," was activated.

*March 27:* President Eisenhower gave his approval to the plans for outer space exploration announced by Secretary of Defense Neil H. McElroy. The Advanced Research Projects Agency (ARPA) was to undertake several space projects including the launching of certain earth satellites and five space probes as a part of this country's contribution to the IGY program. The Air Force Ballistic Missile Division was authorized by ARPA to carry out three lunar probes with a Thor-Vanguard system, and lunar probes utilizing the Jupiter-C rocket were assigned to the Army Ballistic Missile Agency.

*April 2:* In a message to Congress, President Eisenhower proposed the establishment of a National Aeronautics and Space Agency into which the National Advisory Committee for Aeronautics would be absorbed. This agency was to have responsibility for civilian space science and aeronautical research. It would conduct research in these fields in its own facilities or by contract and would also perform military research required by the military departments. Interim projects pertaining to the civilian program which were under the direction of the Advanced Research Projects Agency would be transferred to the civilian space agency. A National Aeronautics and Space Board, appointed by the President and composed of eminent persons outside the Government and representatives of interested Government agencies (with at least one member from the Department of Defense), was to assist the President and the Director of the National Aeronautics and Space Agency.

———: Original budget request of $340 million in new obligational authority for the Advanced Research Projects Agency for fiscal year 1959 was raised to $520 million for advanced research projects in a letter from the Director of the Bureau of the Budget, Maurice H. Stans, which was transmitted to Congress by President Eisenhower.

*April 3:* In a message to Congress on the organization of the Nation's Defense Establishment, President Eisenhower recommended creation of the position of Director of Defense Research and Engineering, which would have a higher rank and replace the present Assistant Secretary of Defense for Research and Engineering.

*April 5:* USAF Atlas ICBM was successfully flown from Cape Canaveral, Fla., to the impact area some 600 miles away.

*April 8:* USAF KC-135 Stratotanker ended a nonstop, nonrefueled record distance jet flight of 10,228 miles, from Tokyo to Lajes Field, Azores.

*April 13:* SPUTNIK II reentered earth's atmosphere.

*April 14:* Proposal for a National Aeronautics and Space Agency drafted by the Bureau of the Budget was submitted to the Congress by the President, and was contained in the following congressional bills:

S. 3609, Senator Lyndon B. Johnson and Senator Styles Bridges
H.R. 11881, Representative John W. McCormack
H.R. 11882, Representative Leslie C. Arends
H.R. 11887, Representative Harry G. Haskell, Jr.
H.R. 11888, Representative Kenneth Keating
H.R. 11946, Representative William H. Natcher
H.R. 11961, Representative Peter Frelinghuysen, Jr.
H.R. 11964, Representative James G. Fulton
H.R. 11996, Representative Gordon L. McDonough

*April 15:* Select Committee on Astronautics and Space Exploration of the House of Representatives opened hearings on outer space leading toward formulation of a national space program.

*April 16:* Grumman F11F-1F Super Tiger flown to world altitude record of 76,828 feet for ground-launched planes, piloted by Cdr. George C. Watkins, at Edwards AFB.

## 1958—Continued

*April 17:* Simulated 7-day trip to the moon made by six Navy men in chamber at Philadelphia Naval Base.

———: British Skylark reached an altitude of 90 miles at Woomera, Australia.

*April 23:* USAF Thor-Able missile was launched from Cape Canaveral in a reentry test; flew short of its goal and the nose cone was not recovered. The nose cone carried a mouse as a biomedical experiment.

*April 24:* Navy rocket sled attained speed of 2,827.5 mph at China Lake, Calif.

*April 25:* First successful launching and erection in space of a 12-foot inflatable sphere for air density measurements, using a Nike-Cajun booster system, by NACA Langley's PARD at Wallops Island, Va.

*April 27: Pravda* reported on Soviet satellite findings that Laika's heartbeat had taken three times as long as expected to return to normal. Weightlessness affecting the nerve centers was suggested as the cause. The Soviet report disclosed that the density and temperature of the atmosphere at a given altitude were not uniform, and that cosmic ray intensity was 40 percent greater at 400 miles than at 135 miles.

*April 28:* Vanguard (TV-5) failed to orbit due to malfunction of minor components in the firing circuit of third stage.

*May 1:* Scientific findings from the two Explorer satellites disclosed an unexpected band of high-intensity radiation extending from 600 miles above earth to possibly an 8,000-mile altitude. The radiation was described by Dr. James A. Van Allen as "1,000 times as intense as could be attributed to cosmic rays."

———: Responsibility for the Project Vanguard portion of the U.S.-IGY scientific satellite program was transferred from Navy to Advanced Research Project Agency monitorship by the Department of Defense.

*May 6–7:* Lt. Comdr. M. Ross (USNR) and A. Mikesell (Naval Observatory) used open gondola STRATO-LAB balloon to reach 40,000-feet altitude from Crosby, Minn.; Mikesell becoming the first astronomer to observe stratosphere, and it was first flight in which crew remained in stratosphere in open basket after sunset.

*May 7:* Flying a Lockheed F-104A Starfighter at Edwards AFB, Calif., Maj. Howard C. Johnson (USAF) set a 91,249-foot world altitude record for ground-launched planes.

*May 11:* Lt. Comdr. Jack Neiman completed 44-hour simulated high altitude flight at between 80,000 and 100,000 feet in pressure chamber at NAS Norfolk.

*May 14–17:* Symposium on "Possible Uses of Earth Satellites for Life Sciences Experiments" held in Washington, D.C., under sponsorship of National Academy of Sciences, National Science Foundation, and American Institute of Biological Science.

*May 15:* SPUTNIK III placed into orbit by the U.S.S.R. with a total payload weight of about 7,000 pounds, and called a "flying laboratory." (See Appendix A.)

*May 16:* In level flight over a 10-mile course at Edwards AFB, Calif., Capt. Walter W. Irwin (USAF), flying a F-104A Starfighter, set a world speed record of 1,404.19 mph.

*May 18:* First U.S. full-size tactical nose cone was recovered from the Atlantic Ocean 4½ hours after launching from Cape Canaveral on a Jupiter missile.

*May 20:* NACA-USAF Memorandum of Understanding signed, "Principles for Participation of NACA in Development and Testing of the Air Force System 464L Hypersonic Boost Glide Vehicle (Dyna-Soar I)."

*May 24:* Gravity load of 83 g's for a fraction of a second withstood by Capt. E. L. Breeding in deceleration of a rocket sled at Holloman AFB.

*May 27:* First USAF Republic F-105 Thunderchief fighter-bomber delivered to the USAF.

———: First launching of production Vanguard satellite vehicle (SLV-1) generally successful with exception of second-stage burnout which prevented achievement of satisfactory orbit.

*During May:* Four-stage rocket launched a 9-pound inflatable sphere to 50-mile altitude at NACA Wallops Island.

——: Dr. Abe Silverstein, Associate Director of Lewis Flight Propulsion Laboratory, was transferred to NACA headquarters to help plan the organization and programs of the National Aeronautics and Space Administration, subsequently becoming Director of the Office of Space Flight Programs.

*June 3:* USAF and NACA jointly announce details on the inertial guidance system to be used on the X-15 research aircraft, a flight instrument system to allow the pilot to prevent the aircraft from reentering dense atmosphere too steeply or too shallow.

*June 4:* USAF Thor flight tested for the first time from a tactical-type launcher at Cape Canaveral.

*June 8:* Test firing of a full-scale upper stage rocket under simulated altitude conditions was made in an engine test cell at the USAF's Arnold Engineering Development Center at Tullahoma, Tenn.

*June 16:* Phase I development contract for Dyna-Soar boost-glide orbital spacecraft awarded by USAF to two teams of contractors headed by Martin Co. (Bell, American Machine & Foundry, Bendix, Goodyear, and Minneapolis-Honeywell) and the Boeing Co. (Aerojet, General Electric, Ramo-Wooldridge, North American, and Chance Vought).

——: Pacific Missile Range, Point Mugu, Calif., officially established under Navy management to provide range support to the Department of Defense and other governmental agencies engaged in missile, satellite, and space vehicle research, development, evaluation and training.

*June 26:* Production Vanguard satellite (SLV-2) failed to orbit due to failure of second stage, but demonstrated structural integrity of tankage which withstood pressure exceeding design values.

*June 27:* First successful launching by NACA Langley's Aircraft Research Division of a Mach 18 five-stage rocket vehicle at Wallops Island, Va.

——: USAF strategic missile squadron successfully completed first military launch of a Snark intercontinental missile at Cape Canaveral.

*June 28:* EXPLORER III reentered the earth's atmosphere.

*June 30:* The NACA reported that 50 percent of its research effort was being devoted to problems associated with missiles and space vehicles.

*During June:* Space Science Board of 16 members established by National Academy of Sciences, with Dr. Lloyd V. Berkner as Chairman, to advise and assist in formulation of U.S. post-IGY space research program and to foster cooperation with space scientists in other nations.

——: NACA-USAF meetings concerning applicability of all solid-propellant launch vehicle (later named Scout) to meet USAF requirements.

——: Recovery of first data capsule at AMR after successful separation from a Thor IRBM at reentry.

*July 1:* Japanese Kappa-6tw two-stage rocket flown to 30-mile altitude over Michikawa Rocket Center, Japan.

*July 8:* First launching of a 10-inch-diameter spherical rocket motor with spin stabilization, at NACA Wallops Island.

*July 9:* Second AF Thor-Able reentry test vehicle was launched, traveling 6,000 miles (no nose cone recovery).

*July 17:* Nose cone of Jupiter missile successfully recovered after intermediate range flight.

*July 21:* Standing Committee on Science and Astronautics established by House of Representatives.

*July 23:* Thor-Able reentry test vehicle made another successful 6,000-mile flight; the nose cone and mouse passenger were not recovered.

*July 23-31:* Feasibility of creating or destroying cloud formations by release of carbon black was established in tests conducted off Florida coast by the Navy Weather Service's Comdr. N. Brango and Dr. Florence Van Straten.

## 1958—Continued

*July 24:* Senate established Standing Committee on Aeronautical and Space Sciences.

*July 26:* EXPLORER IV, fourth U.S.-IGY satellite, successfully launched by Army Jupiter-C. (See Appendix A.)

———: Capt. Ivan C. Kincheloe (USAF) killed when F-104 crashed at Edwards AFB. He had been scheduled to test-fly the X-15.

*July 26-27:* Comdrs. M. Ross and L. Lewis (USN) reached maximum altitude of 82,000 feet in STRATO-LAB HIGH III flight from Crosby, Minn., which set new unofficial record for stratospheric flight of 34.7 hours.

*July 29:* President Eisenhower signed H.R. 12575, making it the National Aeronautics and Space Act of 1958 (Public Law 85-568). In his statement, he said: "The present National Advisory Committee for Aeronautics (NACA) with its large and competent staff and well-equipped laboratories will provide the nucleus for NASA. The NACA has an established record of research performance and of cooperation with the armed services. The coordination of space exploration responsibilities with NACA's traditional aeronautical research functions is a natural evolution . . . [one which] should have an even greater impact on our future."

*July 30:* President Eisenhower requested $125 million to initiate the National Aeronautics and Space Administration (NASA).

———: Successful proof tests subjecting humans to over 20 times the force of gravity were conducted, with NACA's Maxime Faget conceiving concept of the contour couch on centrifuge at Navy AMAL, Johnsville, Pa. This couch became integral part of the Project Mercury concept.

*July 31:* Army Redstone No. 50 successfully fired off Johnson Island in the South Pacific as a part of Project Hardtack.

———: First comprehensive Sputnik data was released by U.S.S.R. to foreign scientists.

*August 1:* AFBMD announced development of a complete inertial guidance system to replace radio inertial system now in use.

*August 2:* First full-powered flight of USAF Atlas ICBM using both the sustainer and booster engines.

*August 6:* Rocketdyne Division of North American announced an Air Force contract for a 1-million-pound thrust engine.

*August 7:* First launching of USAF Bomarc interceptor missile from Cape Canaveral on a signal sent by the SAGE Control Center at Kingston, N.Y.

*August 8:* President nominated Dr. T. Keith Glennan to be Administrator of the National Aeronautics and Space Administration, and Dr. Hugh L. Dryden as Deputy Administrator.

*August 11:* Army Redstone No. 51 successfully fired off Johnson Island in the South Pacific as a part of Project Hardtack.

———: After program review and discussions, NACA drafted specifications of the Scout launch vehicle based upon preliminary designs for a hypervelocity research vehicle and orbiting system.

*August 14:* Nominations of Dr. T. K. Glennan and Dr. H. L. Dryden were approved by the Senate Special Committee on Space and Astronautics.

*August 15:* Saturn Project initiated by ARPA order to Army Ordnance Missile Command, and it was assigned to Redstone Arsenal.

———: Dr. T. Keith Glennan confirmed by the Senate as Administrator of the National Aeronautics and Space Administration.

———: Federal Aviation Agency created with passage by Congress of the Federal Aviation Act.

*August 17:* USAF Thor-Able-1 launch vehicle with first U.S.-IGY lunar payload exploded 77 seconds after launch because of a failure of first-stage engine.

*August 19:* Dr. T. Keith Glennan sworn in as Administrator, and Dr. Hugh L. Dryden as Deputy Administrator, of the

National Aeronautics and Space Administration; 40 days later, as of October 1, 1958, NASA was declared to be ready to function.

———: Navy Tartar surface-to-air missile made successful first flight and interception at NOTS China Lake, Calif.

*August 21:* The National Advisory Committee for Aeronautics held its final meeting, and invited Dr. T. Keith Glennan, newly appointed Administrator of NASA, to receive best wishes for the future.

*August 24:* EXPLORER V successfully launched by ABMA–JPL Jupiter-C and all stages fired, but orbit not achieved because of collision between parts of booster and instrument compartment.

*August 25:* Ninth IAF meeting began at The Hague, which witnessed the first colloquium on space law.

*August 26:* Gen. Thomas D. White, USAF, wrote James H. Doolittle, Chairman, NACA: "There was regret at the passing of an agency that for 43 years has set the world's standard in aeronautical research. . . . There has always been for us in the Air Force, the knowledge that NACA was ready to help in any aerodynamic trouble."

———: Two mice lived 36 days sealed in a chamber and dependent upon oxygen production of algae in an experiment at the University of Texas.

*August 27:* The first Argus experiment (ARPA) was conducted (based upon October 1957 proposal of N. C. Christofilos of the University of California, Livermore), in which a small A-bomb was detonated beyond the atmosphere over the South Atlantic. Launched from the rocketship *Norton Sound*, the initial flash was followed by an auroral luminescence extending upward and downward along the magnetic lines where the burst occurred.

———: Soviet Union reportedly sent two dogs to an altitude of 281 miles and safely returned them to earth, single-stage rocket boosting a total payload of 3,726 pounds.

———: President Eisenhower signed Public Law 85–766 which included $80 million for NASA, including $50 million for research and development, $25 million for construction and expenses, and $5 million for salaries and expenses.

*August 29:* Second full-powered flight of USAF Atlas ICBM traveled 3,000 miles with radio-inertial guidance.

*August 30:* The second Argus small A-bomb detonation beyond the atmosphere was conducted in the South Atlantic.

*During August:* In 3-week period, 19 five-stage Argo E5 sounding rockets were launched in USAF–NACA program to measure radiation caused by Project Argus, rockets reaching 500-mile altitude and were launched from Wallops Island, AMR, and Ramey AFB, Puerto Rico.

———: Experimental "weightlessness" flights in C–131B aircraft begun at Wright Air Development Center.

*September 2:* U.N. Ambassador Henry Cabot Lodge announced that United States would propose a plan for international cooperation in the exploration of outer space to the United Nations.

*September 4:* President Eisenhower appointed Detlev W. Bronk, president of the National Academy of Sciences; William A. M. Burden; James H. Doolittle; and Alan T. Waterman, Director of the NSF, to the National Aeronautics and Space Council. Additionally, the Space Council including the Administrator of NASA, the Secretary of Defense, the Secretary of State, and the Chairman of the AEC as statutory members.

*September 6:* The third of the Argus small A-bomb detonations beyond the atmosphere was conducted over the South Atlantic. Instruments of EXPLORER IV satellite recorded and reported to ground stations resultant electron densities, subsequently reported by James Van Allen.

*September 7:* Black Knight missile of the United Kingdom was launched from the Australian range at Woomera to an altitude of over 300 miles.

*September 8:* Unmanned ONR balloon carried telescope and camera to an altitude of 104,600 feet.

## 1958—Continued

*September 8:* Wearing a Goodrich lightweight full-pressure suit, Lt. R. H. Tabor (USN) completed a 72-hour simulated flight in pressure chamber at NAS Norfolk, in which he was subjected to altitude conditions as high as 139,000 feet.

*September 17:* Joint NASA–ARPA Manned Satellite Panel established to make final recommendation for manned space flight program.

*September 24:* First senior staff meeting of the newly created National Aeronautics and Space Administration (NASA) held, with Dr. T. Keith Glennan as Administrator, and Dr. Hugh L. Dryden as Deputy Administrator.

——: KC–135 jet Stratotanker lifted 77,350-pound payload to an altitude of 1.25 miles.

——: First use of Sidewinder aircraft rocket with heatseeker nose, by Chinese Nationalist F–86's over the Formosa Straits. Chinese Nationalists claimed 10 Communist planes.

——: General Electric delivered first prototype of MIT-developed Polaris guidance system.

*September 25:* Dr. T. Keith Glennan signed proclamation declaring that "as of the close of business September 30, 1958, the National Aeronautics and Space Administration has been organized and is prepared to discharge the duties and exercise the powers conferred on it." Entered upon the Federal Register, this proclamation instituted the National Aeronautics and Space Administration as of October 1, 1958.

——: First launching of an Exos sounding rocket in USAF–NASA joint effort from Wallops Island, Va.

*September 26:* Vanguard (SLV–3) reached 265 miles' altitude and was destroyed 9,200 miles downrange over Central Africa on reentry into the atmosphere.

——: Boeing B–52D set a world distance in a closed-circuit record of 6,233.981 miles, with Lt. Col. V. L. Sandacz at the controls.

*September 28:* Nike-Asp test flight from Navy *LSD Point Defiance* near Puka Island reached 800,000 feet, the highest altitude ever reached by ship-launched rocket, in preliminary test of Nike-Asp for use in IGY solar eclipse studies.

*September 29:* United States announced as policy that all measures to prevent contamination of the moon would be taken in all lunar probes.

*During September:* Saturn design studies authorized to proceed at Redstone Arsenal for development of 1.5-million-pound-thrust clustered first stage.

——: Dr. W. Albert Noyes was appointed chairman of U.S. committee to draft proposals for international cooperation in the space sciences for the consideration of the International Council of Scientific Unions (ICSU).

*October 1:* First official day of the National Aeronautics and Space Administration (NASA). Existing NACA facilities, personnel, policies, and advisory committees were transferred to NASA, and the NACA laboratories were renamed Research Centers.

——: By Executive order of the President, DOD responsibilities for the remaining U.S.–IGY satellite and space probe projects were transferred to the National Aeronautics and Space Administration; included were Project Vanguard, and the four lunar probes and three satellite IGY projects remaining, which had previously been assigned by ARPA to AFBMD and ABMA. Also transferred were a number of engine development research programs.

*October 2:* Executive Board of the International Council of Scientific Unions (ICSU) proposed a plan to establish a Committee on Space Research, which became known as COSPAR.

*October 4:* Vandenberg AFB, first operational ICBM base in free world, was dedicated.

——: Jet transatlantic passenger service inaugurated by British Overseas Airways.

*October 7:* NASA formally organized Project Mercury to: (1) place a manned space capsule in orbital flight around the earth; (2) investigate man's reactions to and capabilities in this environment; and (3) recover capsule and pilot safely. A NASA Space Task Group organized at Langley Research Center drew up specifications for the Mercury capsule, based on studies by the National Advisory Committee for Aeronautics during the preceding 12 months, and on discussions with the Air Force which had been conducting related studies.

*October 8:* U.S.S.R. supplied telemetry code of SPUTNIK III to other IGY members, covering only radiation measurements.

———: In MAN HIGH III balloon launched from Holloman AFB, Lt. Clifton M. McClure attained a near-record altitude of 99,900 feet.

*October 11:* PIONEER I, U.S.-IGY space probe under direction of NASA and with the AFBMD as executive agent, launched from AMR, Cape Canaveral, Fla., by a Thor-Able-I booster. It traveled 70,700 miles before returning to earth, determined radial extent of great radiation belt, first observations of earth's and interplanetary magnetic field, and first measurements of micrometeorite density in interplanetary space.

*October 12:* Naval Research Laboratory rocket firings in Danger Island region of the South Pacific from U.S.S. *Point Defiance*, reached 139, 148, 152, and 150 miles altitude to chart solar spectrum in the ultraviolet and X-ray portion.

*October 14:* NASA requested transfer of Jet Propulsion Laboratory and the space activities of Army Redstone Arsenal to NASA.

*October 15:* First of a series of three X-15 experimental rocket-powered manned research aircraft was rolled out at the Los Angeles plant of North American Aviation, Inc., in the joint USAF-USN-NASA program.

*October 21:* Three weeks after NASA officially began operating, prospective contractors were invited to a briefing at NASA headquarters on development of 1½-million-pound-thrust engine.

———: First launching of two USAF Bomarc missiles within less than 10 seconds of each other at Cape Canaveral; launches signaled from SAGE at Kingston, N.Y., and both missiles scored successful intercepts against different target aircraft.

*October 23:* NASA—with the Army as executive agent—attempted to launch a 12-foot-diameter inflatable satellite of micro-thin plastic covered with aluminum foil known as BEACON. Launched from AMR by a Juno I—a modified Redstone, the payload prematurely separated prior to booster burnout.

*October 26:* Pan American World Airways began regular daily jet service between New York and Paris using Boeing 707's.

*October 30:* William M. Holaday appointed by the President to be Chairman of the NASA–DOD Civilian-Military Liaison Committee (CMLC).

*During October:* Air Force awarded contract to Pratt & Whitney for Centaur vehicle with hydrogen-burning chamber based on research of Lewis Research Center between 1953 and 1957. Centaur project later transferred to NASA.

*November 6:* Army completed Redstone flight testing with a perfect 250-mile shot.

*November 7:* Bidders conference held by NASA on manned-satellite capsule for Project Mercury.

*November 8:* Second U.S.-IGY space probe under direction of NASA with Air Force as executive agent, PIONEER II, was launched from AMR. Unseparated third and fourth stages reached an altitude of about 1,000 miles and flew some 7,500 miles before burning out.

*November 14:* First launch of a 3,750,000-cubic-foot plastic balloon at Holloman AFB; payload was a parachute test vehicle for development of high-Mach parachute systems.

*November 15:* First meeting of COSPAR (Committee on Space Research) proposed bylaws and rules for the approval of the ICSU, at London.

## 1958—Continued

*November 19:* United States and 19 other nations jointly introduced resolution in U.N. General Assembly calling for creation of ad hoc committee to bring about full international cooperation in the peaceful uses of outer space.

*November 21:* NASA formed new Special Committee on Life Sciences to provide advice on human factors, medical, and allied problems on NASA's manned space vehicle program.

*November 26:* Project Mercury, U.S. manned-satellite program, was officially named by NASA.

*November 28:* USAF Atlas made its first successful operational test flight in a 6325 statute-mile flight, landed close to its target.

*During November:* NASA requested DX priority for 1.5-million-pound-thrust F-1 engine project and Project Mercury.

———: Second International Symposium on Physics and Medicine of the Atmosphere and Space was held at San Antonio, Texas.

*December 3:* President transferred the functions and facilities of the Jet Propulsion Laboratory of the California Institute of Technology, Pasadena, Calif., from the Army to NASA. JPL built, designed, and tested upper stages, payloads, and tracking systems for the first IGY Explorer satellites.

———: NASA and the Army reached an agreement whereby ABMA and its subordinate organizations at Redstone Arsenal, Huntsville, Ala., would be responsive to NASA requirements.

———: DOD announced details of Project Discoverer, series of polar orbiting satellites.

*December 5:* Modified Navy Terrier rocket with camera launched to an altitude of 86 miles from Wallops Island, providing a 1,000-mile composite photograph of a frontal cloud formation.

*December 6:* The third U.S.–IGY space probe—the second under direction of NASA and with the Army as executive agent—was launched at 12:45 a.m., from AMR by Juno II rocket. The primary mission of PIONEER III, to place the scientific payload in the vicinity of the moon, was not accomplished although an altitude of 63,580 miles was achieved and it discovered that radiation belt was comprised of at least two bands.

*December 9:* The first meeting of the new NASA Inventions and Contributions Board was held to evaluate scientific or technical contributions and to recommend monetary awards.

*December 10:* First domestic jet airline passenger service, by National Airlines between New York and Miami.

*December 12–16:* SMALL WORLD balloon with four passengers failed in transatlantic attempt, lifting from Canary Islands and landing at sea northeast of Barbados.

*December 13:* U.N. General Assembly adopted resolution bringing into being an 18-member Ad Hoc Committee on the Peaceful Uses of Outer Space.

———: Squirrel monkey Gordo made 1,500-mile flight in nose cone of Army Jupiter with no known adverse effects, but float mechanism failed and nose cone was not recovered.

*December 16:* Two Thor shots, one from Cape Canaveral and one from Vandenberg AFB, were successful. Intermediate range ballistic missile portion of PMR was inaugurated. with successful firing of USAF Thor from Vandenberg AFB.

———: MATS C-133 Cargomaster lifted 117,900 pounds of cargo to 10,000 feet, a weight-lifting record, at Dover AFB, Del.

*December 17:* NASA awarded contract to Rocketdyne of North American to build single-chamber 1.5-million-pound-thrust rocket engine.

———: Project Mercury announced as name of U.S. man-in-space program by NASA.

*December 18:* Plastic balloon flight No. 1,000 launched by the Balloon Branch of the Missile Development Center at Holloman AFB, a series beginning in July 1950.

———: Entire USAF Atlas boosted into orbit communications relay satellite, PROJECT SCORE or the "talking atlas." A total of 8,750 pounds were placed in orbit, of which 150 pounds was payload.

*December 19:* President Eisenhower's Christmas message beamed from PROJECT SCORE satellite in orbit, the first voice beamed in from space.

———: BOLD ORION (WS-199) launched from B-58 Hustler traveling at about 1,100 mph over Cape Canaveral, Fla.

*December 20:* White Sands Proving Ground announced missile range firing record: 2,000 "hot" firings in 1 year.

———: First Titan test launch exploded on the pad at Cape Canaveral.

———: New voice and teletype messages were received and rebroadcast on command by PROJECT SCORE satellite, and a series of experiments were continued in subsequent days.

*December 23:* First Atlas-C fired successfully at AMR.

*December 24:* Dr. Herbert F. York, Chief Scientist of ARPA, was named as Director of Defense Research and Engineering for the Department of Defense by President Eisenhower.

*December 27:* Federal Council for Science and Technology to be headed by Dr. James R. Killian, Jr., was approved by President Eisenhower.

———: PIONEER III data indicated that the earth is surrounded by two bands of radiation.

*December 31:* PROJECT SCORE ceased transmissions, concluding 12 days of operations and 97 successful contacts.

———: IGY scheduled to close, but in October 1958 the International Council of Scientific Unions, meeting in Washington, approved extension of IGY through December 1959 under name of International Geophysical Cooperation—1959 (IGC-59) and also approved establishment of Committee on Space Research (COSPAR) to continue international cooperation in the scientific exploration of space. National Academy of Sciences is U.S. adhering body to COSPAR.

*During December:* National booster program developed by NASA and DOD to provide basis for long-range planning.

———: First vacuum tank for use in ion and plasma electric propulsion research received at NASA Lewis Research Center, three more of which were later put to research, and two large models to be completed by 1962.

*During 1958:* NASA Langley research scientists, Paul Purser and Maxime Faget, conceived Little Joe research rocket; the Scout vehicle system was conceived from PARD's multistage hypersonic solid-propellant rocket program.

———: Twistor and other thin-film semiconductors were developed suitable as memory elements.

———: NASA Lewis Research Center completed 14 years of extensive research on all U.S. turbojet engines.

———: NASA Lewis Center successfully demonstrated first use of fluorine gas to provide reliable ignition for practical hydrogen-oxygen engine (20K thrust); same year first throttling of hydrogen-fluorine thrust chamber demonstrated over wide range.

———: First year that the total number of transatlantic air passengers exceeded the number of sea passengers.

———: Experimental tests for launching satellites via rocket fired from fighter aircraft conducted by Navy Project Pilot.

# 1959

*January 2:* U.S.S.R. launched LUNIK I into a solar orbit, with a total weight of a reported 3,245 pounds, the first man-made object placed in orbit around the sun. It was called MECHTA ("dream") by the Russians. (See Appendix A.)

——: Defense officials indicated fiscal year 1960 budget would begin major integration of long-range missiles into weapons arsenal and replacement of manned aircraft on a large scale.

*January 4:* Vandenberg Air Force Base and the Pacific Missile Range declared officially operational for firings.

*January 5:* LUNIK I transmissions ceased 373,125 miles from earth.

*January 8:* NASA requested eight Redstone-type launch vehicles from the Army to be used in Project Mercury development flights.

*January 9:* NASA–DOD agreement signed for a "National Program To Meet Satellite and Space Vehicle Tracking and Surveillance Requirements" for fiscal year 1959 and fiscal year 1960.

*January 12:* NASA announced selection of McDonnell Aircraft Corp., as source for design, development, and construction of Mercury capsule.

*January 15:* First successful castings of molybdenum made at U.S. Bureau of Mines Laboratory at Albany, Oreg.

*January 19:* The AEC demonstrated a 5-watt radioisotope thermoelectric generator (designated SNAP 3) to President Eisenhower as an example of the potential use of radioisotopes and static thermoelectric conversion for providing long-lived electric power for space.

*January 21:* First Chrysler-made, operational version of Army Jupiter IRBM, successfully launched from AMR.

*January 23:* Dr. T. Keith Glennan, NASA Administrator, announced appointment of chairmen of 13 new research advisory committees to provide technical counsel from industry, universities, and government organizations.

*January 28:* Nike-Cajun successfully launched 12-foot-diameter test inflatable sphere to a height of 75 miles over NASA Wallops Island, the sphere inflating satisfactorily.

——: One hundred ten candidates were selected by NASA in the first screening for Project Mercury astronauts from Air Force, Navy, and Marine Corps test-pilot schools.

*January 29:* First jet passenger service across the United States begun by American Airlines with Boeing 707's.

*During January:* Rocketdyne demonstrated 1-million-pound-thrust liquid-propellant rocket combustion chamber at full power.

*February 2:* First annual report on *Aeronautical and Space Activities*, covering all U.S. activities during the year 1958, was forwarded to the Congress by the President.

*February 6:* First test launch of USAF Titan ICBM (A-3) from Cape Canaveral.

*February 11:* Army announced that a weather balloon, launched at the Signal Research and Development Laboratory, Fort Monmouth, N.J., had established a world altitude record of 146,000 feet.

*February 17:* VANGUARD II (SLV-4), the fifth U.S.–IGY satellite, successfully launched payload containing photocells designed to produce cloud cover images for 2 weeks; precessing or wobbling prevented significant interpretation of data. (See Appendix A.)

——: USAF Committee presided over by Dr. J. Allen Hynek, Associate Director of the Smithsonian Astrophysical Observatory at Cambridge, Mass., recommended that the USAF continue to take a positive approach to UFO's, investigate reported sightings by all scientific means, and keep the public fully informed of existing policy. Of the unknown objects sighted, it reported, no scientific evidence supports the conclusion that the objects were spacecraft.

*February 19:* Monorail two-stage rocket-research sled attained 3,090 mph, or roughly Mach 4.1, at Holloman AFB.

*February 20:* NASA awarded $105 million in contracts for 1959 projects (15 satellites).

*February 23:* Navy revealed development of steerable molybdenum nozzle used in the solid-propellant Polaris missile.

*February 28:* DISCOVERER I, ARPA satellite weighing 1,450 pounds, successfully launched into polar orbit by USAF Thor-Hustler booster from Pacific Missile Range; stabilization difficulties hampered tracking acquisition. (See Appendix A.)

*March 1:* "Poor man's rocket," Scout, was jointly announced by NASA and AF. The concept of Scout originated at Langley Research Center in 1958, based upon extensive experience with staged solid-propellant rockets.

*March 3:* PIONEER IV, fourth U.S.–IGY space probe, a joint ABMA–JPL project under direction of NASA, was launched by a Juno II rocket from AMR and achieved earth-moon trajectory, passing within 37,000 miles of the moon before going into permanent solar orbit. Radio contact was maintained to a record distance of 406,620 miles. It was the first U.S. sun-orbiter. (See Appendix A.)

———: NASA's Langley Research Center launched first in a series of six-stage solid-fuel rocket research vehicles, the world's first, from Wallops Island, Va., to a speed of Mach 26 in a reentry physics program.

*March 4:* British National Committee on Space Research, H. S. W. Massey as chairman, held its first meeting.

*March 6:* Radio signals received from PIONEER IV from a distance of 406,620 miles from earth, a new communications record.

*March 7:* First French Veronique sounding rocket launched from Columb Bechar to an altitude of 155 miles.

*March 10:* First captive flight of X-15 (No. 1) under modified B-52 with A. Scott Crossfield in the cockpit; additional captive flights were made on April 1, April 10, and May 21.

*March 11:* NASA granted $350,000 to National Academy of Sciences-National Research Council for program of research appointments in theoretical and experimental physics to stimulate basic research in the space sciences.

*March 12:* Second British Black Knight rocket reached 350-mile altitude at Woomera, Australia.

*March 12–14:* Second meeting of COSPAR held at The Hague, the Netherlands.

*March 13:* The President announced the establishment of the Federal Council for Science and Technology to promote closer cooperation among Federal agencies in planning their respective research and development programs.

———: From an altitude of 123 miles boosted by an NRL Aerobee-Hi rocket, fired from White Sands, N. Mex., the first ultraviolet photos of the sun were taken and recorded.

*March 14:* National Academy of Sciences delegate to COSPAR transmitted to COSPAR President the offer of NASA to carry experiments by scientists of other nations in U.S. space vehicles.

*March 15:* Army Redstone ejected miniature TV camera which transmitted pictures of its target impact area.

*March 17:* First flight launching of a spin-stabilized 20-inch-diameter spherical rocket, by NASA Langley's PARD at Wallops Station, Va.

———: ARPA announced that DISCOVERER I was no longer in orbit.

*March 18:* Army Signal Corps and RCA announced development of micromodules for electronic devices which ultimately could permit 500,000 components to be packed into a cubic inch of space.

*March 19:* Deputy Secretary of Defense Quarles announced that three atomic blasts were fired in space (Project Argus) in 1958, using modified X-17 rockets.

*March 20:* MIT announced successful radar signal returns from Venus had been performed on February 10 and 12, 1958, return signals being one ten-millionth as strong as transmission signals.

## 1959—Continued

*March 24:* NASA announced that Wallops Station had made over 3,300 rocket firings since 1945.

*April 2:* Seven astronauts were selected for Project Mercury after a series of the most rigorous physical and mental tests ever given to U.S. test pilots. Chosen from a field of 110 candidates, the finalists were all qualified test pilots: Capts. Leroy G. Cooper, Jr., Virgil I. Grissom, and Donald K. Slayton, (USAF); Lt. Malcolm S. Carpenter, Lt. Comdr. Alan B. Shepard, Jr., and Lt. Comdr. Walter M. Schirra, Jr. (USN); and Lt. Col. John H. Glenn (USMC).

———: Lt. Gen. Bernard A. Schriever, Commander AFBMD, was named Commander of Air Research and Development Command.

———: USAF Bold Orion ballistic missile test launched from B-47 jet bomber.

*April 7:* AEC Los Alamos Scientific Laboratory announced development of plasma thermocouple for direct conversion of energy from a nuclear reactor into electricity, offering potential auxiliary power source for space applications.

———: First operational flight of USAF Snark to target on AMR.

*April 8:* Reentry body of USAF Thor-Able recovered at the far end of the Atlantic Missile Range: first recovery after an ICBM range flight by AFMTC task force.

*April 13:* DISCOVERER II satellite successfully placed into polar orbit by Thor-Agena A booster, but capsule ejection malfunctioned causing it to impact in vicinity of Spitsbergen on April 14 instead of vicinity of Hawaii. It was first vehicle known to have been placed in a polar orbit and was the first attempt to recover an object from orbit.

———: VANGUARD (SLV-5) failed to achieve payload orbit because of loss of second-stage pitch attitude control.

*April 16:* First Thor IRBM launched by British crew at Vandenberg AFB.

*April 17:* United States formally requested that the United Nations Committee on the Peaceful Uses of Outer Space convene in New York on May 6.

*April 20:* NASA announced acceptance of proposals by the Canadian Defense Research Telecommunications Establishment for continuing joint rocket and satellite ionospheric experiments of a nonmilitary nature.

*April 23:* Fourth recovery of a data capsule at AMR, USAF Thor 1,500-mile accuracy test flight.

———: President announced the resignation of Richard E. Horner, Assistant Secretary of the Air Force for Research and Development, to become Associate Administrator of NASA effective July 1st.

———: First test flight of USAF GAM-77 Hound Dog at AMR.

*April 24:* Dr. Hugh L. Dryden and Loftus E. Becker appointed to assist Ambassador Henry Cabot Lodge in the forthcoming meetings at the United Nations of the Committee on Peaceful Uses of Outer Space.

*April 27:* Meeting of DOD working group on Project Mercury search and recovery operations was held at Patrick Air Force Base, with major emphasis placed on the first two ballistic Atlas shots, and command relationships.

———: The 1958 Annual Report of the National Advisory Committee for Aeronautics, the 44th and final report of NACA established in 1915, was submitted to Congress by the President. It contained historical sections by Jerome C. Hunsaker and James H. Doolittle.

———: DX priority (highest national priority) assigned to Project Mercury.

*April 28:* NASA announced the signing of a $24 million contract with Douglas Aircraft Co., Inc., for a three-stage Thor-Vanguard launching rocket called Delta.

*April 29-30:* Symposium sponsored by the Space Science Board of the National Academy of Sciences, NASA, and the American Physical Society, held in Washington to review space research findings and the objectives of future research programs in the space sciences.

*During April:* The Tiros meteorological satellite program was transferred from the Department of Defense to the responsibility of NASA for the national meteorological satellite program. At the same time, a Joint Meteorological Satellite Advisory Committee was established.

*May 1:* NASA's Administrator announced the naming of Goddard Space Flight Center under construction near Greenbelt, Md., in commemoration of Robert H. Goddard, American pioneer in rocket research. Dr. Harry J. Goett was appointed Director in September.

———: Smithsonian Optical Tracking Station at Woomera, Australia, successfully photographed VANGUARD I earth satellite at the apogee of its orbit, nearly 2,500 miles from earth. Compared to taking picture of golf ball 600 miles away, this feat was repeated on May 3 and 4.

*May 3:* Dr. Otto Struve of the University of California was appointed Director of the National Radio Astronomy Observatory, to be located at Green Bank, W. Va.

*May 4:* National Bureau of Standards released details on the effect on the ionosphere of the high-altitude nuclear shots called Teak and Orange on August 1 and 12, 1958, over Johnston Island.

*May 6:* NASA created a committee to study problems of long-range lunar exploration to be headed by Dr. Robert Jastrow.

———: ABMA Jupiter IRBM made successful 1,500-mile flight at Cape Canaveral and was declared operational by the USAF.

———: NASA awarded contract to Convair for development of Vega launch vehicle for deep space probes and satellites.

*May 6–June 25:* Ad Hoc Committee on the Peaceful Uses of Outer Space of U.N. met in session at U.N. headquarters in New York.

*May 12:* NASA announced training program for seven Project Mercury astronauts to provide them with technical knowledge and skills required to pilot the Nation's manned orbital capsule.

———: University of Minnesota scientist under ONR contract launched unmanned balloon to 100,000 feet, where first positive measurement of intense solar protons associated with a solar flare was made.

———: USAF Thor launched GE Mark 2 nose cone 1,500 miles down AMF, recovered data capsule contained photograph of the earth from 300-mile altitude.

*May 13:* British plan for launching an earth satellite was revealed by Prime Minister Harold Macmillan before the House of Commons.

*May 14:* Use of moon as relay station for intercontinental transmission made from Jodrell Bank, England, to the USAF Cambridge Research Center at Bedford, Mass.

*May 15:* Lt. Gen. Bernard A. Schriever, Commander of ARDC, unveiled first reentry vehicle ever to be recovered after a full intercontinental range flight.

*May 18:* NASA announced formation of Committee on Long-Range Studies headed by John A. Johnson to fulfill charge of National Aeronautics and Space Act of 1958 (sec. 102), calling for "establishment of long-range studies of the potential benefits to be gained from, the opportunities for, and the problems involved in the utilization of aeronautical and space activities for peaceful and scientific purposes."

*May 26:* ABMA static fired a single H–1 Saturn engine at Redstone Arsenal, Ala.

*May 27:* First flight test of USAF Bomarc B long-range interceptor missile.

*May 28:* Dr. George B. Kistiakowsky of Harvard University named special assistant to the President for science and technology, replacing Dr. James R. Killian, Jr.

———: Army Jupiter IRBM launched a nose cone carrying two living passengers—Able, an American-born rhesus monkey, and Baker, a South American squirrel monkey, to a 300-mile altitude, and both were recovered alive. The medical portions of the experiment were carried out by the Army Medical Service and Army Ballistic Missile Agency, Army

## 1959—Continued

Ordnance Missile Command, with the co-operation of the USN School of Aviation Medicine and the USAF School of Aviation Medicine.

*June 1:* Rhesus monkey Able died from effects of anesthesia given for removal of electrode instrumentation, autopsy revealing no effects from flight on May 28, at Army Research Laboratory, Fort Knox, Ky.

*June 3:* Moon relay transmission of President Eisenhower's voice by recording was made from Millstone Hill Radar Observatory, Westford, Mass., to Prince Albert, Saskatchewan, Canada.

——: DISCOVERER III failed to achieve orbit.

*June 5:* Construction at Cape Canaveral for the Saturn begun.

*June 6:* Army announced that sea urchin eggs fertilized before Jupiter nose cone flight continued to grow normally.

*June 8:* X-15 (No. 1) research airplane made its first glide flight with A. Scott Crossfield as pilot, after being carried by the B-52 mother ship to an altitude of 38,000 feet.

——: Mail carried by missile as 3,000 letters were delivered by a Regulus I from the submarine *Barbero* to NAS Mayport, Fla.

*June 9:* First Polaris-carrier nuclear submarine launched at Groton, Conn., the *George Washington.*

*June 12:* Scientific subcommittee of the U.N. Committee on Peaceful Uses of Outer Space proposed creation of a center to promote international cooperation in outer space research.

*June 17:* First USAF test firing of an experimental escape capsule.

*June 18:* Six U.S. Navy enlisted men began an 8-day experiment in a simulated space cabin at the Air Crew Equipment Laboratory of the Naval Air Material Center at the Philadelphia Naval Base.

*June 22:* VANGUARD (SLV-6) satellite designed to measure the radiation balance of the earth, its atmosphere, and the solar energy flux, failed to go into orbit.

*June 23:* USAF Arnold Engineering Development Center was directed by ARDC to prepare operating and design requirements for a "Large Space Environments Test Facility" for testing and developing military space weapons.

*June 25:* DISCOVERER IV failed to achieve orbit.

*June 29:* NASA welcomed announcement of United Kingdom approval of proposals for cooperative scientific research in space with the United States pending formal arrangements.

*June 30:* Considerable effort devoted to determining the causes of the malfunctions that resulted in the explosion of four out of five Atlas missiles launched before June 30.

*During June:* NASA issued Research Memo (4-17-59L) entitled "Airplane Measurements of Atmospheric Turbulence at Altitudes between 20,000 and 55,000 feet for Four Geographic Areas," analyzing data acquired by Lockheed U-2 aircraft over western United States, England and Western Europe, Turkey, and Japan.

——: Deployment of first USAF operational Thor IRBM squadron to the United Kingdom.

——: Operating velocity of Mach 6 was achieved in AEDC wind tunnel with a 40- by 40-inch test section at Tullahoma, Tenn.

*July 1:* The first experimental reactor (Kiwi-A) in the nuclear space rocket program operated successfully at full temperature and duration at Jackass Flats, Nev.

*July 6:* Comdr. M. Lee Lewis (USN) killed in accident shortly before scheduled launching of high-altitude balloon at St. Paul, Minn. He is credited with originating the Rockoon concept.

*July 7:* Four-stage Argo D4 rocket with an ARDC Javelin payload fired from Wallops Island to an altitude of 750 miles, first in a series of USAF-NASA launchings to measure natural radiation surrounding the earth.

*July 8:* As developmental planning for Project Mercury evolved, NASA notified the Army that to reduce the variety of launching vehicles the Jupiter missile would not be used for Project Mercury tests.

*July 9:* NASA Lewis Research Center operated a research model of an ion rocket in a newly completed electric-rocket test facility designed for basic investigations into the problems associated with a reliable ion rocket with a minimum life of 1 year.

*July 10:* A 10-page report of Soviet, British, and United States scientists recommended that satellites be used to detect nuclear explosions in space.

*July 11:* ONR STRATOSCOPE I balloon with camera to photograph the sun was launched from St. Paul, Minn., to an altitude of 81,250 feet.

*July 13:* Largest plastic balloon to date (6 million cubic feet) launched by Office of Naval Research with 173 pounds of instruments, at Fort Churchill, Canada.

*July 14:* U.N. Assembly Document No. A/4141, Report of the Ad Hoc Committee on the Peaceful Uses of Outer Space, was released.

*July 16:* NASA, with Army as executive agent of a joint ABMA–JPL project, attempted Explorer satellite launch with Juno II booster, but it was destroyed 5½ seconds after launch by range safety officer.

———: Second largest reflector telescope in the world, the 120-inch telescope at the Lick Observatory, was dedicated.

*July 20:* NASA selected Western Electric Co. to build worldwide network of tracking and ground instrument stations to be used in Project Mercury.

*July 21:* A full-scale USAF Atlas ICBM nose cone recovered for the first time after flight down the AMR.

*July 24:* USAF Thor data capsule recovered near Antigua which contained movie film showing nose cone separation.

*July 29:* Two-stage Nike-Asp fired from Naval Missile Facility, Point Arguello, the first of 12 designed to record radiation 150 miles up and also the first ballistic missile fired from this new facility.

*During July:* Project Mercury astronauts completed disorientation flights on three-axis space-flight simulator, the MASTIF (Multiple Axis Space Test Inertia Facility), at NASA Lewis Research Center.

———: Portion of Chincoteague (Va.) Naval Air Station transferred to NASA for use in connection with Wallops Station rocket range.

*August–December:* Conference of the International Telecommunications Union which was held at Geneva, Switzerland, allocated radio frequency bands for space and earth-space use.

*During summer:* Under joint sponsorship of National Science Foundation and the Office of Naval Research, Princeton University scientists successfully photographed sunspots with unprecedented clarity by means of 12-inch solar telescope, STRATOSCOPE I, mounted on a balloon platform at an altitude of near 80,000 feet. (See July 11 and 13.)

*August 3:* First flight test of Navy Subroc antisub missile from NOTS, China Lake, Calif.

*August 7:* EXPLORER VI, popularly called the "Paddlewheel Satellite," launched by NASA Thor-Able 3, contained 14 experiments, and a photocell scanner which transmitted a crude picture of the earth's surface and cloud cover from a distance of 17,000 miles. Placed in highly elliptical orbit (26,000 miles out, 156 miles in), it gave a broad sample of readings. (See Appendix A.)

———: Comdr. M. Ross (USNR) and R. Cooper (High Altitude Observatory) flew STRATO–LAB open gondola balloon to 38,000 feet for solar studies with a coronagraph.

———: USAF launched 39-inch weather balloon with radar reflector (Robin) from rocket at 50-mile altitude.

*August 10:* USAF canceled research program to develop exotic chemical fuels for proposed Mach 3 B–70 bomber and F–108 interceptor.

## 1959—Continued

*August 13:* DISCOVERER V placed into polar orbit by AF Thor-Agena A, but reentry capsule not recovered due to postejection malfunctions.

*August 14:* With Army as executive agent of ABMA–JPL Project, *Beacon* satellite launched by Juno II failed to go into orbit.

——: While EXPLORER VI satellite was passing over Mexico at an altitude of about 17,000 miles, it successfully transmitted a crude picture of a sunlit, crescent-shaped portion of the North Central Pacific Ocean. The area of earth photographed was 20,000 square miles.

*August 17:* First of NIKE-ASP sounding rockets to provide geophysical information on wind activity between 50 and 150 miles high was launched successfully from NASA Wallops Station.

*August 19:* DISCOVERER VI satellite orbited successfully, but reentry capsule not recovered.

*August 21:* Launching of Mercury capsule mockup from Wallops Station to test the escape and recovery systems; emergency escape rocket accidentally fired 30 minutes before scheduled firing of the Little Joe booster.

——: NASA established Bioscience Advisory Committee, headed by Dr. Seymour S. Kety, to study U.S. capability in space-oriented life science research and development and to recommend future NASA role in this area in terms of a national space program.

*August 24:* USAF fired Atlas-C 5,000 miles and recovered nose cone camera with photographs of one-sixth of earth's surface taken from 700 miles up, near Ascension Island.

*August 25:* NASA Western Operations Office, Santa Monica, Calif., made responsible for liaison, administrative, and management support west of Denver, Colo., for rapidly expanding NASA research and development activities.

——: Reflected signals off the moon successfully received at the University of Texas from the Royal Radar Establishment at Malvern, England.

*August 27:* Satellite tracking station at Woomera, Australia, successfully photographed EXPLORER VI at a distance of 14,000 miles.

——: First British Commonwealth Symposium on Space Flight began in London.

*August 29:* Navy technician withstood record 31 g's in centrifuge at AMAL, Johnsville, Pa.

*August 31:* Tenth IAF meeting opened in London.

*September 1:* USAF Atlas ICBM officially declared operational and taken over by the Strategic Air Command, at Vandenberg AFB.

*September 2:* Dr. Theodore von Kármán named chairman of a committee to establish an International Academy on Astronautics.

*September 4:* ONR SKYHOOK unmanned balloon launched from Sioux Falls, S. Dak., by Raven Industries, establishing new unofficial altitude record of 148,000 feet for unmanned balloon.

*September 9:* NASA boilerplate model of Mercury capsule successfully launched on an Atlas (Big Joe) missile from AMR and recovered in South Atlantic after surviving reentry heat of more than 10,000° F.

——: First launch of operational AF Atlas ICBM from Vandenberg AFB was successful, and second Atlas ICBM fired from Cape Canaveral the same day.

*September 12:* Russia's LUNIK II launched with a total payload weight of 858.4 pounds, became the first manmade object to hit the moon on the following day. Its launching coincided with the departure of Premier Nikita Khrushchev for the United States in turboprop Tu-114. (See Appendix A.)

*September 15:* First static test firing of USAF Minuteman, a second generation solid-fuel ICBM.

——: Premier Khrushchev presented President Eisenhower with a replica of the Soviet coat of arms inpacted on the moon on September 13.

*September 16:* Army Jupiter launched with NASA biomedical experiment from Cape Canaveral, destroyed by a range officer after fishtailing.

———: Full-sized USAF Minuteman ICBM model launched from underground silo.

*September 17:* ARPA-Navy TRANSIT IA navigation satellite was successfully launched by Thor-Able booster, but did not orbit due to third-stage malfunction.

———: First powered flight of X-15 (No. 2) research airplane, released from its B-52 mother ship approximately 36 minutes after takeoff (Interim Thiokol-RMD XLR-11 engines), A. Scott Crossfield as pilot.

*September 18:* VANGUARD III, sixth U.S.-IGY satellite, successfully injected into orbit, marking the end of Vanguard launching activities. VANGUARD III provided comprehensive survey of magnetic field, lower edge of radiation belts, and accurate micrometeorite impacts.

———: Secretary of Defense McElroy issued order entitled "Satellite and Space Vehicle Operations," assigning basic responsibilities.

*September 22:* Nuclear submarine *Patrick Henry* launched at Groton, Conn.

*September 23:* Director of Defense Research and Engineering, Dr. Herbert F. York, announced reorganization of military space and missile program, with major role going to Air Force. Four ARPA space projects were to be transferred to the services.

*September 24:* NASA Atlas-Able-4 launch vehicle, minus its payload, undergoing static tests at AMR, exploded while being prepared for the launch of a 375-pound satellite into a lunar orbit in October.

*September 27:* NASA renamed High Speed Flight Station at Edwards, Calif., to be NASA Flight Research Center, consistent with mission responsibility for all but STOL and VTOL flight research at low-speed ranges conducted at NASA Ames Research Center, Moffett Field, Calif.

*September 28:* Pictures taken from satellite EXPLORER VI over Mexico at 19,500 miles altitude on August 14, were released by NASA. Picture showed crescent shape of the sunlit portion of the earth and crude cloud-cover image.

*During September:* Dr. Hugh L. Dryden, Deputy Administrator of NASA, took part in a number of discussions with European scientific community to assess space interest there and to indicate NASA's desire to work out possible cooperative space research programs.

*October 1:* NASA personnel total reached 9,347.

*October 2:* AFMTC Commander Maj. Gen. Donald N. Yates, appointed Department of Defense representative for Project Mercury support operations.

*October 4:* NASA LITTLE JOE launch vehicle carrying a boilerplate Mercury capsule with a dummy escape system successfully launched from Wallops Station, Va.

———: LUNIK III, Russia's translunar earth satellite began photographing trip around the moon, while Premier Khrushchev was visiting Peiping. (See Appendix A.)

*October 6:* EXPLORER VI ceased transmissions.

———: USAF launched an Atlas ICBM and a Thor IRBM at their full range from Cape Canaveral.

*October 8:* PIONEER IV reached first aphelion (estimated 107,951,000 miles) in its orbit around the sun at 8 p.m., e.s.t. Since launch on March 3, PIONEER IV was tracked by JPL's Goldstone tracking station to 407,000 miles from earth.

*October 13:* EXPLORER VIII, the seventh and last U.S.-IGY earth satellite, and now under direction of NASA with the Army as executive agent, launched into an earth orbit by modified Army Juno II. By late December, data from the satellite indicated possible relationships between solar events and geomagnetic storms, and revealed information about trapped radiation and cosmic rays near the earth. With launching of this ABMA-JPL project, all experiments for the U.S.-IGY space program had been successfully placed into orbit.

## 1959—Continued

*October 13:* USAF Bold Orion launched from B-47 near Patrick AFB passed within 4 miles of EXPLORER VI at an altitude of 160 miles in test firing.

*October 14:* First successful flight test of Nike-Zeus at WSPG.

*October 17:* A second powered free flight of the X-15 (No. 2) research airplane accomplished most planned objectives.

*October 18:* LUNIK III provided man's first look at 70 percent of the backside of the moon, 2 weeks after launch, by transmitting automatically taken pictures. Pictures were released on October 26.

*October 21:* The President by Executive Order indicated that the Development Operations Division of ABMA would be transferred to NASA, subject to the approval of Congress.

*October 26:* USSR released photo of the far side of the moon taken by LUNIK III.

*October 28:* 100-foot-diameter inflatable sphere launched on a suborbital test flight from NASA Wallops Station, Va., to an altitude of 250 miles by a first Sergeant-Delta rocket; aluminum-coated Mylar-plastic sphere to be used as passive electronic reflector in Echo was developed by NASA Langley's Space Vehicle Group under the direction of William J. O'Sullivan.

*October 29:* USAF Atlas successfully launched from Cape Canaveral carrying a nose-cone camera which took a series of photographs of the earth's cloud cover from a 300-mile altitude.

*November 2:* President Eisenhower announced his intention of transferring the Saturn project to NASA, which became effective on March 15, 1960.

*November 4:* NASA launched a second LITTLE JOE from Wallops Station, to test the Mercury escape system under severe dynamic pressure; launch vehicle functioned perfectly, but the escape rocket ignited several seconds too late.

*November 5:* Third powered flight of the X-15 (No. 2).

*November 7:* USAF DISCOVERER VII satellite placed into polar orbit, but capsule recovery not achieved.

*November 9:* Entire outer Van Allen radiation belt broke up and disappeared for several days, according to data analysis from EXPLORER VII reported at AAAS meeting in New York, December 29, 1960.

*November 10:* Five-stage sounding rocket launched from NASA Wallops Island to an altitude of 1,050 miles to measure density of electrons in upper atmosphere.

———: The AEC's SNAP 2 Experimental Reactor (SER) achieved initial design power of 50 thermal kilowatts in developmental tests at the Atomics International, Santa Susana, Calif., test site. SER, the first reactor designed for use in space, was being developed for Air Force surveillance satellite systems.

———: Air Force placed contracts for Dyna-Soar project with Boeing and Martin.

*November 11-22:* Under sponsorship of COSPAR, an internationally coordinated program of scientific rocket soundings of the upper atmosphere was conducted. The U.S. contribution included 10 rocket firings.

*November 13:* National Science Foundation and the Office of Naval Research released select photographs from the more than 1,000 taken of the sun on Stratoscope balloon flights over Minnesota on July 11, August 17, and September 4.

*November 14:* World's largest balloon ($10^7$ cubic feet) launched from Stratobowl near Rapid City, S. Dak., by Winzen Research, reaching maximum altitude of near 118,000 feet with a 1-ton payload suspended.

———: New Aerospace Medical Center dedicated at Brooks AFB, Tex.

*November 16:* Capt. Joseph W. Kittinger, Jr. (USAF), made record parachute jump from open balloon gondola at an altitude of 76,400 feet (EXCELSIOR I).

*November 17:* Based on September decision that all Department of Defense satellite and space vehicle programs would be assigned to the military service of primary interest, various projects were assigned. Discoverer, Midas, and Samos were transferred from ARPA to the Air Force.

——: Pending formal transfer of the Saturn project, the Associate Administrator of NASA requested the Director of Space Flight Development to form a study group with membership from NASA, the Directorate of Defense Research and Engineering, ARPA, ABMA, and the Air Force to prepare recommendations for the development, and selection of upper stage configurations.

*November 18:* Nike-Asp sounding rocket fired from NASA Wallops Station emitted sodium vapor at 50-mile altitude to 150 miles, revealing powerful windshear effects.

——: NASA–DOD memorandum of understanding signed providing for interim management of Project Saturn pending its formal transfer to NASA.

*November 19:* Second sodium-vapor-trail experiment in Nike-Asp launch from Wallops Island was not successful.

*November 20:* DISCOVERER VIII satellite successfully placed into polar orbit, but capsule was not recovered.

——: Polaris test missiles successfully launched from launching ship, *Observation Island*, off Cape Canaveral.

*November 26:* Pioneer lunar probe was normally lifted by Atlas-Able 4 launch vehicle, but failure of plastic fairing covering payload (at 45 seconds after launch) caused payload to break away.

*November 27:* Hiller X-18 tilt-wing research transport made first flight at Edwards AFB.

*November 28:* During severe geomagnetic storm, two Geiger tubes on EXPLORER VII found anomalies in the outer radiation zone at about 1,000-km altitude, which appeared to be correlated in space and time with optical emissions from the atmosphere below. Very intense narrow zones of radiation were detected over a visible aurora during one orbit.

*November 28–29:* Comdr. M. Ross and Dr. C. B. Moore flew ONR STRATO-LAB HIGH IV balloon to an altitude of 81,000 feet, using a 16-inch telescope and spectrograph, and observing water vapor in the atmosphere of the planet Venus.

*During November:* Prototype Goodrich full-pressure Mercury astronaut suits (modified Navy Mark IV) were delivered to NASA. Navy Air Crew Equipment Laboratory (NACEL) of Philadelphia fitted suits and indoctrinated the astronauts on their use.

——: Cooperative space efforts were discussed with Soviet scientists attending the American Rocket Society meeting in Washington, D.C.

*December 1:* 12 nations (including United States and U.S.S.R.) signed Antarctic Treaty promoting scientific research and barring any military activity in the area.

——: New Bureau of Naval Weapons, consolidating the Bureau of Ordnance and the Bureau of Aeronautics, began functioning.

——: USAF reduced order for the B-70 bomber to only two prototypes.

*December 2:* Construction of a missile tracking station on Roi Namur Island near Kwajalein in the Central Pacific was announced by DOD.

*December 4:* Third LITTLE JOE successfully launched at NASA Wallops Station as part of Project Mercury development program, carried a monkey named "Sam" 55 miles into space which was recovered safely.

*December 7:* Unofficial altitude record of 98,560 feet set by Navy McDonnell F4H carrier jet at Edwards AFB, Comdr. L. E. Flint as pilot.

——: Administrator of NASA, Dr. T. Keith Glennan, offered services of U.S. worldwide tracking network in support of any manned space flight the U.S.S.R. might plan to undertake, in a speech before the Institute of World Affairs in Pasadena, Calif.

——: Nine nations including the Soviet Union approved a new charter for COSPAR at The Hague, which opened membership in COSPAR to all national academies of science engaged in space research, and created a nine-representative executive board. The U.S.S.R. had not participated in COSPAR deliberations since November 1958.

## 1959—Continued

*December 8:* Maj. Gen. Don R. Ostrander (USAF) named Director of NASA's Office of Launch Vehicle Programs and responsible for launch vehicle development and operations.

——: Brig. Gen. Austin W. Betts (USA) was named Director of ARPA to replace Acting Director, Gen. D. Ostrander (USAF).

*December 9:* USAF Goodyear unmanned balloon launched from Akron, Ohio, to an altitude of 100,000 feet, where radar photographs of the earth's surface were taken.

——: Kaman H-43B established new helicopter altitude record of 30,100 feet.

*December 10:* U.S. Ambassador Lodge presented a resolution to the Assembly of the United Nations recommending that an international conference on the peaceful uses of outer space be convened in 1960 or 1961.

*December 11:* Capt. J. Kittinger (USAF) flew EXCELSIOR II balloon from Holloman AFB to an altitude of 74,700 feet and bailed out, establishing stable free fall for 55,000 feet.

——: New world speed record for 100-km closed course set by Brig. Gen. J. H. Moore (USAF) in F-105B, at 1,216.48 mph.

——: NASA discontinued multistage Vega vehicle program to reduce number of rocket vehicles and to exploit reliability factor in future satellite and space projects.

*December 12:* First Titan ICBM launching testing second stage was unsuccessful at AMR.

——: United Nations created permanent 24-nation committee to study Peaceful Uses of Outer Space and to arrange for an international conference.

*December 14:* Lockheed F-104C piloted by Capt. J. B. Jordan (USAF) climbed to new world's record for jet aircraft of 103,389 feet.

*December 15:* Convair F-106A broke straightaway course record at 1,525.95 mph, piloted by Maj. J. W. Rogers (USAF).

——: NASA released detailed comparison of United States and U.S.S.R. space sciences programs prepared by Dr. Homer E. Newell, which pointed up the importance of leadtime in vehicle technology.

*Mid-December:* NASA team completed study design of upper stages of Saturn launch vehicle.

*December 16:* Transmitters of VANGUARD III, launched on September 18, became silent after providing tracking signals and scientific data for 85 days. Satellite was expected to remain in orbit 40 years.

*December 17:* Launching of NASA-AFBMD Thor-Able space probe designed to boost 90-pound payload to explore space between Earth and Venus was postponed.

*December 18:* Atlas ICBM made second successful 6,325-mile flight at AMR.

*December 19:* The Chairman, AEC, in a letter to the Administrator of NASA, proposed a flight test objective be established for the nuclear rocket program and proposed a technical program and division of agency responsibilities to achieve those objectives.

*December 20:* Dr. Melvin Calvin reported that molecules in meteorites resembled basic constituents of genetic material found on earth.

*December 22:* In a United States-Canadian cooperative project, NASA launched the first four-stage Javelin sounding rocket from Wallops Station to an altitude of 560 miles to measure the intensity of galactic radio noise.

*December 27:* NASA proposed joint space efforts with other nations to promote international cooperation in space research.

*December 30:* U.S.S. *George Washington*, the first fleet Polaris submarine, was commissioned.

——: Scientists associated with EXPLORER VII experiments reported their preliminary findings in a press conference at NASA Headquarters, which indicated sporadic burst of radiation from the sun could influence manned space flight.

*December 31:* Mercury astronauts completed basic and theoretical studies in their training program and started practical engineering studies.

——: More than 100 drop tests of boilerplate Mercury capsules had been completed from aircraft to test and develop the parachute system.

——: Approximately 300 U.S. research rockets were launched during the 30-month IGY/IGC–59 period: 221 of these were launched during the IGY. This compared with the some 400 U.S. research rockets fired during the entire preceding 12-year period from the beginning of high-altitude rocket research circa 1945 to July 1, 1957.

——: The IGY/IGC–59 program ended, but international cooperation in geophysics was to continue without a formal name under the sponsorship of International Council of Scientific Unions. NASA continued to make data from scientific satellites and space probes available to the world scientific community utilizing COSPAR and World Data Centers established during the IGY.

*During December:* National Radio Astronomy Observatory at Green Bank, W. Va., placed its 85-foot equatorially mounted radio telescope in full operation and continued construction of its 140-foot telescope which was planned for operation in 1961. All qualified U.S. astronomers have access to these facilities sponsored by the National Science Foundation, with priorities determined by the scientific merit of their respective projects.

——: USAF Test Pilot School at Edwards AFB proposed curriculum for Space Research Pilot Course in defining training needs for 1960–65.

——: Briefing on the orbiting Astronomical Observatory Satellite (AOS) program was given for interested members of industry at NASA headquarters.

*During 1959:* Lewis Research Center developed general method for automatic computation of theoretical rocket performance for propellant combinations involving up to 10 chemical elements; method permitting rapid performance calculation for virtually any conceivable fuel-oxidant combination.

——: Pratt & Whitney conducted thrust chamber tests of high-energy upper stage rocket engine using liquid hydrogen (RCIO).

——: Previous experience led NASA Lewis Research Center to design and construct experimental high-temperature jet engine which demonstrated feasibility of gas turbine operation at inlet gas temperatures up to 2,500° F, almost 1,000° above conventional gas-turbine engine. This test engine had a cooled turbine.

——: Aeromedical Laboratory completed development and testing of the full-pressure pilot suit for use by pilots of the X–15.

——: The National Science Foundation sited a national observatory on Kitt Peak, Ariz., 40 miles southwest of Tucson, for construction of a 36-inch reflector and an 80-inch telescope, and a 60-inch solar telescope. The solar telescope is scheduled for completion in 1961 and will be several times larger than the largest instrument of its kind in existence.

——: NASA Lewis Research Center first operated hydrogen fluorine thrust chambers at simulated high-altitude conditions obtaining unusually high performance.

——: Aeromedical Field Laboratory at Holloman AFB began training of chimpanzees for flights in ballistic and orbital flights for Project Mercury.

——: School of Aviation Medicine undertook to evolve a system for maintaining animals in sealed, self-contained ecological systems under a variety of physical conditions, such as weightlessness, acceleration, vibration, and spinning.

——: Transatlantic air passengers totaled 1,367,000 persons on scheduled flights and 173,000 on charter and special flights for the year, as compared to 884,000 sea passengers.

# 1960

*January 1:* NASA headquarters reorganization became effective, including a new Office of Launch Vehicle Programs.

*January 6:* Second full-range Atlas ICBM impacted target area 6,325 miles down AMR.

*January 7:* In his State-of-the-Union message, President Eisenhower requested revision of the National Aeronautics and Space Act of 1958, to abolish the National Aeronautics and Space Council and the Civilian-Military Liaison Committee (CMLC).

———: Polaris test vehicle achieved first fully guided flight from AMR.

*January 8–16:* First International Space Science Symposium held at Nice, France, under sponsorship of COSPAR. U.S. delegation from the National Academy of Sciences participated.

*January 11:* Skybolt air-launched ballistic missile announced by USAF, prototypes of which had already been launched from subsonic and supersonic aircraft.

*January 14:* The President formally asked Congress to amend the National Aeronautics and Space Act of 1958, "to clarify management responsibilities and to streamline organizational arrangements concerning the national program of space exploration."

———: The President directed NASA Administrator to examine need for additional money for high-thrust launching vehicles, which resulted ultimately in NASA's request for $113 million additional for the fiscal year 1961 budget.

*January 15:* Research Division created in USAF Research and Development Command to coordinate basic research.

*January 16:* Second Sergeant-Delta launched 100-foot-diameter inflatable sphere to an altitude of 250 miles from Wallops Station, a development flight of Project Echo.

———: House Science and Astronautics Committee announced members of its Panel on Science and Technology to provide consultation on major problems.

*January 20:* NASA presented its 10-year plan of space activities to Congress; the plan included 25 major vehicle launchings per year of increasing mission capability as research and development programs proceed.

———: Navy Polaris successfully tested in 900-mile flight from Cape Canaveral.

———: U.S.S.R. fired long-range ballistic missile into Pacific.

*January 21:* Fourth LITTLE JOE fired Mercury capsule in successful test of emergency-escape system to an altitude of 9 miles from Wallops Station; rhesus monkey passenger, Miss Sam, successfully recovered after 20-g and 48,900-foot-altitude flight.

*January 22:* Director of USIA, George V. Allen, stated that the United States was facing a loss of prestige in world opinion because of Soviet successes in space, in testimony before the House Science and Astronautics Committee.

*January 25:* NASA's Bioscience Advisory Committee submitted its report recommending establishment of an office to concern itself with the role of life sciences in space exploration.

———: United States and Britain announced cooperative satellite project using a Scout launch vehicle and several British experiments.

*January 26:* Javelin four-stage sounding rocket reached an altitude of 600 miles from Wallops Station.

———: Navy 173-foot-diameter balloon launched from USS *Valley Forge* east of Puerto Rico, carrying 1,630-pound payload to record cosmic ray and secondary particles. Payload film packs were recovered the next day by the destroyer *Hyman*.

*January 28:* NRL Communications Moon Relay (CMR) system using the moon as a reflector of radio signals between Hawaii and Annapolis, Md., was first publicly demonstrated.

*January 29:* In agreement with the desire of the Department of State, NASA established the Office for the United Nations Conference to prepare for U.S. participation in an international scientific conference on the peaceful uses of outer space. Dr. John P. Hagen was named as Director of OUNC.

*January 30:* USS *Valley Forge* launched second cosmic ray research balloon south of Puerto Rico, the payload of which was recovered the next day.

——: Airman B. Barwise (USAF) completed 72-hour test afloat in survival capsule designed for B-70.

*January 31:* U.S.S.R. fired second long-range missile into the Pacific.

*During January:* SPACETRACK (National Space Surveillance Control Center) at Bedford, Mass., began operations.

——: Initial experimental investigation of the plug-nozzle rocket engine was completed by General Electric, a NASA contractor.

*February 1:* NASA Administrator requested another $113 million for fiscal year 1961 to increase large launch vehicle program based on study directed by the President on January 14.

——: University of Chicago Project ICEF (International Cooperative Emulsion Flights), sponsored by the NSF and ONR, launched 10-million-cubic-foot SKYHOOK balloon to 21.4-mile altitude, capturing ultra-high-energy cosmic ray particles for analysis by international groups of physicists.

*February 2:* Titan ICBM fired from AMR, successfully achieving separation and ignition of second stage.

*February 3:* Simulated weightlessness experiment at USAF Aerospace Medical Laboratory ended, in which Dr. Duane E. Graveline was submerged in a liquid in centrifuge with a 5-g spin, and which demonstrated muscle deterioration without exercise.

*February 4:* Stanford University scientists reported on successful reflection of radar signals off the sun's corona on April 7, 10, and 12, 1959.

——: DISCOVERER IX failed to orbit from Vandenberg AFB.

*February 7:* New EXPLORER VII data showed that outer Van Allen belt rim moved north and south as much as 500 miles in latitude and varied in intensity tenfold within a few hours.

*February 9:* X-15 (No. 1) rocket research airplane was delivered by North American to NASA for further testing.

——: Test-stand construction progress announced for the development of large F-1 rocket engine, near Edwards AFB, Calif.

——: Air Force dedicated the National Space Surveillance Control Center (SPACETRACK) at Bedford, Mass.

*February 10:* Department of Defense announced that "mystery satellite" in near-polar orbit since last January may be ejected DISCOVERER V recovery capsule launched August 1959.

——: President Eisenhower toured Cape Canaveral.

*February 11:* X-15 (No. 2) ascended to 86,700 feet in powered flight.

*February 13:* France became fourth nuclear power with explosion of plutonium bomb in the Sahara Desert.

*February 16:* Reaction Motors Division of Thiokol Chemical reported successful completion of a series of 36 tests on the final XLR-99 engine for the X-15, at the Arnold Engineering Development Center, Tullahoma, Tenn.

——: First color photographs of the earth taken at high altitude, secured with recovery of data capsule of Thor launched on December 1, 1959.

*February 19:* DISCOVERER X launched but did not attain orbit.

——: Exos four-stage rocket launched reached 68-mile altitude at Eglin AFB, Fla.

*February 24:* Titan ICBM successfully fired 5,000 miles from Cape Canaveral, its longest flight to date.

——: NASA plans were outlined for the conduct of a nuclear rocket flight test program in a letter from the Administrator to the Chairman, AEC.

## 1960—Continued

*February 25:* First test launch of Army's Pershing tactical missile from Cape Canaveral.

*February 26:* First USAF Midas test launch with Atlas-Agena from AMR failed when a malfunction at staging damaged Agena.

——: Establishment of Project Mercury tracking networks in Australia was sanctioned by joint agreement.

*February 27:* 100-foot-diameter inflatable sphere successfully launched on third suborbital test to an altitude of 225 miles, from NASA Wallops Station, Va. Radio transmissions were reflected via the sphere from Holmdel, N.J., to Round Hill, Mass.

——: Atmosphere entry simulator at NASA Ames Research Center completed first successful launch and recovery of test model launched at satellite speed of 17,000 mph. First proposed by A. Eggers in 1955, it had previously provided important information at ballistic speeds. Throughout 1959–60, Ames scientists contributed to understanding of flight characteristics at altitudes over 100 miles, using low density research apparatus.

*During early 1960:* NASA Lewis Research Center completed flight safety research program involving over 30 full-scale experimental crashes and laboratory studies leading to improved criteria for survivability.

*March 1:* NASA announced establishment of the Office of Life Sciences to provide focal point for broad-based scientific study of life processes provided by the space exploration program, not to duplicate existing effort in military laboratories. Dr. Clark T. Randt was named as Director.

——: House Science and Astronautics Committee voted $915 million for NASA in fiscal year 1961.

*March 8:* First USAF Atlas flight using inertial guidance system.

*March 9:* Navy fired Polaris 900 miles in successful test of flight control equipment.

*March 10:* Office of Reliability and Systems Analysis was established in NASA Headquarters to conduct program design to evaluate and improve operational reliability of launch vehicles and payloads. Landis S. Gephardt was named as Director.

*March 11:* PIONEER V, NASA space probe, successfully launched by Thor-Able-4, the start of a historic flight to measure radiation and magnetic fields between Earth and Venus, and to communicate over great distances. Managed by AFBMD and Space Technology Laboratories for NASA, PIONEER V carried experiments designed by various civilian and governmental scientists. (See Appendix A.)

*March 13:* PIONEER V transmitted radio signals from a distance of more than 409,000 miles, a new communications record.

——: Lunar atlas published by the USAF, representing a comprehensive collection of high-quality photographs of the visible surface of the moon prepared by G. P. Kuiper.

*March 15:* Saturn project officially transferred to NASA from ABMA.

——: George C. Marshall Space Flight Center at Huntsville, Ala., named by Executive Order of the President.

*March 16:* Ban on nuclear weapons being placed in orbit around the earth in the future proposed by the representatives of the Western nations at the Geneva Disarmament Conference.

*March 17:* VANGUARD I still in orbit and transmitting on its second anniversary after traveling 131,318,211 miles. NASA reported that VANGUARD I orbit was being altered by solar pressure.

——: X-15 (No. 2) passed stress flight test.

*March 18:* PIONEER V reported on command to NASA Headquarters at 2 a.m. from 1,002,700 miles away and transmitting seven kinds of scientific readings.

——: Princess Margaret of England commanded PIONEER V 1,040,000 miles away and received answer 25 seconds later.

*March 19:* United States-Spanish agreement on Project Mercury tracking station in Canary Islands was announced (1 of 16 similar agreements with other nations).

*March 22:* USAF Titan fired 5,000 statute miles and data capsule recovered.

*March 23:* Explorer satellite launched by Juno II but did not orbit.

———: Los Alamos Scientific Laboratories disclosed controlled thermonuclear fusion was achieved by Scylla II device for less than a millionth of a second at about 13 million degrees centigrade.

*March 25:* Aerobee 150–A, a new type, fired from new launch tower at Wallops Station, reached an altitude of 150 miles and achieved rocket performance objectives as well as micrometeorite impact counts.

———: First flight and first powered flight of the X–15 (No. 1) in the NASA/USAF research program, NASA's Joseph A. Walker as pilot.

———: First launch of missile from a nuclear submarine when a Regulus I was fired from the *Halibut* off Oahu, Hawaii.

———: DOD formally announced high priority for Midas project.

———: Signals received from a distance of 2 million miles from PIONEER V.

*March 28:* Two of Saturn's first-stage engines passed initial static firing test of 7.83 seconds duration at Huntsville, Ala.

———: NASA announced selection of Aerojet-General to build the power conversion equipment for the SNAP–8 (System for Nuclear Auxiliary Power), and to integrate the reactor into an operational system. SNAP–8 is a joint NASA–AEC project.

*March 29:* Naval Weapons Annex, Charleston, S.C., was opened, providing capability for missile final assembly and loading of submarines.

———: First fully guided flight of Polaris from *Observation Island.*

*During March:* NASA let contract with Naval Ordnance Test Station, Inyokern, Calif., to study the feasibility of controlling the direction of thrust from a nozzle by injecting gas or liquid into the nozzle expansion cone.

*April 1:* First known weather observation satellite, TIROS I (Television Infra-Red Observation Satellite), launched into orbit by Thor-Able, and took pictures of earth's cloud cover on a global scale from 450 miles above until June 29. TIROS I was hailed as ushering in "a new era of meteorological observing." (See Appendix A.)

———: Fourth suborbital Shotput test of the 100-foot-diameter sphere later known as Echo was launched from NASA Wallops Station to an altitude of 235 miles and inflated successfully.

*April 2:* LUNIK I completed first orbit around the sun.

*April 4:* Project Ozma initiated to listen for possible signal patterns from outer space other than natural "noise," at the National Radio Astronomy Observatory at Green Bank, W. Va.

*April 6:* Four Saturn's first-stage engines successfully tested at Huntsville, Ala.

———: SPUTNIK III reentered the earth's atmosphere.

*April 7:* Maj. Gen. Donald N. Yates (USAF) named Deputy Director of Defense Research and Engineering for Ranges and Space Ground Support.

*April 12:* First production model of McDonnell-built Mercury capsule was delivered to NASA.

*April 13:* Navy TRANSIT I–B launched into orbit by Thor-Able-Star with navigation payload experiment at Cape Canaveral. Flight demonstrated the first engine restart in space and the feasibility of using satellites as navigational aids. (See Appendix A.)

*April 14:* First underwater launch of Polaris missile, from an underwater tube off San Clemente Island, Calif.

———: William M. Holaday's resignation as Chairman of the Civil-Military Liaison Committee accepted by the President.

## 1960—Continued

*April 14:* One week in self-sustained simulated space capsule environment concluded by C. A. Metzgen at USAF Aerospace Medical Laboratory.

*April 15:* DISCOVERER XI launched from Vandenberg AFB and stayed in orbit, reentry capsule was not recovered. (See Appendix A.)

*April 17:* PIONEER V transmitted telemetry a distance of 5 million miles from earth.

*April 18:* Scout test vehicle, with live first and third stages, fired from Wallops Station, but vehicle broke up after first-stage burnout.

———: NASA selected Avco Manufacturing and General Electric to conduct engineering and development studies on an electric rocket engine.

*April 19:* NASA announced negotiation of a contract for development of a spacecraft solar powerplant, Sunflower I, with Thompson-Ramo-Woolridge.

———: ONR Aerobee-Hi made series of X-ray photographs of the sun from an altitude of 130 miles.

*April 20:* Spin of TRANSIT I-B was reduced from 170 to 4 rpm by ground control.

*April 22:* Radar beam transmitted along electron lines of the earth's magnetic field extending into the exosphere, first confirmation of theory and work of Roger M. Gallet of the National Bureau of Standards and Henry G. Booker of Cornell University. Echo reflected from the earth successfully received 0.2 of a second later after traveling 37,000 miles, perhaps offering a new tool to study the effect of solar eruptions on the earth's magnetic field and a new long-range surveillance method using radar.

*April 23:* NASA fired first of five Aerobee-Hi sounding rockets from Wallops Station in program to measure ultraviolet radiation.

———: NASA announced that Robert E. Gottfried of GSFC had successfully "repaired" faulty diode in PIONEER V (5.5 million miles from earth) by reworking of telemetry.

*April 26:* IRAC Table of Frequency Allocations (official allocation of frequency table for United States and possessions) was approved, relating to frequency assignments for space research based on 1959 ITU Conference in Geneva, Switzerland.

———: NASA announced selection of Douglas Aircraft for construction of second (S-4) stage of initial C-1 Saturn launch vehicle.

*April 27:* Completion of technical review of Dyna-Soar program announced by the Air Force.

———: NASA signed contract with Aeronutronic, a division of Ford Motor Co., for development and production of the first survivable capsule for landing instruments on the moon.

*April 29:* Milestone achieved in completion of interim or formal agreements concluded for all oversea Mercury tracking stations.

———: NASA press conference with participating scientists reporting on the correlation of data received from EXPLORERS VI and VII, and PIONEER V during the solar storm on March 21.

*April 29:* All eight engines of the Saturn engine were fired for the first time at Huntsville, Ala.

*During April:* Seven Mercury astronauts completed training session at the Navy Aviation Medical Acceleration Laboratory, Johnsville, Pa.

*May 4:* Lewis Research Center began testing of high-energy hydrogen-oxygen engines in an altitude test facility capable of subjecting an entire propulsion system to a space environment. On June 17, LRC began similar testing of hydrogen-fluorine engines.

*May 5:* NASA held a press conference on high-altitude weather research using Lockheed U-2 aircraft, one of which was reportedly lost on May 1 over Turkey.

*May 8:* 150-watt transmitter on PIONEER V interplanetary spacecraft was commanded at 5:04 a.m. e.d.t., and operated satisfactorily while it was 8,001,000 miles from earth, another communications record.

*May 9:* First production model of Project Mercury spacecraft was successfully launched from NASA Wallops Station to test escape, landing, and recovery systems. Known as the "beach abort" shot, the Mercury capsule reached 2,540 feet before parachute landing and pickup by Marine helicopter returned it to Wallops' hangar 17 minutes after launch.

*May 10:* Submarine U.S.S. *Triton* completed 41,519-mile submerged cruise around the world.

*May 12:* Speed of Mach 3.2 and 78,000-foot altitude attained in X-15 (No. 1) with interim engines by NASA's Joseph A. Walker. This was the first remote-launch operation (100 miles from release from "mother" aircraft to landing site at Edwards AFB).

*May 13:* Echo satellite, a 100-foot passive reflector sphere, failed to orbit with first complete three-stage Thor-Delta launch vehicle.

*May 14:* Founding of the International Academy of Astronautics announced by the IAF and the Daniel and Florence Guggenheim Foundation.

*May 15:* SPACECRAFT I weighing 10,000 pounds launched into orbit by the U.S.S.R., the first successful effort to orbit a vehicle large enough to contain a human passenger, although efforts to recover the space capsule failed. (See Appendix A.)

*May 19:* TIROS I weather satellite spotted a tornado storm system in the vicinity of Wichita Falls, Tex.

———: X-15 (No. 1) flown to 107,000 feet, its highest altitude to date, by Maj. Robert M. White (USAF), at Edwards AFB.

*May 20:* Atlas ICBM fired 9,040 statute miles from AMR to Indian Ocean, longest known flight of an ICBM to date. Missile attained an apogee of about 1,000 miles.

*May 21:* First public showing of F-1 engine mockup.

*May 24:* MIDAS II test satellite successfully launched into orbit from AMR by an Atlas-Agena launch vehicle, a test of an USAF surveillance system designed to provide warning of long-range missile launching, the first anti-missile early-warning satellite.

*May 27:* Rate of spin of TIROS I satellite was increased by ground command.

———: ONR Aerobee-Hi launched to 135-mile altitude carrying eight telescopes to map sky by means of ultraviolet light, from WSPG.

*May 30:* NASA established Office of Technical Information and Educational Programs (OTIEP) in Headquarters to carry out pertinent requirements of the National Aeronautics and Space Act of 1958 and related functions. Shelby Thompson of AEC was named as Director.

*May 31:* 100-foot inflatable sphere launched from NASA Wallops Station to an altitude of 210 miles to test payload configuration carrying two beacon transmitters, a development flight of Project Echo.

———: NASA disseminated telemetry calibration for EXPLORER VII to members of the Committee on Space Research (COSPAR).

———: NASA selected Rocketdyne Division of North American Aviation to develop a 200,000-pound-thrust engine utilizing hydrogen and oxygen propellants. This engine is second only to the F-1 in single-thrust chamber level.

*June 1:* Navy assumed operational responsibility for PMR.

*June 2–3:* Panel on Science and Technology of the House Committee on Science and Astronautics held its second meeting in Washington.

*June 5:* Winzen Research launched $10^7$-cubic-foot balloon from NAS Glynco, Ga., for cosmic ray studies; after 10 days of flight the balloon disappeared over the Pacific on a westerly heading.

*June 7:* Contract for ion engine development was awarded by NASA to Hughes Aircraft.

## 1960—Continued

*June 8:* Complete eight-engine static firing of Saturn successfully conducted for 110 seconds at MSFC, the longest firing to date.

———: XLR-99 engine mounted in X-15 (No. 3) during test-stand runs by the contractor exploded, which damaged aircraft but did not injure contractor's test pilot in the cockpit.

*June 14:* AEC's SNAP-2 Experimental Reactor (SER) reached 147,300 kilowatt-hours of operation at design temperatures and power during which 1,000 hours of continuous operation was attained.

———: NASA announced creation of Launch Operations Directorate (LOD) to become operational on July 1, to be headed by Dr. Kurt Debus of Marshall Space Flight Center, who headed the Army launch operations of EXPLORER I and the first American payload to orbit the sun, PIONEER IV.

———: Small explosive charge ignited flare package on side of Titan ICBM at AMR, causing first missile fatality (J. G. Sibole) in 10 years of missile launchings at Cape Canaveral.

*June 15:* Saturn static test firing of 121 seconds successful at MSFC.

*June 22:* Navy TRANSIT II-A, an experimental navigation satellite with two payloads (navigation and radiation measurement), successfully launched into orbit by Thor-Able-Star vehicle. This was the first time that two instrumented satellites have been placed into orbit at the same time. (See Appendix A.)

*June 24:* 500-w SNAP mercury-Rankin cycle-turbine alternator package endurance test was successfully terminated at 2,500 hours of operation at design conditions, by AEC.

*June 25:* Aerospace Corp., a nonprofit civilian organization to manage engineering, research, and development aspects of missile and military space programs, was established by the USAF.

*June 26:* Six-minute message received by Jodrell Bank, England, was last communication received from PIONEER V, then 22.5 million miles from earth moving at a relative velocity of 21,000 mph. Since March 11 when launched, PIONEER V traveled some 180 million miles, and it would fly 18 million miles closer to the sun than any manmade object.

*June 28:* The Smithsonian Institution awarded its highest honor, the Langley Medal, to Robert H. Goddard posthumously.

———: U.S.S.R. announced that it would conduct new series of long-range missile shots into the Pacific, July 5–31, 1960.

*June 29:* DISCOVERER XII failed to go into polar orbit.

———: TIROS I ended its operational lifetime, transmitting a total of 22,952 picture frames of the earth's cloud cover and completing 1,302 orbits since launch on April 1.

*July 1:* NASA George C. Marshall Space Flight Center, with Dr. Wernher von Braun as its Director, officially opened with formal transfer to NASA from ABMA, at Redstone Arsenal, Huntsville, Ala.

———: First complete Scout launch vehicle fired from NASA Wallops Station, but fourth stage separation and firing was not accomplished.

———: Pacific Missile Range (PMR) Facility established at Eniwetok, Marshall Islands.

———: First operational version of Titan ICBM failed to launch at Cape Canaveral.

*July 4:* Soviet *Tass* announced that Russia last month successfully launched a new 4,400-pound-thrust rocket carrying a rabbit and two dogs to a reported altitude of 124.8 miles.

———: Piper Comanche set a world distance record in a closed circuit of 6,921.28 miles, Max Conrad as pilot.

*July 8:* Second experimental reactor (Kiwi-A Prime) in the Project Rover nuclear rocket program was successfully tested at full power and duration at Jackass Flats, Nev.

*July 11:* NASA selected Hughes, North American, Space Technology Laboratory, and McDonnell to study designs for the first lunar soft-landing spacecraft.

———: Dr. Ivan A. Getting of Raytheon was named first president of the Aerospace Corp.

———: Bell Telephone outlined to FCC a plan for worldwide service based upon a network of 50 satellites in polar orbit at 3,000-mile altitude.

*July 12:* Mistran (Missile trajectory measurement system) for AFMTC initiated by USAF contract with General Electric.

*July 17:* First of three NASA experiments carried by USAF balloons, carrying a NASA capsule containing 12 mice to 130,000-foot altitude for 11½ hours, in support of study of effects of heavy primary cosmic ray particles.

*July 18:* Dr. Robert C. Seamans, Jr., formerly chief engineer of RCA Missile Electronics and Control Division, was named Associate Administrator of NASA to replace Richard E. Horner.

*July 20:* Two Polaris (A–1X) test missiles successfully launched from submerged submarine, the *George Washington*, marking a major milestone in the Navy ballistic missile program.

*July 21:* NASA fired a Nike-Cajun sounding rocket from Fort Churchill, Manitoba, Canada, containing an instrumented payload to measure data on energetic particles during a period of low solar activity.

*July 22:* First flight of NASA's Iris sounding rocket successful, designed for 100-pound payloads to altitudes of about 200 miles, from Wallops Station.

*July 23:* Second of USAF–NASA balloon flights carrying NASA life science experiment to an altitude of over 130,000 feet for 11½ hours.

*July 24:* Donald Piccard established Class I world altitude record of 3,740 feet in plastic balloon HOLIDAY, from Minneapolis, Minn.

*July 26:* End of series of Army Bell HU–1 Iroquois helicopter flights which established four new world records.

*July 28–29:* First NASA-Industry Program Plans Conference held in Washington, D.C.

*July 29:* Project Apollo, advanced manned spacecraft program, was first announced at NASA's Industry Conference.

———: Atlas launch vehicle carrying unmanned Mercury capsule exploded 65 seconds after launch from AMR.

———: The 300-kw(e) static reactor electric power system attained first criticality. SNAP 10, utilizing thermoelectric conversion with no moving parts, was being developed for satellite application.

*July 31:* Dr. John F. Victory, the first employee of NACA hired in 1915 and recently Assistant to the Administrator of NASA, retired after 52 years of continuous Government service, including many important contributions in formulating national air policies and in establishing aeronautical research facilities and programs.

*August 2:* NRL Aerobee reached 90-mile altitude from WSPG with instruments to measure ultraviolet spectrum of the sun.

———: Army Ordnance five-stage Strongarm sounding rocket launched from Wallops Station, reaching an altitude of 300 miles, although fifth stage did not function.

*August 3:* First Sparrowbee sounding rocket launched from Wallops Station, lifting 56-pound University of Michigan payload to 260-mile altitude.

*August 4:* X–15 (No. 1) rocket airplane with interim engines established new unofficial world speed record of 2,196 mph, with Joseph Walker, NASA test pilot, at the controls. This topped Captain Apt's speed of 2,094 mph attained in the X–2 on September 27, 1956.

*August 5:* NASA and the Department of Defense announced the settlement of patent infringement claim by the estate of the late Robert H. Goddard, which had been pending since 1951, for $1 million ($765,000 by USAF, $125,000 by USA, $100,000 by NASA, and $10,000 by USN).

## 1960—Continued

*August 5:* IGY data released indicated that upper atmosphere's density becomes twice as great in December as in June.

*August 6:* While over Blossom Point, Md., simultaneously with a Class 1 solar flare, TRANSIT II-A satellite transmitted 6 minutes of clear reception showing history of development of ultraviolet and X-ray emission in relation to ionospheric behavior and to solar-radio noise.

*August 10:* DISCOVERER XIII launched successfully into a polar orbit.

*August 11:* First manmade object recovered from an orbiting satellite, the 85-pound instrumented capsule of DISCOVERER XIII recovered from the ocean off Hawaii after 16 orbits. Silken 50-star American flag it carried was presented to the President on August 15.

*August 12:* X-15 (No. 1) with interim engines and with Maj. Robert M. White (USAF) at the controls, established a new altitude record for a manned vehicle of 136,500 feet. This topped Captain Kincheloe's record altitude of 126,200 feet attained on September 7, 1956, in the X-2 rocket research aircraft.

——: NASA's ECHO I, the first passive communications satellite, successfully launched into orbit by a Thor-Delta. It reflected a radio message from the President across the Nation, thus demonstrating the feasibility of global radio communications via satellites. The 100-foot-diameter aluminized Mylar-plastic sphere was the most visible and largest satellite launched to date. (See Appendix A.)

——: USAF Atlas carrying radiation experiments in nose cone was fired 5,000 miles from Cape Canaveral, but nose cone was not recovered.

——: Navy Polaris missile fired 1,000 miles down AMR.

*August 13:* Army announced completion of a project for mapping lunar landing sites.

*August 15:* NASA announced selection of Plasmadyne Corp. for contract negotiations on a 1-kilowatt electric arc jet rocket engine.

——: Two pilots sealed in "space cabin" for 17-day simulated flight to the moon, at SAM, Brooks AFB, Tex.

*August 16:* Capt. Joseph W. Kittinger, Jr. (USAF), parachuted from EXCELSIOR III balloon at 103,000 feet, falling 17 miles before chute was employed at 17,500 feet, a new parachute record.

——: 11th Congress of the IAF opened in Stockholm.

*August 17:* ECHO I visible to skywatchers and provided reflection for numerous long-range radio transmissions by private and governmental research agencies in the United States.

*August 18:* ECHO I utilized for transatlantic communications when carrier signal was received by the French Telecommunications Establishment (CNET). Subsequently, voice and music transmissions were received by the University of Manchester at Jodrell Bank and other British installations.

——: USAF DISCOVERER XIV launched into polar orbit from Vandenberg AFB, Calif. (See Appendix A.)

——: USAF-Army COURIER IA communications satellite failed to orbit due to premature shutdown of first stage of Thor-Able-Star.

*August 19:* SPACECRAFT II launched into orbit by the U.S.S.R. weighing 5 tons and carrying a biological payload including two dogs. (See Appendix A.)

——: Second time a manmade object was recovered intact from earth orbit and the first midair recovery of an object from space, when USAF C-119 transport snared the 300-pound capsule of DISCOVERER XIV at 10,000 feet, Capt. H. F. Mitchell (USAF) as pilot of the C-119.

——: Wirephoto of President Eisenhower transmitted from Cedar Rapids, Iowa, to Dallas, Tex., via ECHO I satellite.

*August 21:* U.S.S.R. announced safe recovery of biologic payloads of SPACECRAFT II after 17 orbits, and reported that 2 dog passengers were in excellent physical condition. This was the first successful recovery of life forms from orbit.

*August 22:* Smithsonian Astrophysical Observatory reported that solar pressure was pushing ECHO I's perigee 1½ miles closer to the earth every 24 hours.

*August 23:* Bell Laboratory technicians successfully transmitted a voice and music message from New Jersey to Jodrell Bank, England, via ECHO I.

——: Aerobee-Hi with 208-pound payload launched from NASA Wallops Station 118-mile altitude.

*August 24:* ECHO I first went into earth's shadow with its two tracking beacons still operating. Since going into orbit on August 12, it had relayed hundreds of telephonic experiments and transmissions.

*August 26:* Construction begun on the world's largest radar at Arecibo, P.R., capable of bouncing signals off Venus, Mars, and Jupiter, with Cornell University as the prime contractor under direction of ARPA and USAF.

*August 30:* First Industry Conference conducted by NASA Goddard Space Flight Center.

*August 31:* Joint NASA-AEC Nuclear Propulsion Office (NPO) created at Germantown, Md., with Harold B. Finger as Manager.

*During August:* NASA suspended work on geodetic satellite program pending determination of whether it was to be a civilian or military program.

——: USAF Atlas squadrons became operational at Warren AFB, Wyo.

*September 5:* McDonnell F4H-1 Phantom II Navy fighter flown 1,216.78 mph on 500-km closed course for a new record, Lt. Col. T. H. Miller (USMC) as pilot.

*September 8:* ONR announced that radio signals had been received from the planet Saturn and a star 3,000 light-years away by the University of Michigan's 85-foot radio telescope.

——: President Eisenhower formally dedicated the NASA George C. Marshall Space Flight Center at Huntsville, Ala.

*September 10:* X-15 flown at more than 2,100 mph and to 80,000 feet.

*September 13:* NASA and DOD announced the creation of the Aeronautics and Astronautics Coordinating Board "to review planning, avoid duplication, coordinate activities of common interest, identify problems requiring solution either by NASA or the Department of Defense and insure a steady exchange of information." Dr. Hugh L. Dryden, Deputy Administrator of NASA, and Dr. Herbert F. York, Director of Research and Engineering of DOD, were named co-chairmen of the Board.

——: DISCOVERER XV placed into polar orbit. (See Appendix A.)

——: NASA gave bidders briefing to industry representatives on Project Apollo study contract at Space Task Group, Langley AFB, Va.

——: Bilateral agreement with South Africa ratified providing for construction of new tracking station in South Africa.

*September 13-14:* First meeting of the NASA Advisory Committee on Space Biology, chaired by Dr. Melvin Calvin.

*September 14:* Recovery capsule of DISCOVERER XV located from aircraft, but bad weather prevented surface pickup before it sank.

*September 15:* Two USAF pilots, Capt. W. D. Habluetzel and Lt. J. S. Hargreaves, completed a 30-day, 8-hour, and 24-minute simulated round trip to the moon in the space cabin simulator at the School of Aviation Medicine, Brooks AFB, Tex.

*September 16-22:* 27 research rockets were launched by U.S. scientists as a part of the COSPAR International Rocket Interval for 1960.

*September 19:* Atlas ICBM fired 9,000 miles from Cape Canaveral to the Indian Ocean in 50 minutes, the second record distance flight.

——: NERV (Nuclear Emulsion Recovery Vehicle) experiment successfully launched from Point Arguello, Calif., by an Argo D-8 rocket, the first NASA launching at PMR. NERV instrumented capsule reached an altitude of 1,260 miles before landing 1,300 miles downrange where it was picked up by Navy ships 3 hours later. It reached the highest

## 1960—Continued

known altitude that any manmade object had attained to be recovered successfully from space.

*September 20:* Aero Commander 680F set a world class altitude record of 36,932 feet for light aircraft, Jerrie Cobb as pilot.

*September 21:* USAF Blue Scout rocket fired from Cape Canaveral placed instrumented payload 16,600 miles above the earth, the first of 11 such tests, but no data were received due to radio malfunction.

*September 22:* President Eisenhower's address to the General Assembly of the United Nations pointed to the importance of international agreement on measures to "enable future generations to find peaceful and scientific progress not another fearful dimension in the arms race, as they explore the universe."

*September 25:* Atlas-Able 3 lunar orbital probe of NASA failed to achieve trajectory because of malfunction in one of the upper stages.

——: McDonnell F4H-1 Phantom II Navy fighter flown record 1,390.21 mph over 100-km closed course at Edwards AFB, Comdr. J. F. Davis as pilot.

*September 26:* NASA and Weather Bureau issued joint invitation to scientists of 21 nations to participate in meteorological research connected with future Tiros satellite.

——: Heat balance between atmospheric pressure areas near the earth's surface and temperature readings in space, reported as a result of experiments in EXPLORER VII launched October 13, 1959, by Dr. Verner E. Suomi of the University of Wisconsin.

——: Formal meeting of the DOD–NASA Aeronautics and Astronautics Coordinating Board (AACB).

*September 27:* Parachute designed to slow reentry speed of space capsules successfully tested at a speed of 2,000 mph after rocket boost to 30-mile altitude, over Eglin AFB, Fla.

——: Massive Soviet news buildup for this as "a day in the history of the world," while Premier Khrushchev was at the U.N. General Assembly meeting in New York. Rumored space spectacular did not apparently take place.

*September 30:* To date, the Smithsonian Astrophysical Observatory had photographed approximately 17,200 satellite passages with the Baker-Nunn Optical Network, and had recorded 17,000 visual observations by Moonwatch.

——: Soviet test pilot K. K. Kokkinaki established world speed record of 2,148.3 km/hr in delta-wing E–66 jet aircraft over 100-km closed course.

——: Formal agreements for all NASA tracking stations, planned at present, were either concluded or near conclusion.

*October 1:* First BMEWS (Ballistic Missile Early Warning System) station went into operation, at Thule, Greenland.

*October 2:* JPL announced that 85-foot receiving antenna for space tracking at Woomera, southwestern Australia, would be operational by November 1.

*October 3:* ONR STRATOSCOPE balloon carrying equipment to photograph the halo around the sun was launched at 80,000 feet in a series of high-altitude coronascope flights.

*October 4:* COURIER I–B active communications satellite successfully placed into orbit by Thor-Able-Star launch vehicle from Cape Canaveral. After completing one orbit it received and recorded a transcribed message to the United Nations by President Eisenhower transmitted from Fort Monmouth, N.J., and retransmitted it to another earth station in Puerto Rico. This marked the 100th launch of the Douglas Thor, military and scientific combined, and a Thor record of 60 percent of the U.S. satellites boosted into orbit.

——: Second complete NASA Scout vehicle fired successfully to its predicted 3,500-mile altitude and 5,800-mile impact range, from Wallops Station.

*October 7:* AEC briefing held at the Nevada Test Site at Jackass Flats, Nev., for representatives of 26 companies for proposals to study the requirements for a

National Nuclear Rocket Engine Development Facility. Existing test facilities are fully committed to the development of nuclear reactors.

———: Fédération Aéronautique Internationale meeting at Barcelona, Spain, accepted first rules to govern establishment of official records for manned spacecraft. The first record to be recognized must be at least 100 km, and later records must exceed existing record by 10 percent. Four categories for records are duration of flight, altitude without orbiting earth, altitude in orbit, and mass lifted above 100 km.

*October 10:* Interagency meeting on the establishment of an operational meterological satellite system was held at NASA Headquarters.

*October 11:* USAF SAMOS I launched from Vandenberg AFB, but failed to orbit.

*October 12:* Dr. T. Keith Glennan, NASA Administrator, announced that communications satellites developed by private companies on a commercial basis would be launched by NASA at cost to assist private industry in developing a communications network.

———: Heavy-equipment parachute drop record of 41,740 pounds, from Lockheed C-130 Hercules transport to ground at El Centro, Calif.

*October 13:* USAF Atlas launched at AMR placed nose cone containing three black mice 650 miles up and 5,000 miles downrange at 17,000 mph. Nose cone was recovered in target area near Ascension Island, the three mice surviving the flight in "good condition."

———: Transmitter of EXPLORER VII failed to stop as programed.

———: Camera mounted in nose of Atlas photographed stars at 700-mile altitude, providing first color picture of the earth from 600-mile altitude.

*October 15–18:* Four operational-type Polaris missiles successfully launched from submerged *Patrick Henry* off the Florida coast.

*October 17:* Project Mercury weather support group established at NASA's request in the Office of Meteorological Research of the Weather Bureau.

*October 18:* Second Iris rocket rose to 140 miles with a payload of 125 pounds from Wallops Station.

*October 19:* Kiwi-A No. 3 static test of nuclear rocket propulsion was successfully conducted at AEC Nevada test site, resulting in NASA–AEC call for bids for industrial development phase of Project Rover on November 1, 1960.

———: NASA announced award of preliminary design contracts for solid-fuel rockets with thrusts between 2 and 15 million pounds to Aerojet-General, Grand Central, and Thiokol.

———: Dr. Hugh L. Dryden received the Elliott Cresson Medal of the Franklin Institute.

*October 21:* FCC received formal application of American Telephone & Telegraph for authority to operate a communications satellite.

*October 23:* COURIER I–B stopped transmitting, but radio tracking beacon continued to function. In 18 days it had transmitted 118 million words.

*October 24:* Titan ICBM fired 6,100 miles, 100 miles longer than any previous shot, with tactical-type nose cone.

*October 25:* NASA selected Convair, General Electric, and Martin to conduct individual feasibility studies of an advanced manned spacecraft as part of Project Apollo.

*October 26:* USAF DISCOVERER XVI successfully launched with new payload, but failed to go into polar orbit.

*October 27:* Institute of the Aeronautical Sciences (IAS) changed its name to the Institute of the Aerospace Sciences.

*October 31:* DOD ordered a stepup in development of the mach 3 B–70 supersonic bomber.

———: USAF announced consideration of proposals for "aerospace plane" capable of scooping up tons of oxygen in upper atmosphere before space flight, then re-entering for landing as an airplane.

## 1960—Continued

*During October:* Construction of space simulator began at Rye Canyon Research Center of Lockheed for study of disintegration of materials at simulated 800,000 feet at temperature of −320° F.

———: Structures Research Division of NASA Langley continued ablation studies begun in 1956 with electric arc-powered jet, achieving 9,000° F for 105 seconds on an illustrative test.

*November 2:* Lunar atlas prepared for USAF by group under technical direction of G. P. Kuiper was released, an "Orthographic Atlas of the Moon" charted 5,000 base points combined with best available photos and grids.

*November 3:* EXPLORER VIII launched into an elliptical orbit from AMR by four-stage Juno II, containing instrumentation for detailed measurements of the ionosphere. This was the 10th time that JPL-developed upper stage rocket clusters had successfully placed satellites or deep space probes into orbit. (See Appendix A.)

*November 4:* New results in sustaining hydrogen fusion for 1 millisecond at 60° F reported by University of California scientists.

*November 5:* Operational date of first Minuteman squadron advanced a full year to July 1962 by USAF.

*November 6:* U.S.S.R. published atlas on the far side of the moon based on LUNIK III photographs.

———: Japanese Space Development Council recommended initiation of basic studies for launching an earth satellite.

*November 8:* USAF Blue Scout Junior with radiation-study payload reached 24,500-mile altitude, but second stage did not burn full program.

———: NASA LITTLE JOE test flight of Mercury capsule, capsule did not separate from booster.

*November 9:* NRL Aerobee-Hi collected data on ultraviolet radiation in the night sky 131 miles above WSPG.

———: Post Office Department transmitted a "speed mail" letter from Washington to Newark, N.J., by bouncing microwave transmission off ECHO I.

*Early November:* NASA–DOD Aeronautics and Astronautics Coordinating Board (AACB) and cognizant members agreed that NASA could drop the tracking light geodetic satellite and utilize other space projects to obtain geodetic data for the scientific community.

*November 10:* Advanced Polaris (A-2) successfully launched on record 1,600-mile flight at AMR.

———: Department of Defense placed Navy's SPASUR (Space Surveillance Detection Net) and the Air Force's SPACETRACK (National Space Surveillance Control Center) under the North American Air Defense Command for military functions. NASA would assume SPACETRACK's function of passing on information on space vehicles to the world's scientific community.

*November 12:* DISCOVERER XVII placed into polar orbit from Vandenberg AFB, restartable Agenda B second stage successfully flown for the first time.

———: Navy announced development of techniques for low-cost satellite-launching facilities from airplanes, barges, ships, or from underwater.

*November 14:* Capsule of DISCOVERER XVII ejected after 31 orbits and successfully snared at 9,000 feet by USAF C-119 aircraft, the second such recovery in midair of a space object.

———: IGY Warning Center reported that solar flares were causing "extremely severe" magnetic disturbance of the earth's atmosphere, an event detected by EXPLORER VII and later analyzed as greatest burst of solar radiation in the satellite's 13 months of operation.

———: DOD announced that NASA, USAF, USA, and USN were jointly building a geodetic satellite to map the earth accurately.

———: USAF reported that printed messages and weather maps had been sent up to 900 miles by bouncing radio signals off meteor trails.

———: First letter carried by satellite mail (31 orbits and a distance of about a million miles), a letter from USAF Chief of Staff to the Secretary of Defense carried in capsule recovered from DISCOVERER XVII.

*November 15:* X–15 (No. 2) with new XLR–99 engine (57,000-pound thrust) flown to nearly 80,000 feet and 2,000 mph on first test flight by A. Scott Crossfield at Edwards AFB, Calif. Earlier interim engine, XLR–11 with one-quarter of the thrust of the XLR–99, had pushed the X–15 to new world speed and altitude records of 2,196 mph and 136,500 feet.

———: Prof. A. Gib DuBusk, geneticist at Florida State University, reported that bread mold specimens, rocketed to 1,200-mile altitude on Argo D–8 capsule on September 19, had shown 30 times as many changes as control cells.

———: USAF Mace-B flight tested 1,000 miles.

———: Data capsule fired 5,000 miles downrange from AMR by Atlas ICBM, which was recovered 1 hour later.

———: Aerobee-Hi launched to 145-mile altitude from NASA Wallops Station, Va.

*November 17:* NASA established Test Support Office at the Pacific Missile Range (PMR), to function under Launch Operations Directorate, Marshall Space Flight Center.

———: Last test of Polaris (A–1, 1,300-mile series) from AMR unsuccessful.

———: U.S. proposed upper atmosphere rocket probes from Woomera Rocket Range in Australia.

*November 19:* Albert Hibbs of JPL reported that EXPLORER I had also discovered clouds of cosmic dust in its orbit, information found by continued examination of data obtained during 4 months of payload transmission after launch on January 31, 1958. EXPLORER I remained in orbit.

*November 21:* Mercury Redstone flight test (MR–I) at AMR terminated prior to liftoff because of faulty ground-support circuitry which had not been noted on some 60 previous Redstone firings.

———: 500-pound capsule of USAF launched to 32-mile altitude and recovered intact by means of drag balloon and parachute known as the "Ballute" system.

*November 22:* India and United States announced joint program of some 40 high-altitude balloon flights from India, starting in December.

———: Aerobee-Hi fired to 105-mile altitude from NASA Wallops Station with four stellar spectrometers developed for an experiment by the University of Rochester's Institution of Optics.

———: National Science Foundation announced that the National Center for Atmospheric Research, operated by a group of universities, would be sited at Table Mountain, near Boulder, Colo. Walter Orr Roberts was named as Director of this NSF Center which will do fundamental research and serve as a coordinating center for a network of atmospheric investigations.

———: ARPA technical advisory group established to facilitate exchange of information between technical management and research personnel on Project Defender.

*November 23:* X–15 (No. 2) flown on second test flight with XLR–99 engine by A. Scott Crossfield, restarting the engine in flight for the first time.

———: TIROS II weather satellite launched by Thor-Delta at AMR, the 14th successful U.S. satellite launched to date in 1960. (See Appendix A.)

———: In a letter to the chairman of the Senate Committee on Aeronautical and Space Sciences, NASA Administrator Glennan defined low-altitude (orbits of 2,000 to 6,000 miles) active communications satellite development to "stimulate those developments which promise early benefits to our citizens."

———: Plastic balloon launched from Sioux Falls, S. Dak., with University of Michigan instrument package designed to take cloud pictures to compare with those taken by cameras in TIROS II.

*November 25:* NASA scientists increased the speed of spin of TIROS II by means of ground radio command.

## 1960—Continued

*November 27:* Report of the President's Commission on National Goals was released, which stated that the United States "should be highly selective in our space objectives and unexcelled in their pursuit. Prestige arises from sound accomplishment, not from the merely spectacular, and we must not be driven by nationalistic competition into programs so extravagant as to divert funds and talents from programs of equal or greater importance . . . ."

*November 28:* Discussions on creation of an European space research organization undertaken by scientific representatives of Belgium, Denmark, France, West Germany, Italy, the Netherlands, Norway, Sweden, Switzerland, and the United Kingdom, with an observer from Spain.

———: TIROS II had successfully transmitted 998 pictures to receiving stations at Fort Monmouth, N.J., and San Nicholas Island, Calif., 85 percent of narrow angle and 5 to 10 percent of the wide-angle pictures having some value.

*November 28–December 3:* Space Research Symposium sponsored by Argentina in which Dr. Hugh Dryden and other U.S. scientists participated.

*November 30:* TRANSIT III–A navigation satellite, with two instrumented payloads, was destroyed 40 minutes after launch from AMR by Thor-Able-Star booster.

*During November:* Under arrangements of the AACB (Aeronautics and Astronautics Coordinating Board), NASA will utilize existing NASA tracking stations for initial Centaur development vehicles and switch to the Advent network (which is to be planned, funded, and constructed by DOD) when Centaur is operational, perhaps as early as the fourth of 10 development launchings of Centaur.

*December 1:* SPACECRAFT III launched by U.S.S.R., weighing over 5 tons and carrying a biological payload in its "space cabin." (See Appendix A.)

———: USAF delivered to JPL the first 1:1,000,000 scale map of the lunar landing site selected by NASA, the second in a continuing series of 1:1,000,000 charts prepared on USAF contract in response to NASA requirements.

———: Army Nike-Zeus A–ICBM missile with guidance successfully test fired from WSPG.

———: Delegates of 11 Western European nations approved an agreement aimed at establishing an organization for space research. Proposed intergovernmental agency would concentrate on satellites rather than rockets for launch vehicles.

*December 2:* First of new series of static firings of Saturn considered only 50 percent successful in 2-second test at MSFC.

———: Human tissues exposed to heavy radiation during 50-hour flight of recovered DISCOVERER XVII capsule according to USAF.

*December 3:* Moscow Radio reported that SPACECRAFT III descended along an "uncalculated trajectory" and burned up in the dense atmosphere.

———: Titan ICBM exploded in its silo at Vandenberg AFB during night fueling operations.

———: Senate Committee on Science and Astronautics issued staff study entitled "Policy Planning for Space Communications," which stated that the United States "must have a unified policy which effectively coordinates all our diverse and extensive resources in this area."

*December 4:* American Bar Association's "Report to NASA on the Law of Outer Space" was released, which contained collation of legal opinion on the broad spectrum of space activities.

———: Attempt to launch a Beacon satellite with a four-stage solid-propellant Scout from Wallops Station did not succeed due to failure of second stage.

*December 5:* Polaris A–2 successfully test fired 1,400 nautical miles down AMR.

———: USAF completed Snark R&D program with a 5,000-mile flight from Cape Canaveral.

*December 6:* Civil Service Commission approved new examination for career professional positions in aerospace technology, part A covering work in the physical sciences, engineering, and mathematics, and part B covering work in the life sciences and related systems.

*December 7:* DISCOVERER XVIII launched into polar orbit by new Thor-Agena B from Vandenberg AFB, carrying surveillance-system equipment and human tissue in recovery capsule.

———: X–15 (No. 2) flown on final contractor's test flight by A. Scott Crossfield, making two midair engine shutdowns and restarts.

*December 7–10:* Series of upper atmosphere sounding rockets from NASA Wallops Station, sodium vapor being ejected at about 212 miles altitude and a lithium flare released near peak altitude of about 450 miles to measure wind velocities and temperatures.

*December 9:* X–15 made first flight with ball-shaped "hot nose," reaching 50,000 feet and 1,254 mph, NASA's Neil Armstrong making his second familiarization flight.

———: Tory IIA reactor, part of AEC-USAF Pluto program to demonstrate feasibility of nuclear ramjet propulsion, achieved criticality of 1-watt nominal power, and later in day was run up to 200 watts.

*December 10:* 300-pound capsule of DISCOVERER XVIII caught at 14,000 feet by USAF C–119 crew, after making 48 polar orbits. Capsule contained human eye-lid tissue and blood and bone marrow to study effect of radiation in space. This was the second DISCOVERER capsule catch by C–119 crew headed by Capt. Gene Jones, while precision of the entire operation beginning with launch 3 days previous was considered the most successful to date.

*December 11:* SAM scientists reported that human tissue recovered from the capsule of DISCOVERER XVII after about 50 hours on 31 orbits (November 14), survived radiation in space, including that generated by one of the largest solar storms ever observed.

*December 12:* SAM scientists at Brooks AFB reported that biological specimens including human tissue recovered from the capsule of DISCOVERER XVIII two days ago, showed far less radiation effects than specimens recovered from DISCOVERER XVII in November.

———: Initial flight test of new guidance system of Army Pershing missile successful.

*December 13:* North American A3–J Vigilante set a world altitude record with 1,000 kilogram payload of 91,450.8 feet, Comdr. Leroy Heath (USN) as pilot.

———: *Palaemon,* a 180-foot barge built to transport the Saturn launch vehicle from MSFC to Cape Canaveral by water, was formally accepted by MSFC Director from Maj. Gen. Frank S. Besson, Chief of Army Transportation.

*December 14:* USAF B–52G completed 10,000-mile nonstop flight without refueling in 19 hours and 45 minutes, at Edwards AFB, which broke world and jet distance records over a closed course without refueling.

*December 15:* Atlas-Able launch vehicle with NASA cislunar spacecraft exploded 70 seconds after launch from Cape Canaveral.

*December 16:* Scientists from Great Britain and NASA completed a series of meetings leading to planning for British scientific satellite to be flown on a Scout vehicle.

———: Atlas-D with Mark 3 nose cone fired 4,384 nautical miles into Eniwetok Atoll in first SAC launching from Vandenberg AFB.

———: AEC–NASA Nuclear Propulsion Office announced selection of TALANT industrial team proposal to conduct study of the requirements for a National Nuclear Rocket Engine Development Facility.

*December 17:* National Science Foundation announced grants totaling $22.7 million to support summer institutes for 20,000 teachers of science, mathematics, and engineering in high schools and colleges.

## 1960—Continued

*December 19:* Unmanned Project Mercury spacecraft launched by modified Redstone booster (MR-1) in a suborbital trajectory, impacting 235 miles downrange after reaching an altitude of 135 miles and a speed of near 4,200 mph. Capsule was recovered about 50 minutes after firing.

——: Secretariat of COSPAR released official Soviet data on 27 U.S.S.R. rockets launched in a series of high-altitude experiments from a research ship in the Pacific, and a total of 73 rocket launchings in the first half of the year 1960.

*December 20:* USAF DISCOVERER XIX successfully launched into polar orbit from PMR carrying Project Midas test payload. (See Appendix A.)

——: President-elect Kennedy announced that Vice President-elect Lyndon B. Johnson would chair the National Aeronautics and Space Council.

——: Founded in 1912 by Glenn L. Martin, the Martin Co. delivered its last airplane, a P5M-2, to the Navy, having produced more than 12,000 aircraft and entering the missile/space business with the NRL Viking research rocket in 1948.

——: Second stage of near-operational Titan ICBM failed to ignite over Cape Canaveral.

*December 21:* Space Technology Laboratories was selected by NASA for contract negotiations for an orbiting geophysical observatory (OGO) satellite program. To be managed by GSFC, OGO will be NASA's first standardized satellite, often referred to as the "streetcar" satellite, capable of placing 50 different geophysical experiments on any one flight.

——: Eight-engine cluster of Saturn successfully static fired for 65 seconds at MSFC, the firing generating 1,300,000 pounds of thrust.

*December 22:* Nuclear submarine *Robert E. Lee* fired Polaris A-1 IRBM 1,300 miles in an Atlantic shot.

*December 23:* Goddard Space Flight Center scientists, Robert Jastrow and Robert Bryant, reported that atmospheric drag acting on ECHO I during the severe solar storm of November 12, was increased by about a factor of two. Scientists had previously noted the rise and fall of the density of the upper atmosphere, and the heating effect of a solar flare had been noted on the orbit of SPUTNIK III in 1959.

*December 25:* Jet Propulsion Laboratory announced selection of Blaw-Knox Equipment, Hughes Aircraft, North American Aviation, and Westinghouse Electric to study feasibility of a large space-tracking antenna.

*December 26:* Successful firing of a solid-propellant rocket motor using "building block" method was announced by NASA.

*December 27:* EXPLORER VIII ceased transmitting ionospheric measurement data acquired in 20,866,706 miles and 694.3 orbits, which produced more than 700 miles of magnetic tape since launch on November 3.

*December 28:* U.S. Weather Bureau sent TIROS II cloud-cover picture to Australia, which was taken over the Indian and South Pacific Oceans and served as a basis for forecasting a break in severe heat wave.

*December 29:* Dr. T. Keith Glennan offered his resignation as Administrator of NASA, to be effective January 20, 1961.

*December 31:* To date, the United States had successfully launched 31 earth satellites (9 of 16 still in orbit were still transmitting) and 2 deep space probes into orbit around the sun. The U.S.S.R. had launched seven satellites (one of which remained in orbit) and one deep space probe. The U.S.S.R. had also launched one lunar impact mission (LUNIK II), while LUNIK III had passed once around the moon and then went into earth orbit before decaying.

*During 1960:* World Data Center A, Rockets and Satellites, of the National Academy of Sciences, continued to provide a means for international exchange of scientific data.

——: JPL turned the Army Sergeant missile over to Sperry Gyroscope Co. as production contractor.

——: Through a contract with the University of Chicago, the USAF's Aeromedical Field Laboratory and Missile Test Center developed a system for ascertaining the types and intensities of primary cosmic particles in space.

——: NASA launching record for the year: 22 major space flight attempts, over two-thirds of which were fully successful.

——: World's scheduled airlines (excluding U.S.S.R. and Communist China) carried 108 million passengers during the year according to ICAO, the first year air passenger traffic exceeded 100 million persons.

# Appendices

APPENDIX A

# Chronicle of Earth Satellites and Space Probes

1957–1960

THIS APPENDIX was compiled from statistics prepared by the NASA Office of Public Information, Washington, D.C. It does not include description of spent rocket casings, etc., that have gone into orbits or trajectories along with payloads. For more detail on instrumentation and other items, the reader is advised to consult "Space Activities Summary," prepared and issued by the NASA Office of Public Information. Russian data are unofficial.

## CHRONICLE OF EARTH SATELLITES AND SPACE PROBES, 1957–60

| Launch date | Name | International designation | Launch vehicle | Payload data | Apogee (st. mi.) | Perigee (st. mi.) | Period (minutes) | Inclination | Remarks |
|---|---|---|---|---|---|---|---|---|---|
| Oct. 4, 1957 | SPUTNIK I (U.S.S.R.) | 1957 ALPHA: I (case), II (payload), III (nose cone). | Not disclosed | Total weight in orbit: About 4 tons (unofficial). Scientific instrumentation: 184 lbs. 2 chemical battery transmitters (ceased 10/27/57). | 588 | 142 | 96.2 | 65° to Equator. | 1st manmade earth satellite. Status: Payload down on 1/4/58 (92 days). |
| Nov. 3, 1957 | SPUTNIK II (U.S.S.R.) | 1957 BETA | Not disclosed | Total weight in orbit: About 4 tons (unofficial). Scientific instrumentation: 1,120 lbs. Experiments: Cosmic rays; solar ultraviolet and X-radiation; test animal "Laika" (dog); temperature; pressures. Transmitters ceased 11/10/57. | 1,038 | 140 | 103.7 | 65° to Equator. | 1st biomedical experiment and data disclosed significant solar influence on upper atmosphere densities. Status: Down on 4/14/58 (162 days). |
| Jan. 31, 1958 | EXPLORER I (United States) | 1958 ALPHA | Jupiter-C | Total weight: 30.8 lbs. Scientific instrumentation: 18.13 lbs. Experiments: Cosmic rays, micrometeorites, and temperatures. Transmitter stopped 5/23/58 (intermittent operation 2/11–2/28). | 1,573 ¹(1,171) | 224 ¹(217) | 114.8 ¹(107.5) | 33.34° to Equator. | Made most important discovery of the IGY, the radiation belt around the earth identified by Van Allen. (2d belt later discovered by Pioneer III.) Status: Still in orbit 10/25/60. |

| Date | Name | Designation | Vehicle | Remarks | Weight (lbs.) | Apogee/Perigee | Period (min.) | Inclination | Status/Data |
|---|---|---|---|---|---|---|---|---|---|
| Mar. 17, 1958 | VANGUARD I (United States) | 1958 BETA: I (case); II (payload). | Vanguard (TV-4) | Scientific payload and total weight: 3.25 lbs. (50-lb. 3d-stage rocket casing also in orbit.) 1st solar-powered batteries: 1 transmitter ceased on 4/5/58, the other will operate indefinitely. | 2,453 [1] (2,451) | 409 [1] (408) | 107.9 [1] (133.9) | | Stability of orbit provided geodetic observations including determination that the earth is slightly pear shaped. Status: Still in orbit and still transmitting with estimated orbit lifetime of 200 to 1,000 years. |
| Mar. 26, 1958 | EXPLORER III (United States) | 1958 GAMMA | Jupiter-C | Total weight: 31 lbs. Instrumentation: 18.56 lbs. Experiments: Cosmic rays with tape recorder, temperatures, micrometeor gauges. Telemetering and beacon ceased 5/14/58, erratic 5/22–6/5/58. | 1,746 | 121 | 115.87 | 33.4° | Yielded data on radiation belt discovered by Explorer I, on micrometeorite impacts and temperature. Status: Down 6/28/58 (93 days). |
| May 15, 1958 | SPUTNIK III (U.S.S.R.) | 1958 DELTA | Not disclosed | Total weight: About 7,000 lbs. (unofficial). Instrumentation: 2,925 lbs. Experiments: Atmospheric pressure; ions; and earth's electrostatic and magnetic fields; micrometeors, etc. | 1,167 | 135 | 106 | 65.3° to Equator. | Provided data on radiation belts, etc. Status: Down 5/6/60. |
| July 26, 1958 | EXPLORER IV (United States) | 1958 EPSILON | Jupiter-C | Total weight: 38.4 lbs. Instrumentation: 25.8 lbs. Experiments: 2 Geiger-Mueller counters and 2 scintillation counters. 2 transmitters with 5 concurrent channels (channels 2 and 5 ceased 9/3/58). All transmission ceased 10/6/58. | 1,380 | 163 | 110.27 | 50.29° to Equator. | Provided data on radiation belts, etc. Status: Down 10/23/59. |

[1] As of 10/25/60.

# CHRONICLE OF EARTH SATELLITES AND SPACE PROBES, 1957-60—Continued

| Launch date | Name | International designation | Launch vehicle | Payload data | Apogee (st. mi.) | Perigee (st. mi.) | Period (minutes) | Inclination | Remarks |
|---|---|---|---|---|---|---|---|---|---|
| Oct. 11, 1958 | PIONEER I (United States) | None | Thor-Able I | Total weight: 84.4 lbs. including 43.7 lbs. venier and retrorockets. Instrumentation: 39 lbs. Experiments: Radiation in space measurements; magnetic field of earth; density of micrometeors; internal temperatures; electronic scanner. | Attained altitude of 70,700 statute-miles before reentering atmosphere over South Pacific 43 hrs. 17½ min. | | | | Determined: Radial extent of radiation bands (first observation that radiation is a band); total ionizing flux; first observation of hydromagnetic oscillation of earth's magnetic field; discovery of departure of magnetic field from predictions; first determinations of density of micrometeors in space; first measurements of interplanetary magnetic field. |
| Nov. 8, 1958 | PIONEER II (United States) | None | Thor-Able I | Total weight: 86.4 lbs. Instrumentation: 34.3 lbs. Experiments: Radiation, cosmic ray flux, magnetic fields of Earth and Moon, etc. | Attained altitude of 963 miles before 3d stage failed to ignite. | | | | Acquired some data that equatorial region has higher flux and radiation energy before reentering atmosphere 42.4 minutes after launch. |
| Dec. 6, 1958 | PIONEER III (United States) | None | Juno II | Total weight including instrumentation: 12.95 lbs. Experiments: Radiation measurements. | Attained altitude of 63,580 miles before reentering atmosphere over French Equatorial Africa. | | | | Attained 24,000-mph velocity. Discovered 2d radiation belt around the earth. Status: Down 12/7/58 (38 hrs. 6 min.). |

142

| Date | Name | Designation | Launch vehicle | Weight & Experiments | Apogee/Distance | Perigee | Period (min) / Inclination | Remarks |
|---|---|---|---|---|---|---|---|---|
| Dec. 18, 1958 | PROJECT SCORE (United States) | None | Atlas | Total weight in orbit including booster: 8,750 lbs. Instrumentation in payload: 150 lbs. Experiments: Twin package of transmitters, recording and receiving apparatus (12 days). | 920 | 110 | 101.46 / 32.3° | 1st transmittal of human voice from space. Status: Down 1/21/59 (34 days). |
| Jan. 2, 1959 | LUNIK I—Mechta or "Dream" (U.S.S.R.) | None | T-3 | Total weight: 3,245 lbs. (unofficial). Instruments: 800 lbs. Experiments: Study gas components of interplanetary matter and corpuscular radiation of sun; magnetic fields of earth and moon; meteoric particles; heavy nuclei in primary cosmic radiation. | Aphelion 123 million miles from sun. | Perihelion 91 million miles from sun. | 443 days / 1° to ecliptic. | 1st successful deep space probe. Status: In orbit around sun in 15-month cycle. |
| Feb. 17, 1959 | VANGUARD II (United States) | 1959 ALPHA | Vanguard (SLV-4) | Total weight in orbit: 70.74 lbs. (including 3d stage). Experiments: Cloud cover; 2 photocells designed to produce images for 2 weeks. | 2,050 [1] (2,042) | 346 [1] (350) | 125.4 [1] (125.3) / 32.86° | Precession of satellite prevented interpretation of data. Status: Still in orbit.[1] |
| Feb. 28, 1959 | DISCOVERER I (United States) | 1959 BETA | Thor-Agena | Total weight: 1,300 lbs. Instruments: 245 lbs. | 519 | 176 | 95.57 / −3° off N.-S. axis. | 1st polar orbit. Status: Down early March 1959. |
| Mar. 3, 1959 | PIONEER IV (United States) | None | Juno II | Total weight: 13.40 lbs. Experiments: Radiation measurements in space on earth-moon trajectory; photoelectric scanner for use in vicinity of moon. | Came within 37,300 miles of moon on 3/4/59. 107.9 million from sun. | 91.7 million from sun. | 406.9 days / 0.127° to ecliptic. | Yielded data on radiation in space, tracked for 82 hours to a distance of 407,000 miles. Status: In sun orbit. |
| Apr. 13, 1959 | DISCOVERER II (United States) | 1959 GAMMA | Thor-Agena | Total weight: 1,600 lbs. Instruments: 245 lbs. for communications and performance, 195 lbs. data capsule. Experiments: Recovery of ejected data capsule. | 225 | 156 | 90.4 / 0.2° off N.-S. axis. | First satellite to carry recoverable instrument package. Data capsule impacted near Spitsbergen in 4/14/60, payload decayed on 4/26/60. |

[1] As of 10/25/60.

CHRONICLE OF EARTH SATELLITES AND SPACE PROBES, 1957-60—Continued

| Launch date | Name | International designation | Launch vehicle | Payload data | Apogee (st. mi.) | Perigee (st. mi.) | Period (minutes) | Inclination | Remarks |
|---|---|---|---|---|---|---|---|---|---|
| Aug. 7, 1959 | EXPLORER VI (United States) | 1959 DELTA | Thor-Able | Total weight including paddlewheels: 142 lbs. Experiments: Measure specific levels of earth's radiation belt; TV scanner to relay cloud cover; micrometeor detection; 2 magnetometers. | 26,357 | 156 | 12½ hrs. | 46.9° | Acquired valuable data on radiation levels, transmitted crude cloud cover image, detected ring of electrical current circling the earth. Status: Position uncertain as of 10/25/60. |
| Aug. 13, 1959 | DISCOVERER V (United States) | 1959 EPSILON | Thor-Agena | Same as Discoverer II | 450 | 136 | 94 | 78.9° | Reentry capsule not recovered. Status: Down 9/28/59 (47 days). |
| Aug. 19, 1959 | DISCOVERER VI (United States) | 1959 ZETA | Thor-Agena | Same as Discoverer II | 537 | 139 | 95.3 | 84° | Separation of capsule occurred 8/20, but no recovery was made. Payload orbit decayed 10/20/59. |
| Sept. 12, 1959 | LUNIK II (U.S.S.R.) | None | "Multistage rocket." | Weight: 858.4 lbs. (estimated). Experiments: Measure external and internal temperatures; study magnetic fields of earth and moon; meteoric particles; heavy nuclei and other properties of cosmic rays. | Lunar impact on 9/13/59 | | | 65° | 1st lunar impact, probe traveled 236,875 miles in 35 hours. |

| Date | Name | Designation | Vehicle | Payload description | (a) | (b) | (c) | (d) | Remarks |
|---|---|---|---|---|---|---|---|---|---|
| Sept. 18, 1959 | VANGUARD III (United States) | 1959 Eta | Vanguard (SLV-7) | Total weight around 100 lbs. Instruments: 50 lbs. Experiments: elliptical orbit study of magnetic fields, micrometeorites, X-rays, and temperatures. | 2,329 [1] (2,318) | 319 [1] (2,320) | 130 [1] (129.8) | 33.3° | Provided comprehensive survey of earth's magnetic field; detailed location of lower edge of radiation belts; accurate count of micrometeorite impacts. Status: Still in orbit 10/25/60. |
| Oct. 4, 1959 | LUNIK III (U.S.S.R.) | 1959 Theta | Not disclosed | Total weight: 614 lbs. scientific satellite; last stage weighing 3,423 lbs. without fuel also went into orbit which contained 345 lbs. of instruments—a total of 4,037 lbs. Experiments: Photographed back side of moon. | 292,000 | 24,840 Payload passed 4,372 miles from the moon on 10/6/59, before perigee on 10/18/59. | 15 days pictures transmitted | 80° to Equator after passing moon. | Produced photograph of 70 percent of moon's far side. Status: Reentered earth's atmosphere about 4/20/60. |
| Oct. 13, 1959 | EXPLORER VII (United States) | 1959 Iota | Juno II (19A) | Total weight: 91.5 lbs. Experiments: Earth-Sun heat budget; Lyman-alpha and X-ray counters, cosmic ray counters. | 680 [1] (671) | 342 [1] (344) | 101.33 [1] (101.2) | 50.3° | Provided significant geophysical data on radiation and magnetic storms; first penetration of a sensor in flight by micrometeorite. Status: Tracking beacon inoperative on 12/5/59. As of 6/16/60, information transmitted on 3 of 7 experiments not intelligible after 8 months. Data still being received on remaining experiments as of September 1960. |

[1] As of 10/25/60

## CHRONICLE OF EARTH SATELLITES AND SPACE PROBES, 1957-60—Continued

| Launch date | Name | International designation | Launch vehicle | Payload data | Apogee (st. mi.) | Perigee (st. mi.) | Period (minutes) | Inclination | Remarks |
|---|---|---|---|---|---|---|---|---|---|
| Nov. 7, 1959 | DISCOVERER VII (United States) | 1959 KAPPA | Thor-Agena | Contained telemetry equipment to measure its performance. | 550 | 104 | 95 | 82° | Electrical malfunction prevented stabilization in orbit and separation of capsule. Status: Down on 11/26/59 (19 days). |
| Nov. 20, 1959 | DISCOVERER VIII (United States) | 1959 LAMBDA | Thor-Agena | Capsule contained telemetry equipment to measure its performance. | 1,000 | 120 | 103 | 81° | Capsule ejected but not located. Status: down 3/8/60 (110 days). |
| Mar. 11, 1960 | PIONEER V (United States) | 1960 ALPHA | Thor-Able IV | Total weight: 94.8 lbs., including 4 vanes covered by 4,800 solar cells and 40 lbs. of instruments. Experiments: High-energy radiation; plasma, cosmic radiation and solar particle counters; magnetometer; aspect indicator; temperatures. | Aphelion 92.3 million from sun. | Perihelion 74.9 million from sun. | 311.6 days | 3.35° to ecliptic. | Highly successful exploration of interplanetary space between orbits of Earth and Venus; established communication record of 22.5 million miles on 6/26/60; made 1st measurements of solar flare effects, particle energies and distribution, and magnetic field phenomena in interplanetary space. Status: In orbit around sun. |
| Apr. 1, 1960 | TIROS I (United States) | 1960 BETA | Thor-Able | Total weight: 270 pounds. Objective: Test of experimental TV techniques leading to worldwide meteorological system. | 467 1 (468) | 429 1 (429) | 99.1 1 (99.1) | 48.327° to Equator. | Provided 1st global cloud-cover photographs (22,500 total) from near circular orbit. Status: In stable orbit. |

| Date | Name | Designation | Launch vehicle | Description | (col) | (col) | (col) | (col) | Remarks |
|---|---|---|---|---|---|---|---|---|---|
| Apr. 13, 1960 | TRANSIT I-B (United States) | 1960 GAMMA | Thor-Able-Star | Total weight: 265 pounds. Objective: Determine feasibility of navigation satellite. | 479 [1] (433) | 233 [1] (238) | 96 [1] (95.4) | 51° | Demonstrated feasibility of navigation satellite. Status: In orbit 10/25/60. |
| Apr. 15, 1960 | DISCOVERER XI (United States) | 1960 DELTA | Thor-Agena | Total weight: 1,700 pounds, including 2d stage casing and 300-pound capsule. Objective: Gather data on propulsion, communications, orbital performance, stabilization and recovery techniques. | 380 | 109 | 92.25 | 80.1° | Data capsule ejected but not observed. |
| May 15, 1960 | SPACECRAFT I (U.S.S.R.) | 1960 EPSILON | Not disclosed | Total weight: 10,008 lbs. Pressure cabin: 5,512 lbs. Instruments: 3,250 lbs. Objective: Place space cabin into orbit and test life support systems, and recover cabin from orbit. Tape of voice transmitted to ground stations. Cabin contained "dummy space man." | 288.7<br>429 [1] (384) | 188.5<br>191 [1] (175) | 91.1<br>94.25 [1] (93.6) | 64.9° | After firing of retrorockets on May 19, 1960, went into eccentric orbit. Space cabin placed into near circular orbit, but recovery operations malfunctioned. Status: In orbit. |
| May 24, 1960 | MIDAS II (United States) | 1960 ZETA | Atlas-Agena | Total weight including entire 2d stage: 5,000 lbs. Objective: Test of system for detection of missile launchings with satellite-borne infrared sensors. | 322 [1] (316) | 292.1 [1] (300) | 94.3 [1] (94.4) | 33° | Data link telemetry ceased functioning 2 days after launch. Status: In orbit 10/25/60. |

[1] As of 10/25/60.

CHRONICLE OF EARTH SATELLITES AND SPACE PROBES, 1957–60—Continued

| Launch date | Name | International designation | Launch vehicle | Payload data | Apogee (st. mi.) | Perigee (st. mi.) | Period (minutes) | Inclination | Remarks |
|---|---|---|---|---|---|---|---|---|---|
| June 22, 1960 | TRANSIT II-A (United States) | 1960 Eta I and II. | Thor-Able-Star. | Weight of 2 payloads (a) 223 lbs.; (b) 42 lbs. Objectives: Demonstrate navigation satellite; increase accuracy of geodetic measurements; provide accurate time standards. Payload: (a) 2 oscillators, infrared scanner, electronic clock and galactic noise receiver; (b) instruments to read solar radiation. | (a) 665 (b) 657 | (a) 389 (b) 382 | (a) 101.7 (b) 101.6 | (a) 66.7° (b) 66.8° | 1st time 2 active satellites placed in orbit in single firing. Status: Still active. |
| Aug. 10, 1960 | DISCOVERER XIII (United States) | 1960 Theta | Thor-Agena | Total weight: 1,700 lbs. including 2d-stage casing and 300-lb. capsule. Objective: Gather data on propulsion, communications, orbital performance, stabilization and recovery techniques. | 436 ¹ (267) | 161 ¹ (153) | 94.1 ¹ (91.3) | 82°51′ | 1st successful recovery man-made object from space when data capsule recovered at sea on 8/11/60. Telemetry check points increased by almost 400 percent over previous Discoverers. Status: Payload minus capsule in orbit 10/25/60. |
| Aug. 12, 1960 | ECHO I (United States) | 1960 Iota | Thor Delta | Total weight: 132 lbs. Objective: Place 100-foot inflatable sphere in orbit. | 1,049 ¹ (1,192) | 945 ¹ (781) | 118.3 117.8 | 47.2° | Demonstrated use of radio wave reflection for global communications, numerous successful transmissions; visible to naked eye. Status: Affected by solar pressure, still in orbit 10/25/60. |

| Date | Name | Designation | Vehicle | Objective/Description | | | Period | Inclination | Status |
|---|---|---|---|---|---|---|---|---|---|
| Aug. 18, 1960 | DISCOVERER XIV (United States) | 1960 KAPPA | Thor-Agena | Total weight: 1,700 lbs, including 2d-stage casing and 300-lb. capsule. Objective: Gather data on propulsion, communications, orbital performance, stabilization and recovery techniques. Also carried 10-lb. tracking experiment as part of Navy Transit satellite. | 502 | 116 | 94.5 | 79.6° | 1st successful snatch of space object 8,000 feet in air by C-119 aircraft, 360 miles southwest of Hawaii. Status: Payload minus capsule down 9/15/60. |
| Aug. 19, 1960 | SPACECRAFT II (U.S.S.R.) | 1960 LAMBDA | Not disclosed | Satellite weight without final stage of rocket: 10,120 lbs. Objective: Test capsule and recovery system for ultimate development of manned space flight. Experiments: Capsule contained 2 dogs, rats, mice, flies, plants, fungi, seeds, etc., with TV cameras and transmitter. | 211 | 190 | 90.72 | 64.57° | 1st biologic payload recovered, on 18th orbit on 8/20/60, after traveling 437,500 miles. Capsule and carrier reportedly landed less than 7 miles from predetermined point. |
| Sep. 13, 1960 | DISCOVERER XV (United States) | 1960 MU | Thor-Agena | Total weight: 1,700 lbs, including 2d-stage casing and 300-lb. capsule. Objective: Gather data on propulsion, communications, orbital performance, stabilization and recovery techniques. Instrumentation included reentry capsule, retrorocket and parachute for recovery, radio beacon and aluminum radar chaff for recovery. | 472 | 130 | 94.24 | 80.93° | Capsule ejected and landed in sea south of Hawaii, sighted but not recovered due to rough seas on 9/15/60. Carrier satellite down on 10/17/60. |

[1] As of 10/25/60

## CHRONICLE OF EARTH SATELLITES AND SPACE PROBES, 1957–60—Continued

| Launch date | Name | International designation | Launch vehicle | Payload data | Apogee (st. mi.) | Perigee (st. mi.) | Period (minutes) | Inclination | Remarks |
|---|---|---|---|---|---|---|---|---|---|
| Oct. 4, 1960 | COURIER I-B (United States) | 1960 Mu 1 | Thor-Able-Star | Total weight: 500 lbs., including 300 lbs. of electrical equipment. Objective: Test feasibility of active global communications using delayed repeater transmission equipment capable of handling 68,000 coded words per minute. | 658 | 501 | 107 | 28.3° | Transmissions successfully received and transmitted. Status: Active 10/25/60. |
| Nov. 3, 1960 | EXPLORER VIII (United States) | 1960 XI | Juno II | Total weight: 90.14 lbs. Objective: Study ionosphere by direct measurement of positive ion and electron composition; gather data on frequency, momentum and energy of micrometeorite impacts. | 1,423 | 258 | 112.7 | 49.9° | Orbit achieved and data received is being analyzed. Transmitter ceased functioning on 12/27/60. |
| Nov. 12, 1960 | DISCOVERER XVII (United States) | 1960 OMICRON | Thor-Agena B | Total weight: 1,700 lbs. including 2d-stage casing and 300-lb capsule. Objective: Gather data on propulsion, communications, orbital performance, stabilization and recovery techniques. Instrumentation included reentry capsule, retrorocket and parachute for recovery, radio beacon and aluminum radar chaff for recovery. | 615 | 118 | 96.45 | | Polar orbit achieved. After 31 orbits, capsule ejected and successfully recovered in air by C-119 aircraft on 11/14/60, the 2d midair recovery of a space object. |

| Date | Name | Designation | Launch vehicle | Weight (lbs) / Objective | Apogee | Perigee | Period (min) | Inclination | Remarks |
|---|---|---|---|---|---|---|---|---|---|
| Nov. 23, 1960 | TIROS II (United States) | 1960 Pi 1 | Thor-Delta | Total weight: 280 lbs. Objective: Provide optical and infrared photos of global cloud cover. | 431 | 406 | 98.2 | | 14th U.S. satellite orbited in 1960. Satellite data to be correlated with ground acquired data; estimated lifetime 3 months. |
| Dec. 1, 1960 | SPACECRAFT III (U.S.S.R.) | 1960 Rho | Not disclosed | Payload weight: 10,060 lbs., a 2-part craft. Objective: Test of equipment for manned space flight. | 165 | 117 | 88.6 | 65° | Referred to as Sputnik VI. Spacecraft carrying biological payload burned up on unprogramed re-entry on 12/2/60. |
| Dec. 7, 1960 | DISCOVERER XVIII (United States) | 1960 SIGMA | Thor-Agena B | | 391 | 143 | 93.2 | | Spacecraft carried surveillance test equipment while capsule contained human tissue. Capsule caught by C-119 aircraft after 48 orbits on 12/10/60. |
| Dec. 19, 1960 | DISCOVERER XIX (United States) | 1960 | Thor-Agena B | Objective: To test Midas system equipment. | 400.3 | 119.6 | 93 | 83.4° | Polar orbit achieved; the 31st successful U.S. satellite to date. |

APPENDIX B

# Chronicle of World Airplane Records

THE FOLLOWING official records are those compiled by the Fédération Aéronautique Internationale, Paris, France, as supplied by its American representative, the National Aeronautic Association. It does not include seaplane records or discontinued categories.

1. Airplanes: Distance (1925–46).

2. Airplanes: Distance in closed circuit (1906–60).

3. Airplanes: Maximum speed over straightaway course (1906–59).

4. Airplanes: Altitude (1909–59).

## 1. AIRPLANES: DISTANCE IN A STRAIGHT LINE

| Record holder and country; date and place | Miles |
|---|---|
| Capts. Ludovic Arrachart and Henri LeMaitre. France. | |
| Feb. 3–4, 1925. Etampes to Cisneros | 1,967.257 |
| Capts. Ludovic Arrachart and Paul Arrachart. France. | |
| June 26–27, 1926. Le Bourget to Shaibah | 2,674.998 |
| Capt. Andre Girier and Lt. Francis Dordilly. France. | |
| July 14–15, 1926. Le Bourget to Omsk | 2,930.319 |
| Lt. Leone Challe and Capt. Rene Weiser. France. | |
| Aug. 31–Sept. 1, 1926. Le Bourget to Bender Abbas | 3,214.968 |
| Lt. Dieudonne Costes and Capt. Georges Rignot. France. | |
| Oct. 28–29, 1926. Le Bourget to Jask | 3,352.912 |
| Charles A. Lindbergh. United States. | |
| May 20–21, 1927. New York to Paris | 3,609.538 |
| Clarence Chamberlin and Charles A. Levine. United States. | |
| June 4–6, 1927. New York to Isleben, Germany | 3,910.902 |
| Majs. A. Ferrarin and C. P. Del Prete. Italy. | |
| July 3–5, 1928. Rome to Touros | 4,466.569 |
| Dieudonne Costes and Maurice Bellonte. France. | |
| Sept. 27–29, 1929. Le Bourget to Moulant | 4,911.929 |
| Russell N. Boardman and J. Polando. United States. | |
| July 28–30, 1931. Brooklyn to Istanbul | 5,011.349 |
| O. R. Gayford and C. E. Nicholetts. Great Britain. | |
| Feb. 6–8, 1933. Cranwell to Walvis Bay | 5,308.985 |
| Maurice Rossi and Paul Codos. France. | |
| Aug. 5–7, 1933. New York to Rayack, Syria | 5,656.932 |
| Mikhail Gromov, Andre Youmachev, and Sergei Danilin. U.S.S.R. | |
| July 12–14, 1937. Moscow to San Jacinto, Calif. | 6,305.662 |
| R. Kellett, R. T. Gething, and M. L. Gaine, one flight. A. N. Combe, B. K. Burnett, and H. B. Gray, second plane. Great Britain. | |
| Nov. 5–7, 1938. Ismalia, Egypt, to Darwin, Australia | 7,158.440 |
| Col. C. S. Irvine, Lt. Col. G. R. Stanley and crew, United States Army Air Force. United States. | |
| Nov. 19–20, 1945. Guam to Washington, D.C. | 7,916.000 |
| Comdrs. T. D. Davies, E. P. Rankin, W. S. Reid; Lt. Comdr. R. A. Tabeling, United States Navy. United States. | |
| Sept 29–Oct. 1, 1946. Perth, Australia to Columbus, Ohio | 11,235.600 |

## 2. AIRPLANES: DISTANCE—IN CLOSED CIRCUIT RETURNING TO POINT OF DEPARTURE (WITHOUT PAY LOAD)

| Record holder and country; date and place | Miles |
|---|---|
| A. Santos Dumont. France. | |
| Nov. 12, 1906. Bagatelle | 0.136 |
| Henri C. Farman. France. | |
| Oct. 26, 1907. Issy-les-Moulineaux | .478 |
| Henri C. Farman. France. | |
| Jan. 13, 1908. Issy-les-Moulineaux | .621 |
| Henri C. Farman. France. | |
| Mar. 21, 1908. Issy-les-Moulineaux | 1.242 |

| Record holder and country; date and place | Miles |
|---|---|
| Leon Delagrange. France. | |
| Apr. 11, 1908. Issy-les-Moulineaux | 2.438 |
| Leon Delagrange. Italy. | |
| May 30, 1908. Cantocelle | 7.922 |
| Leon Delagrange. France. | |
| Sept. 16, 1908. Issy-les-Moulineaux | 14.989 |
| Wilbur Wright. France. | |
| Sept. 21, 1908. Auvours | 41.382 |
| Wilbur Wright. France. | |
| Dec. 18, 1908. Auvours | 62.012 |
| Wilbur Wright. France. | |
| Dec. 31, 1908. Auvours | 77.485 |
| Louis Paulhan. France. | |
| Aug. 25, 1909. Betheny | 83.263 |
| Hubert Latham. France. | |
| Aug. 26, 1909. Betheny | 96.064 |
| Henri C. Farman. France. | |
| Aug. 27, 1909. Betheny | 111.846 |
| Henri C. Farman. France. | |
| Nov. 4, 1909. Mourmelon | 145.531 |
| Rene Labouchere. France. | |
| July 9, 1910. Rheims | 211.265 |
| Jan Olieslagers. France. | |
| July 20, 1910. Rheims | 244.043 |
| Maurice Tabuteau. France. | |
| Oct. 20, 1910. Etampes | 289.414 |
| Georges Legagneux. France. | |
| Dec. 11, 1910. Pau | 320.564 |
| Maurice Tabuteau. France. | |
| Dec. 30, 1910. Buc | 363.344 |
| Jan Olieslagers. Belgium. | |
| July 16, 1911. Kiewit | 388.356 |
| Georges Fourny. France. | |
| Sept. 1, 1911. Buc | 449.209 |
| Andre Gobe. France. | |
| Dec. 24, 1911. Pau | 459.998 |
| Georges Fourny. France. | |
| Sept. 11, 1912. Etampes | 628.142 |
| Augustin Seguin. France. | |
| Oct. 13, 1913. Buc-le-Barp | 634.541 |
| Lucien Bossoutrot and Jean Bernard. France. | |
| May 3–4, 1920. Villesauvage | 1,189.992 |
| Lts. Oakley C. Kelly and John A. Macready, USAS. United States. | |
| Apr. 16–17, 1923. Dayton, Ohio | 2,516.548 |
| Maurice Drouhin and Jules Landry. France. | |
| Aug. 7–9, 1925. Etampes-Chartres | 2,734.026 |
| Cornelius Edzard and Johann Risztics. Germany. | |
| Aug. 3–5, 1927. Dessau | 2,895.970 |
| Arturo Ferrarin and C. P. Del Prete. Italy. | |
| May 31–June 1–2, 1928. Casale del Prati | 4,763.798 |
| Dieudonne Costes and Paul Codos. France. | |
| Dec. 15–17, 1929. Istres | 4,988.969 |

| Record holder and country; date and place | Miles |
|---|---|
| Maj. Umberto Maddalen and Lt. Fausto Cecconi. Italy. May 31–June 1–2, 1930. Montecelio-Stazion Ladispoli | 5,088.267 |
| Lucien Bossoutrot and Aime Rossi. France. Feb. 26–28, 1931. Oranie | 5,481.927 |
| Anthoine Paillard and Jean Mermoz. France. Mar. 30–Apr. 2, 1931. Oran | 5,567.475 |
| Joseph LeBrix and Marcel Doret. France. June 7–10, 1931. Istres | 6,444.881 |
| Lucien Bossoutrot and Maurice Rossi. France. Mar. 23–26, 1932. Oran | 6,587.441 |
| Yuzo Fujita and F. Takahashi. Japan. May 13–15, 1938. Kisarasu | 7,239.588 |
| Angelo Tondi, Roberto Dagasso and Ferrucio Vignoli. Italy. July 30–Aug. 1, 1939. Rome | 8,037.899 |
| Lt. Col. O. F. Lassiter, Capt. W. J. Valentine, and crew, USAAF. United States. Aug. 1–2, 1947. Tampa, Fla | 8,854.308 |
| Lt. Col. J. R. Grissom and crew, USAF. United States. Dec. 14, 1960. Edwards, Calif | 10,078.84 |

## 3. AIRPLANES: MAXIMUM SPEED OVER STRAIGHTAWAY COURSE

| Record holder and country; date and place | Miles per hour |
|---|---|
| A. Santos Dumont. France. Nov. 12, 1906. Bagatelle | 25.66 |
| Henri C. Farman. France. Oct. 26, 1907. Issy-les Moulineaux | 32.75 |
| Paul Tissandier. France. May 20, 1909. Pau | 34.06 |
| Glenn H. Curtiss. France. Aug. 23, 1909. Rheims | 43.38 |
| Louis Bleriot. France. Aug. 24, 1909. Rheims | 46.18 |
| Louis Bleriot. France. Aug. 28, 1909. Rheims | 47.84 |
| Hubert Latham. France. Apr. 23, 1910. Nice | 48.21 |
| Leon Morane. France. July 10, 1910. Rheims | 66.18 |
| Alfred Leblanc. United States. Oct. 29, 1910. New York | 68.20 |
| Alfred Leblanc. France. Apr. 12, 1911. Pau | 69.47 |
| Edouard Nieuport. France. May 11, 1911. Chalons | 74.42 |
| Alfred Leblanc. France. June 12, 1911. Etampes | 77.67 |
| Edouard Nieuport. France. June 16, 1911. Chalons | 80.81 |

| Record holder and country; date and place | Miles per hour |
|---|---|
| Edouard Nieuport. France. | |
| June 21, 1911. Chalons | 82.73 |
| Jules Vedrines. France. | |
| Jan. 13, 1912. Pau | 90.20 |
| Jules Vedrines. France. | |
| Feb. 22, 1919. Pau | 100.22 |
| Jules Vedrines. France. | |
| Feb. 29, 1912. Pau | 100.94 |
| Jules Vedrines. France. | |
| Mar. 1, 1912. Pau | 103.66 |
| Jules Vedrines. France. | |
| Mar. 2, 1912. Pau | 104.33 |
| Jules Vedrines. France. | |
| July 13, 1912. Rheims | 106.12 |
| Jules Vedrines. United States. | |
| Sept. 9, 1912. Chicago | 108.18 |
| Marcel Prevost. France. | |
| June 17, 1913. Rheims | 111.73 |
| Marcel Prevost. France. | |
| Sept. 27, 1913. Rheims | 119.24 |
| Marcel Prevost. France. | |
| Sept. 29, 1913. Rheims | 126.67 |
| Sadi Lecointe. France. | |
| Feb 7, 1920. Villacoublay | 171.04 |
| Jean Casale. France. | |
| Feb. 28, 1920. Villacoublay | 176.14 |
| Bernard de Romanet. France. | |
| Oct. 9, 1920. Buc | 181.86 |
| Sadi Lecointe. France. | |
| Oct. 10, 1920. Buc | 184.36 |
| Sadi Lecointe. France. | |
| Oct. 20, 1920. Villacoublay | 187.98 |
| Bernard de Romanet. France. | |
| Nov. 4, 1920. Buc | 192.01 |
| Sadi Lecointe. France. | |
| Dec. 12, 1920 Buc | 194.52 |
| Sadi Lecointe. France. | |
| Sept. 26, 1921. Villesauvage | 205.22 |
| Sadi Lecointe. France. | |
| Sept. 21, 1922. Villesauvage | 211.90 |
| W. E. Mitchell. United States. | |
| Oct. 13, 1922. Detroit, Mich | 222.97 |
| Sadi Lecointe. France. | |
| Feb. 15, 1923. Istres | 233.01 |
| Lt. R. L. Maughan, USAS. United States. | |
| Mar. 29, 1923. Dayton, Ohio | 236.59 |
| Lt. Harold J. Grow, USN. United States. | |
| Nov. 2, 1923. Mineola, N.Y | 259.15 |
| Lt. A. J. Williams, USN. United States. | |
| Nov. 4, 1923. Mineola, N.Y | 266.58 |
| Adj. A. Bonnet. France. | |
| Dec. 11, 1924. Istres | 278.48 |

| Record holder and country; date and place | Miles per hour |
|---|---|
| J. H. Doolittle. United States. | |
| Sept. 5, 1932. Cleveland, Ohio | 294.38 |
| J. R. Wedell. United States. | |
| Sept. 4, 1933. Glenview, Ill | 304.98 |
| Raymond Delmotte. France. | |
| Dec. 25, 1934. Istres | 314.319 |
| Howard R. Hughes. United States. | |
| Sept. 13, 1935. Santa Anna, Calif | 352.388 |
| Herman Wurster. Germany. | |
| Nov. 11, 1937. Augsburg, Germany | 379.626 |
| Hans Dieterle. Germany. | |
| Mar. 30, 1939. d'Orianenburg, Germany | 463.917 |
| Fritz Wendel. Germany. | |
| Apr. 26, 1939. Augsburg, Germany | 469.220 |
| Grp. Capt. H. Wilson. Great Britain. | |
| Nov. 7, 1945. Herne Bay | 606.255 |
| Grp. Capt. E. M. Donaldson. Great Britain. | |
| Sept. 7, 1946. Little Hampton | 615.778 |
| Col. A. Boyd, USAAF. United States. | |
| June 19, 1947. Muroc, Calif | 623.738 |
| Comdr. T. F. Caldwell, USN. United States. | |
| Aug. 20, 1947. Muroc, Calif | 640.663 |
| Maj. Marion Carl, USMC. United States. | |
| Aug. 25, 1947. Muroc, Calif | 650.796 |
| Maj. R. L. Johnson, USAF. United States. | |
| Sept. 15, 1948. Muroc, Calif | 670.981 |
| Capt. James S. Nash, USAF. United States. | |
| Nov. 19, 1952. Salton Sea, Calif | 698.505 |
| Lt. Col. Wm. F. Barnes. United States. | |
| July 16, 1953. Salton Sea, Calif | 715.745 |
| Sqd. Ldr. N. F. Duke. Great Britain. | |
| Sept. 7, 1953. Little Hampton | 727.624 |
| Michael J. Lithgow. Great Britain. | |
| Sept. 25, 1953. Azizia, Tripoli | 735.702 |
| Lt. Comdr. J. B. Verdin, USN. United States. | |
| Oct. 3, 1953. Salton Sea, Calif | 752.943 |
| Lt. Col. F. K. Everest, Jr., USAF. United States. | |
| Oct. 29, 1953. Salton Sea, Calif | 755.149 |
| Col. H. A. Hanes, USAF. United States. | |
| Aug. 20, 1955. Palmdale, Calif | 822.266 |
| L. Peter Twiss. Great Britain. | |
| Mar. 10, 1956. Ford-Chichester | 1,132.136 |
| Maj. Adrian E. Drew, USAF. United States. | |
| Dec. 12, 1957. Edwards, Calif | 1,207.6 |
| Capt. Walter W. Irwin, USAF. United States. | |
| May 16, 1958. Edwards, Calif | 1,404.09 |
| Gueorgui Mossolov. U.S.S.R. | |
| Oct. 31, 1959. Joukovski-Petrovskoe | 1,483.85 |
| Maj. Joseph W. Rogers, USAF. United States. | |
| Dec. 15, 1959. Edwards, Calif | 1,525.965 |

## 4. AIRPLANES: ALTITUDE

| Record holder and country; date and place | Feet |
|---|---|
| Hubert Latham. France. | |
| Aug. 29, 1909. Rheims | 509 |
| Comte Charles de Lambert. France. | |
| Oct. 18, 1909. Paris | 984 |
| Hubert Latham. France. | |
| Dec. 1, 1909. Chalons | 1,486 |
| Hubert Latham. France. | |
| Jan. 7, 1910. Chalons | 3,281 |
| Louis Paulhan. United States. | |
| Jan. 12, 1910. Los Angeles | 3,967 |
| Walter R. Brookins. United States. | |
| June 14, 1910. Indianapolis | 4,380 |
| Hubert Latham. France. | |
| July 7, 1910. Rheims | 4,541 |
| Walter R. Brookins. United States. | |
| July 10, 1910. Atlantic City, N.J | 6,234 |
| Anthony Drexel. United States. | |
| Aug. 11, 1910. Lamark | 6,601 |
| Leon Morane. France. | |
| Sept. 3, 1910. Deauville | 8,471 |
| Georges Chavez. France. | |
| Sept. 8, 1910. Issy-les-Moulineaux | 8,488 |
| Henri Wijnmalen. France. | |
| Oct. 1, 1910. Mourmelon | 9,121 |
| Anthony Drexel. United States. | |
| October 1910. Philadelphia | 9,449 |
| Ralph Johnstone. United States. | |
| Oct. 31, 1910. Belmont Park | 9,711 |
| Georges Legagneaux. France. | |
| Dec. 8, 1910. Pau | 10,171 |
| Marcel Loridan. France. | |
| July 8, 1911. Chalons | 10,423 |
| Capt. Julien Felix. France. | |
| Aug. 9, 1911. Etampes | 10,466 |
| Roland Garros. France. | |
| Sept. 4, 1911. Saint-Malo | 12,828 |
| Roland Garros. France. | |
| Sept. 6, 1912. Houlgate | 16,076 |
| Georges Legagneaux. France. | |
| Sept. 17, 1912. Corbaulieu | 17,881 |
| Roland Garros. France. | |
| Dec. 11, 1912. Tunis | 18,405 |
| Edmond Perreyon. France. | |
| Mar. 11, 1913. Buc | 19,291 |
| Georges Legagneaux. France. | |
| Dec. 28, 1913. Saint-Raphael | 20,079 |
| Maj. R. W. Schroeder, USAS. United States. | |
| Feb. 27, 1920. Dayton, Ohio | 33,113 |
| Lt. J. A. Macready, USAS. United States. | |
| Sept. 18, 1921. Dayton, Ohio | 34,508 |

| Record holder and country; date and place | Feet |
|---|---|
| Sadi Lecointe. France. | |
|    Sept. 5, 1923. Villacoublay | 35,243 |
| Sadi Lecointe. France. | |
|    Oct. 30, 1923. Issy-les-Moulineaux | 36,565 |
| Lt. C. C. Champion, USN. United States. | |
|    July 25, 1927. Washington, D.C. | 38,419 |
| Lt. Apollo Soucek, USN. United States. | |
|    May 8, 1929. Washington, D.C. | 39,140 |
| Willi Neuenhofen. Germany. | |
|    May 26, 1929. Dessau | 41,795 |
| Lt. Apollo Soucek, USN. United States. | |
|    June 4, 1930. Washington, D.C. | 43,166 |
| Capt. C. F. Uwins. Great Britain. | |
|    Sept. 16, 1932. Filton, Bristol | 43,976 |
| G. Lemoine. France. | |
|    Sept. 28, 1933. Villacoublay | 44,819 |
| Comdr. Renato Donati. Italy. | |
|    Apr. 11, 1934. Rome | 47,352 |
| Georges Detre. France. | |
|    Aug. 14, 1936. Villacoublay | 48,697 |
| F. R. D. Swain. Great Britain. | |
|    Sept. 28, 1936. South Farnborough | 49,944 |
| Mario Pezzi. Italy. | |
|    May 8, 1937. Montecello | 51,361 |
| Fl. Lt. M. J. Adam. Great Britain. | |
|    June 30, 1937. Farnborough | 53,937 |
| Col. Mario Pezzi. Italy. | |
|    Oct. 22, 1938. Montecello | 56,046 |
| John Cunningham (Jet). Great Britain. | |
|    Mar. 23, 1948. Hatfield | 59,445 |
| Walter F. Gibb. Great Britain. | |
|    May 4, 1953. Bristol | 63,668 |
| Walter F. Gibb. Great Britain. | |
|    Aug. 29, 1955. Bristol | 65,889 |
| George E. Watkins. United States. | |
|    Apr. 18, 1958. Edwards, Calif | 76,932 |
| Maj. Howard C. Johnson, USAF. United States. | |
|    May 7, 1958. Palmdale, Calif | 91,243 |
| Vladimir Iljiuchin. U.S.S.R. | |
|    July 14, 1959. Podmoskounoie | 94,635 |
| Comdr. Lawrence E. Flint, USN. United States. | |
|    Dec. 6, 1959. Edwards, Calif | 98,557 |
| Capt. Joe B. Jordan, USAF. United States. | |
|    Dec. 14, 1959. Edwards, Calif | 103,389 |

APPENDIX C

# Chronicle of Select Balloon Flights

1927–1961

THIS CHRONICLE provides appreciation of balloon flights and operations in man's conquest of the air, and helps illustrate the more recent use of the balloon as a research tool in the scientific exploration of space. It includes both official and unofficial record flights. The guidance of Comdr. Malcolm Ross (USNR) of the Office of Naval Research and Maj. Gen. Orvil A. Anderson (USAF retired) of the Air Force Historical Foundation is gratefully acknowledged.

*March 9, 1927:* Capt. H. C. Gray (AAC) ascended to 28,910 feet in a free balloon for an American altitude record. World record was held by Suring and Berson of Germany who ascended to 35,433 feet on June 30, 1901.

*May 4, 1927:* Record balloon flight by Capt. H. C. Gray (AAC) reached 42,470 feet over Scott Field, Ill., but he was forced to bail out so that record was not official.

*November 4, 1927:* Capt. H. C. Gray (AAC) ascended to 42,470 feet, the identical altitude of his May 4 flight, but he did not survive the flight and thereby again failed to achieve official world record.

*May 27, 1931:* Auguste Piccard and Paul Kipfer made first successful manned ascent into stratosphere from Augsburg, Germany, establishing new world altitude record of 51,777 feet in Belgian FNRS balloon.

*August 18, 1932:* Auguste Piccard and Max Cosyns flew FNRS balloon from Zurich, Switzerland, to a record altitude of 53,152 feet.

*September 30, 1933:* Russian balloon USSR launched at Moscow reached altitude of 60,695 feet which never became an official record, crew including Prokovief, Birnbaum and Godrenow.

*November 29, 1933:* Lt. Comdr. T. G. W. Settle (USN) and Maj. C. Fordney (USMC) established official world altitude record of 61,237 feet over Akron, Ohio, in balloon CENTURY OF PROGRESS.

*January 30, 1934:* Russian balloon OSA-VIAKHIM reached 73,000-foot altitude but crew perished when gondola fell free, personnel including Fedosienko, Wasienko, and Vsyskin.

*July 28, 1934:* EXPLORER I balloon launched from Stratobowl near Rapid City, S. Dak., failed at 60,613 feet, Maj. W. Kepner (AAC) and Capts. O. A. Anderson and A. W. Stevens (AAC) parachuted safely.

*August 18, 1934:* Jeanette and Jean Piccard flew CENTURY OF PROGRESS balloon from Dearborn, Mich., to an altitude of 57,579 feet. Jeanette Piccard setting an unofficial record for women and becoming first woman to enter the stratosphere.

*July 26, 1935:* Russian balloon USSR reached 52,000 feet, crew including Warigo, Christofil, and Prilucki.

*November 11, 1935:* Army Air Corps-National Geographic EXPLORER II established new official world record of 72,395 feet in ascent from Stratobowl, crew personnel were Capts. Orvil A. Anderson and A. W. Stevens (AAC).

*During 1935:* Dr. Jean Piccard, in collaboration with Dr. Thomas Johnson of Bartol Research, flew first unmanned plastic balloon (cellophane) at Swarthmore, Pa.

*During summer-fall 1936:* Dr. Jean Piccard and John Ackerman flew first constant-level plastic (cellophane) balloons from the University of Minnesota stadium.

*During 1937:* First manned cluster balloon flight (PLEIADES) made by Dr. Jean Piccard using rubber meteorological balloons from Rochester, Minn.

*During 1944–45:* Japan launched approximately 10,000 unmanned *Fugo* balloons (30-foot diameter). They floated at 30,000 feet, carried incendiaries, and were aimed at the North American continent.

*During March 1947:* First test flights of plastic balloons conducted by General Mills for ONR Project Helios.

*June 5, 1947:* First AAF research balloon launch (cluster of rubber balloons) at Holloman, by New York University team under AMC contract.

*July 3, 1947:* Start of polyethylene balloon operations at Holloman, a 10-balloon cluster launched by New York University staff with a payload of less than 50 pounds, which reached an altitude of 18,500 feet.

*September 25, 1947:* First successful ONR SKYHOOK polyethylene plastic balloon (220,000 cubic feet) launched from St. Cloud, Minn., carrying 63-pound cosmic ray emulsions to 100,000 feet.

*September 28, 1948:* Army Signal Corps balloon set a 140,000-foot altitude unmanned record, at Belmar, N.J.

*January 30, 1949:* U.S. Navy launched first polyethelene SKYHOOK balloons from ship, USS *Saipan* off the coast of Cuba, in a series of 12 flights.

*November 3, 1949:* Charles B. Moore (General Mills) made first manned flight in a plastic balloon, over Minneapolis, Minn.

*During May 1950:* New York University research balloon released from Holloman AFB drifted 7,000 miles and was recovered in Myrdal, Norway, several days later.

*July 21, 1950:* First polyethylene balloon launched by USAF personnel at Holloman AFB, N. Mex.

*August 28, 1952:* First successful *Rockoon* (balloon-launched rocket) launched from icebreaker off Greenland by University of Iowa team headed by J. A. Van Allen and on ONR contract, rocket was launched from balloon at 70,000 feet and reached maximum altitude of 37.9 miles.

*February 19–26, 1953:* First of the MOBY DICK balloon flights to 60,000–100,000 feet to study high-altitude winds and carrying capsule containing fruit flies, launched from NAS Vernalles, Calif., by USAF Cambridge Research Center.

*May 18, 1954:* *Super Skyhook*, largest polyethylene balloon built to date (3,000,000 cubic feet) was successfully launched by General Mills for ONR and carried emulsions to 115,000 feet.

*July 18, 1955:* First Aeromedical Laboratory's ($2^7$ cubic feet) plastic balloons manufactured by Winzen Research, launched at Fleming Field, Minn., attained an altitude of over 120,000 feet; the second launched on the next day attained a record altitude of 126,000 feet.

*August 10, 1956:* Lt. Comdrs. Malcolm D. Ross (USNR) and M. Lee Lewis (USN) made first stratospheric manned flight on a polyethylene balloon from Minneapolis as a part of ONR Project Strato-Lab, flying in open basket and reaching an altitude of 40,000 feet.

*September 7, 1956:* University of Minnesota launched ONR Mylar plastic balloon from Minneapolis, establishing an unofficial world altitude record of 145,000 feet for unmanned balloon.

*September 24, 1956:* H. Froehlich and K. Long (General Mills) flew ONR STRATO-LAB balloon to new altitude record of 42,000 feet for an open basket gondola.

*November 8, 1956:* Lt. Comdrs. M. D. Ross (USNR) and M. L. Lewis (USN) set unofficial world altitude record of 76,000 feet in STRATO-LAB HIGH I ascent from Stratobowl, bringing balloon down safely although it failed.

*June 2, 1957:* Capt. Joseph Kittinger (USAF) made first solo balloon flight into statosphere in MANHIGH I, setting a new unofficial world altitude record of 96,000 feet after launch from St. Paul, Minn.

*June 28, 1957:* First phase of Project Far Side completed, with the lifting by world's largest plastic balloon (3,700,000 cubic feet) of a load of 2,306 pounds of military equipment and instruments to a height of more than 104,000 feet.

*August 19–20, 1957:* Maj. David G. Simons (USAF MC) set official world altitude record of 101,516 feet in MANHIGH II balloon, setting unofficial duration record of 32 hours, and becoming first person to remain in stratosphere overnight, over Crosby, Minn.

*August 19, 1957:* Stratoscope I, an unmanned balloon-telescope system, launched by General Mills under Navy contract for Princeton University astronomers, produced first "clear" photos of the sun taken at 80,000-foot altitude with 12-inch telescope.

*September 18, 1957:* Donald Piccard made first successful manned low-level flight on a cluster of plastic balloons, from Swarthmore, Pa.

*October 18, 1957:* Lt. Comdrs. M. D. Ross (USNR) and M. L. Lewis (USN) ascended to unofficial two-man altitude record of 85,700 feet in STRATO-LAB HIGH II balloon.

*May 6-7, 1958:* Lt. Comdr. M. D. Ross (USNR) and A. Mikesell (Naval Observatory) used open-gondola STRATO-LAB to ascend to 40,000 feet from Crosby, Minn., Mikesell becoming first astronomer to observe from stratosphere and first flight in which crew remained in stratosphere in open basket after sunset.

*July 26-27, 1958:* Comdrs. M. D. Ross (USNR) and M. L. Lewis (USN ret.) reached maximum altitude of 82,000 feet in STRATO-LAB HIGH III flight from Crosby, Minn., which set new unofficial record for stratospheric flight of 34.7 hours at −65° C.

*October 8, 1958:* Lt. D. McClure (USAF) flew MANHIGH III balloon to an altitude of 99,000 feet from Holloman AFB, N. Mex.

*December 12-16, 1958:* Balloon SMALL WORLD with four passengers failed in transatlantic attempt, lifting from Canary Islands and landing at sea northeast of Barbados.

*December 18, 1958:* Balloon Flight No. 1,000 launched at Holloman AFB, a series of USAF plastic balloon flights begun in July 1950.

*February 11, 1959:* Army weather balloon, launched at Signal Research and Development Laboratory, Fort Monmouth, N.J., established world altitude record of 146,000 feet.

*April 22, 1959:* A. Dollfus flew from Paris, France, on a cluster of 100 weather balloons to an altitude of 42,000 feet.

*May 12, 1959:* University of Minnesota scientists under Navy contract launched unmanned balloon to 100,000 feet, where first positive measurement of intense solar protons associated with a solar flare were measured.

*July 13, 1959:* Largest plastic balloon to date ($6^7$ cubic feet) launched by ONR at Fort Churchill, Canada.

*August 7, 1959:* Comdr. M. D. Ross (USNR) and R. Cooper (High Altitude Observatory) flew Navy STRATO-LAB open gondola balloon from Stratobowl to 38,000 feet for solar studies with a coronagraph.

*September 4, 1959:* Raven Industries launched Office of Naval Research *Skyhook* unmanned balloon from Sioux Falls, So. Dakota, establishing new unofficial altitude record of 148,000 feet for unmanned balloon.

*November 14, 1959:* Winzen Research successfully launched world's largest balloon ($10^7$ cubic feet) from Stratobowl, reading maximum altitude of near 118,000 feet with a ton load suspended.

*November 16, 1959:* Capt. J. Kittinger (USAF) flew EXCELSIOR I balloon from Holloman AFB to an altitude of 76,400 feet and successfully bailed out of open gondola.

*November 28-29, 1959:* Comdr. M. Ross (USNR) and C. Moore (Arthur D. Little, Inc.) flew Navy STRATO-LAB HIGH IV from the Stratobowl to an altitude of 81,000 feet using a 16-inch telescope and spectrograph, observing water vapor in the atmosphere of the planet Venus.

*December 16, 1959:* Capt. J. Kittinger (USAF) flew EXCELSIOR II balloon from Holloman AFB to an altitude of 74,700 feet and successfully bailed out and established stable free fall for 55,000 feet.

*January 26, 1960:* Navy 292-foot-diameter balloon launched from U.S.S. *Valley Forge* east of Puerto Rico, carrying 1,630-pound payload to record cosmic ray and secondary particles, film packs being recovered by a destroyer the next day.

*February 1, 1960:* University of Chicago Project ICEF (International Cooperative Emulsion Flights) sponsored by the National Science Foundation and the Office of Naval Research launched SKYHOOK ($10^7$ cubic feet) balloon to 21.4-mile altitude.

*June 5, 1960:* Navy launched $10^7$ SKYHOOK balloon from NAS Glencoe, Ga., which disappeared over the Pacific Ocean 2 weeks later carrying 3,300-pound payload.

*July 24, 1960:* Donald Piccard established Class I world altitude record of 3,740 feet in plastic balloon HOLIDAY, from Minneapolis, Minn.

*July 17, 1960:* First of three NASA experiments carried by USAF balloons to an altitude of 130,000 feet for 11½ hours from Minneapolis, Minn., NASA capsule containing 12 mice for study of effects of heavy primary cosmic ray particles.

*August 12, 1960:* First "balloon satellite," NASA ECHO I (a 100-foot diameter inflatable sphere) was launched into orbit from AMR and inflated; while not a balloon in sense of being dependent upon atmospheric displacement for lift and its orbit altitude, it became popularly known as the "balloon satellite" because of its bright star intensity as viewers observed it around the world. ECHO I provided basis for numerous passive communications experiments as well as significant data on solar pressure and atmospheric drag. (See Appendix A.)

*August 16, 1960:* Capt. J. Kittinger (USAF) flew EXCELSIOR III from Holloman AFB and successfully bailed out of open gondola at 102,800 feet, establishing unofficial new altitude record for manned balloon and parachute descent record.

*May 4, 1961:* ONR STRATO-LAB HIGH V flight from carrier *Antietam* reached 113,740-foot altitude (21.5 miles), a record ascent for manned balloon, Comdr. Malcolm Ross (USNR) and Lt. Comdr. Victor G. Prather (MC USN) as crew.

APPENDIX D

# Select Awards and Honors in Aeronautics and Astronautics

THE FOLLOWING LISTING of recipients of major awards and honors over the years provides one index to some of the individuals considered to have contributed to the advancement of aeronautics and astronautics and related fields. This listing is focused upon those honors most related to research and development activities, and it does not necessarily include the "elder statesman" type of an award.

*Contents*

| | Page |
|---|---|
| 1. Applied Meteorology Award (AMS), 1956– | 169 |
| 2. Astronautics Award (ARS), 1954– | 169 |
| 3. Astronautics Engineer Achievement Award (NRC), 1959– | 169 |
| 4. Thurman H. Bane Award (IAS), 1943–56 | 170 |
| 5. Laura Taber Barbour Award, 1956– | 170 |
| 6. Louis H. Bauer Award (AMA), 1961– | 171 |
| 7. Grover E. Bell Award (IAS, AHS), 1957– | 171 |
| 8. Bioclimatology Outstanding Achievement Award (AMS), 1960– | 171 |
| 9. Melbourne W. Boynton Space Medicine Award (AAS), 1957– | 172 |
| 10. Frank G. Brewer Trophy (NAA), 1943– | 172 |
| 11. Octave Chanute Award (IAS), 1939– | 172 |
| 12. Robert J. Collier Trophy (NAA), 1911– | 173 |
| 13. Fédération Aéronautique Internationale Gold Medal, 1925– | 175 |
| 14. Flight Achievement Award (AAS), 1958– | 176 |
| 15. Robert H. Goddard Memorial Award (ARS), 1948–. | 176 |
| 16. Daniel Guggenheim Medal, 1929– | 176 |
| 17. Harmon International Trophies, 1926– | 177 |
| 18. Louis W. Hill Space Transportation Award (IAS), 1958– | 180 |

| | Page |
|---|---|
| 19. Nelson P. Jackson Aerospace Award (NRC), 1961– | 180 |
| 20. John Jefferies Award (IAS), 1940– | 180 |
| 21. Alexander Klemin Award (AHS), 1951–. | 181 |
| 22. The Langley Medal, 1909– | 181 |
| 23. Eric Liljencrantz Award (AMA), 1957–. | 182 |
| 24. Raymond F. Longacre Award (AMA), 1947– | 182 |
| 25. Robert M. Losey Award (IAS), 1940– | 182 |
| 26. Theodore C. Lyster Award (AMA), 1947– | 183 |
| 27. The Mackay Trophy (USAF), 1912– | 184 |
| 28. Charles M. Manly Memorial Award (SAE), 1928–. | 185 |
| 29. Clarence L. Meisinger Award (AMS), 1938– | 186 |
| 30. Harry G. Mosley Award (AMA), 1961–. | 187 |
| 31. NACA/NASA Distinguished Service Medal, 1956– | 187 |
| 32. NACA/NASA Exceptional Service Medal, 1956– | 188 |
| 33. NASA Invention and Contribution Award, 1961–. | 188 |
| 34. National Air Council Research Award, 1948–51 | 189 |
| 35. National Rocket Club Award, 1961–. | 189 |
| 36. G. Edward Pendray Award (ARS), 1951– | 189 |
| 37. Propulsion Award (ARS), 1948–. | 190 |
| 38. Sylvanus A. Reed Award (IAS), 1934–. | 190 |
| 39. Carl-Gustaf Rossby Award (AMS), 1951–. | 192 |
| 40. David C. Schilling Award (AFA), 1948– | 192 |
| 41. Science Trophy (AFA), 1948– | 193 |
| 42. Space Flight Award (AAS), 1955– | 193 |
| 43. Lawrence B. Sperry Award (IAS), 1936– | 193 |
| 44. Spirit of St. Louis Medal (ASME), 1929– | 195 |
| 45. Thomas L. Thurlow Navigation Award, 1945–. | 195 |
| 46. Arnold D. Tuttle Memorial Award (AMA), 1952– | 196 |
| 47. Wright Brothers Medal (SAE), 1928– | 196 |
| 48. Wright Brothers Lecturers (IAS), 1937– | 197 |
| 49. Wright Brothers Memorial Trophy Award (NAA), 1948–. | 198 |
| 50. James H. Wyld Memorial Award (ARS), 1954– | 198 |
| 51. Miscellaneous NASA Award winners: | |
|     Flemming Award | 199 |
|     National Civil Service League Award | 199 |
|     Rockefeller Public Service Award | 199 |

## 1. APPLIED METEOROLOGY AWARD (AMS)

The award for Applied Meteorology is administered by the American Meteorological Society.

1956. Joseph J. George, for contributions to the improvement of practical forecasting for airways operations.

1957. Vincent J. Schaefer, for contributions in the field of experimental and physical meteorology, particularly his pioneering work in artificial nucleation.

1959. Carl-Gustaf Arvid Rossby (posthumously), for convincing industry of the importance of meteorology and for persuading the U.S. Weather Bureau to make weather information over teletype facilities available to private meteorologists without charge. In 1927 under the Daniel Guggenheim Fund for Aeronautics he established the first model airway weather service in the United States.

1960. Henry T. Harrison, for pioneering work in aviation meteorology and continued key role in the development of modern aeronautical meteorology and its application to the jet era.

## 2. ASTRONAUTICS AWARD (ARS)

This award is given annually for outstanding contribution to the advancement of space flight. It is administered by the American Rocket Society.

1954. Theodore von Kármán.

1955. Wernher von Braun.

1956. Joseph Kaplan.

1957. Krafft Ehricke.

1958. Iven C. Kincheloe, Jr.

1959. Walter Dornberger.

1960. A. Scott Crossfield.

## 3. ASTRONAUTICS ENGINEER ACHIEVEMENT AWARD (NRC)

Established in 1959, the Astronautics Engineer Achievement Award is given annually to an accredited engineer who has made an outstanding contribution to the advancement of space technology, an award based on personal accomplishment. It is administered by the National Rocket Club.

1959. Rudolf F. Hoelker.

1960. Richard C. Canright.

1961. William J. O'Sullivan, Jr.

## 4. THURMAN H. BANE AWARD (IAS)

The Thurman H. Bane Award was given "to an officer or civilian of the Air Research and Development Command of the U.S. Air Force, for an outstanding achievement in aeronautical development during the year." It was administered by the Institute of Aeronautical Sciences until suspended in 1956.

1943. Lt. Col. Hollingsworth F. Gregory, for his contribution to the military and commercial development of the helicopter.

1944. Col. Donald J. Keirn, for his contribution to the development and utilization of the jet propulsion engine.

1945. Capt. Myron Tribus, for reducing the icing hazards of high-speed flying through research and flight testing.

1946. Col. Leighton I. Davis, for gyro computing sights for aiming guns, bombs, and rockets from fighter aircraft.

1947. Adolph L. Berger, for development and applications of high-temperature ceramic materials to military aircraft engines.

1948. Col. James M. Gillespie, who was responsible for the preparation of the first long-range automatic "Push Button" flight.

1949. Capt. Harold W. Robbins, for his contribution to the development of jet-assisted takeoff for cargo aircraft.

1950. Col. George W. Goddard, for his development of a new system of night aerial photography at low altitudes.

1951. J. B. Johnson, for outstanding contributions in the field of high-temperature materials used in aircraft propulsion units.

1952. Maj. Patrick L. Kelly, for developing techniques providing for artificial stability and control of aircraft during high-performance phases of operation.

———. Henry Seeler, for developing and standardizing a resuscitator, capable of operating at varying altitudes, for use in combat or evacuation aircraft.

1953. Benjamin F. Greene, Jr., for the development of an electronic system for airport traffic control which greatly increases the landing rate of aircraft under high-density traffic or poor visibility conditions.

1954. Helmut G. Heinrich, for his outstanding accomplishments in the field of personnel parachute development.

1955. Gottfried Guderley, for his outstanding contributions to the development of transonic aerodynamics essential to the engineering design of supersonic aircraft.

1956. Award suspended.

## 5. LAURA TABER BARBOUR AWARD

The Laura Taber Barbour Award was established in 1956 and is given annually for notable achievement which shall tend to advance safety in aeronautics and which contributes toward a method of avoiding or minimize suffering or loss of life in air travel. It is administered by the Flight Safety Foundation.

1956. I. Irving Pinkel and associates, NACA.

1957. Harry F. Guggenheim.

1958. James Martin.

1959. Al Morse.

1960. Melvin Gough.

## 6. LOUIS H. BAUER AWARD (AMA)

Established in 1960 in honor of the founder of Aerospace Medical Association, the Louis H. Bauer Award will be awarded for the first time in 1961 for the most significant contribution in space medicine. It is administered by the Aerospace Medical Association.

## 7. GROVER E. BELL AWARD (IAS)

Established under the terms of the will of the late Lawrence D. Bell in memory of his brother Grover E. Bell, aviation pioneer, the Grover E. Bell Award is given jointly by the Institute of the Aerospace Sciences, the American Helicopter Society, and the Helicopter Council of the Aircraft Industries Association. It is presented "for the purpose of fostering and encouraging research and experimentation in the important and relatively new field of helicopter development to the person or persons making an outstanding contribution to helicopter development during the preceding calendar year in the United States." It is administered by the IAS.

1957. Kurt H. Hohenemser, McDonnell Aircraft Corp., for outstanding achievement in solving the complex dynamic problem required in the development of the world's first successful convertiplane, the McDonnell XV-1.

1958. Engineers of the Vertol Aircraft Corp., for the development of the Vertol 76, the first tilt-wing aircraft to perform successful transition from vertical to horizontal flight and back.

1959. Igor I. Sikorsky, for his foresight and personal contributions to the successful development and demonstration of the helicopter crane concept with the Sikorsky S-60 Skycrane.

1960. Combat Development Office, U.S. Army Aviation School, Fort Rucker, Ala., for the development of the helicopter as a close air-support combat vehicle, thereby significantly increasing its military effectiveness.

## 8. BIOCLIMATOLOGY OUTSTANDING ACHIEVEMENT AWARD (AMS)

Established in 1959, the Bioclimatology Outstanding Achievement Award is administered by the American Meteorological Society.

1960. Frederick Sargent II, for organizing the field of bioclimatology in this country and for outstanding studies on the effects of weather on man.

## 9. MELBOURNE W. BOYNTON SPACE MEDICINE AWARD

Awarded annually for outstanding contributions to aeronautics through medical research, this award is administered by the American Astronautical Society.

1957. Lt. Col. David G. Simon, USAF (MC).

1958. Capt. Charles F. Gell, MC USN.

1959. Lt. Col. Stanley C. White, USAF.

1960. Brig. Gen. Don Flickinger, USAF (MC).

## 10. FRANK G. BREWER TROPHY (NAA)

Established in 1943, the Frank G. Brewer Trophy is awarded annually "for the greatest achievement in the field of Air Youth Education and Training, accomplished by any individual, group of individuals, or organization." It is administered by the National Aeronautic Association.

1943. Civil Aeronautics Administration.

1944. Edgar Fuller.

1945. H. W. Hurt.

1946. Frank E. Sorenson, University of Nebraska.

1947. Nickolaus L. Englehardt, Jr.

1948. Philip S. Hopkins, Link Aviation.

1949. Elsie W. Adams.

1950. Lt. John H. Burton, USN.

1951. Harold E. Mehrens, CAA.

1952. Civil Air Patrol.

1953. Leslie A. Bryan, University of Illinois.

1954. John H. Furbay, TWA.

1955. Willis C. Brown, U.S. Office of Education.

1956. Ray C. Mertes.

1957. Edwin A. Link, Link Aviation.

1958. Evan Evans, NAEC.

1959. Paul E. Garber, Curator, National Air Museum.

1960. George N. Gardner, PAA.

## 11. OCTAVE CHANUTE AWARD (IAS)

Established by the Institute of the Aeronautical Sciences in 1939 in honor of Octave Chanute, American aeronautical pioneer, it is awarded "for a notable contribution made by a pilot to the aeronautical sciences." It is administered by the Institute of the Aerospace Sciences.

1939. Edmund T. Allen, Boeing Aircraft.

1940. Howard Hughes.

1941. Melvin N. Gough, NACA.

1942. Albert Lewis MacClain, Pratt & Whitney.

1943. William H. McAvoy, NACA.

1944. Col. Benjamin S. Kelsey, USAAF.

1945. A. Elliott Merrill and Robert T. Lamson, Boeing Aircraft.

1946. Ernest A. Cutrell, Air Transport Command.

1947. Lawrence A. Clousing, NACA.

1948. Herbert H. Hoover, NACA.

1949. Capt. Frederick M. Trapnell, USN.

1950. Comdr. Donald B. MacDiarmid, USCG.

1951. Lt. Col. Marion E. Carl, USMC.

1952. John Clifford Seal, Cornell Aeronautical Laboratory.

1953. William T. Bridgeman, Douglas Aircraft.

1954. George E. Cooper, NACA.

1955. Maj. Gen. Albert Boyd, WADC.

1956. A. M. Johnston, Boeing Airplane.

1957. Lt. Col. Frank K. Everest, Jr., USAF.

1958. A. Scott Crossfield, North American.

1959. John P. Reeder, NASA.

1960. Joseph J. Tymczyszyn.

1961. Joseph A. Walker, NASA

## 12. THE ROBERT J. COLLIER TROPHY (NAA)

Donated in 1911 by Robert J. Collier, to be awarded each year for the greatest achievement in aviation in America, it is administered by the National Aeronautic Association. It is considered a premier award and it is presented by the President of the United States.

1911. Glenn H. Curtiss, for development of the hydroaeroplane.

1912. Glenn H. Curtiss, for the flying boat.

1913. Orville Wright, for development of the automatic stabilizer.

1914. Elmer A. Sperry, for gyroscopic control.

1915. W. Sterling Burgess, for the Burgess-Dunn hydroaeroplane.

1916. Elmer A. Sperry, for drift indicator.

1917–20. No awards because of the war.

1921. Grover Loening, for development of the aerial yacht.

1922. Personnel of the U.S. Air Mail Service.

1923. Personnel of the U.S. Air Mail Service, for night flying.

1924. The U.S. Army, for its round-the-world flight.

1925. S. Albert Reed, for development of the metal propeller.

1926. Maj. E. L. Hoffman, for development of a practical parachute.

1927. Charles L. Lawrance, for his radial air-cooled engine.

1928. Aeronautics Branch of the Department of Commerce, for development of airways and air navigation facilities.

1929. The National Advisory Committee for Aeronautics, for the NACA cowling.

1930. Harold Pitcairn and his staff, for their autogiro.

1931. The Packard Motor Car Co., for the aircraft Diesel engine.

1932. Glenn L. Martin, for development of an outstanding bi-engined, high-speed weight-carrying airplane.

173

1933. The Hamilton Standard Propeller Co. and chief engineer, Frank W. Caldwell, for the controllable pitch propeller.

1934. Maj. Albert F. Hegenberger, for blind landing experimentation.

1935. Donald Douglas and his staff, for development of the DC–2.

1936. Pan American Airways, for its transpacific and overwater operations.

1937. The Army Air Corps, for design and equipment of a substratosphere airplane.

1938. Howard Hughes and his crew, for his round-the-world flight.

1939. Airlines of the United States for their record of safety in air travel.

1940. Dr. Sanford Moss and the Army Air Corps, for development of the turbosupercharger.

1941. The Air Forces and the airlines, for the worldwide operations typified in the routes of the Air Transport Command.

1942. Gen. H. H. Arnold, for organization and leadership of the mightiest air force in the world.

1943. Capt. Luis de Florez, USNR, for development of synthetic training devices for flyers.

1944. Gen. Carl A. Spaatz, for demonstrating the airpower concept through employment of American aviation in the war against Germany.

1945. Dr. Luis W. Alvarez, for development of the ground-controlled approach radar landing system.

1946. Lewis A. Rodert, NACA, for development of thermal ice-prevention system.

1947. John Stack (NACA), Lawrence D. Bell and Capt. Charles E. Yeager (USAF), for supersonic flight.

1948. Radio Technical Commission for Aeronautics, for the development of a system of air traffic control to permit safe and unlimited operations under all weather conditions.

1949. William P. Lear, for development of the Lear F–5 automatic pilot and automatic control coupler system.

1950. The helicopter industry, the military services, and the Coast Guard, for development and use of rotary-wing aircraft for air-rescue operations.

1951. John Stack and associates at the Langley Aeronautical Laboratory, NACA, for the conception, development, and practical application of the transonic wind tunnel throat.

1952. Leonard S. Hobbs of United Aircraft Corp., for design, development, and production of the J–57 jet engine.

1953. James H. Kindleberger and Edward H. Heinemann, for development of the first supersonic airplanes in service.

1954. Richard Travis Whitcomb, NACA research scientist, for discovery and experimental verification of the area rule, yielding higher speed and greater range with same power.

1955. Willian M. Allen and the Boeing Airplane Co. and to Gen. Nathan F. Twining and the U.S. Air Force, for development and operational use of the B–52.

1956. Charles J. McCarthy and associates of Chance Vought Aircraft, Inc., and Vice Adm. James S. Russell and associates of the U.S. Navy Bureau of Aeronautics, for conception, design, and development of the F8U Crusader.

1957. Edward P. Curtis, for his report entitled "Aviation Facilities Planning," developed while he was special assistant to the President of the United States.

1958. The U.S. Air Force and industry team responsible for the F–104 Interceptor: Clarence L. Johnson of Lockheed Aircraft Corp. for design of the airframe; Neil Burgess and Gerhard Neumann of the Flight Propulsion Division, General Electric Co., for development of its J–79 turbojet engine; Maj. Howard C. Johnson, USAF, for establishing a world landplane altitude record; and Capt. Walter W. Irwin, USAF, for establishing a world straightaway speed record.

1959. The U.S. Air Force, Convair, and Space Technology Laboratories for developing, testing, producing, and putting into operation the Atlas, America's first intercontinental ballistic missile, so vital to the security and space exploration needs of the United States and the free world.

## 13. FÉDÉRATION AÉRONAUTIQUE INTERNATIONALE GOLD MEDAL

The FAI Gold Medal is awarded to those who have contributed highly to the development of aeronautics. It is administered by the Fédération Aéronautique Internationale, U.S. representative of which is the National Aeronautic Association.

1925. Gen Francesco de Pinedo, Italy.

1926. Sir Alan Cobham, Great Britain.

1927. Col. Charles A. Lindbergh, United States.

1928. H. J. L. Hinkler, Great Britain.

1929. Dieudonne Costes, France.

1930. Gen. Italo Balbo, Italy.

1931. Dr. Hugo Eckener, Germany.

1932. Juan de la Cierva, Spain.

1933. Wiley Post, United States.

1934. C. W. A. Scott, Great Britain.

1935. No award.

1936. Jean Mermoz, France.

1937. Miss Jean Batten, Great Britain.

1938. No award (tie on third vote).

1939-45. No awards.

1946. Igor I. Sikorsky.

1947. No award made.

1948. Capt. Charles E. Yeager, United States.

1949. No award made.

1950. Frank Whittle.

1951. Dr. Edward P. Warner, United States.

1952. No award made.

1953. Jacqueline Cochran, United States.

1954. Lt. Gen. James H. Doolittle, United States.

1955. Maurice Hurel, France.

1956. L. Peter Twiss, Great Britain.

1957. Maj. David G. Simons, United States.

1958. Audrey Micholaevich Tupolev, U.S.S.R.

1959. Pierre Satre, France.

## 14. FLIGHT ACHIEVEMENT AWARD (AAS)

Established in 1958, this award is "presented annually to the pilot or crew who, by outstanding ability and courage, has personally extended the frontiers of flight from the earth's environment into space." The Flight Achivement Award is administered by the American Astronautical Society.

    1958. Capt. Iven C. Kincheloe, USAF (posthumously).

    1959. A. Scott Crossfield.

    1960. No award.

## 15. ROBERT H. GODDARD MEMORIAL AWARD (ARS)

Established in 1947, the Robert H. Goddard Memorial Award is presented annually for outstanding work in liquid propellant rockets. It is the highest award of the American Rocket Society.

| | |
|---|---|
| 1948. John Shesta. | 1955. E. N. Hall. |
| 1949. Calvin M. Bolster. | 1956. Chandler C. Ross. |
| 1950. Lovell Lawrence, Jr. | 1957. Thomas F. Dixon. |
| 1951. Robert C. Truax. | 1958. Richard B. Canright. |
| 1952. Richard W. Porter. | 1959. Samuel K. Hoffman. |
| 1953. David A. Young. | 1960. Theodore von Kármán. |
| 1954. A. M. O. Smith. | |

## 16. DANIEL GUGGENHEIM MEDAL

Established by the Daniel Guggenheim Fund for the Promotion of Aeronautics in 1928, this medal is given for notable achievement in the advancement of aeronautics. It is administered by the Institute of Aeronautical Sciences, the Society of Automotive Engineers, the American Society of Engineers, and the United Engineering Trustees, Inc.

| | |
|---|---|
| 1929. Orville Wright. | 1934. William E. Boeing. |
| 1930. Ludwig Prandtl. | 1935. William F. Durand. |
| 1931. Frederick W. Lanchester. | 1936. George W. Lewis. |
| 1932. Juan de la Cierva. | 1937. Hugo Eckener. |
| 1933. Jerome C. Hunsaker. | 1938. Alfred H. R. Fedden. |

1939. Donald W. Douglas.

1940. Glenn L. Martin.

1941. Juan T. Trippe.

1942. James H. Doolittle.

1943. Edmund Turney Allen.

1944. Lawrence D. Bell.

1945. Theodore P. Wright.

1946. Frank Whittle.

1947. Lester Durand Gardner.

1948. Leroy Randle Grumman.

1949. Edward P. Warner.

1950. Hugh L. Dryden.

1951. Igor I. Sikorsky.

1952. Sir Goeffrey de Havilland.

1953. Charles A. Lindbergh.

1954. Clarence D. Howe.

1955. Theodore von Kármán.

1956. Frederick B. Rentschler (posthumous).

1957. Arthur E. Raymond.

1958. William Littlewood.

1959. Sir George R. Edwards.

1960. Grover Loening.

## 17. HARMON INTERNATIONAL TROPHIES

Established in 1926 by Clifford B. Harmon, early balloonist and aviator, trophies are awarded for outstanding achievements in the arts and/or sciences of aeronautics for the preceding year, with the art of flying receiving first consideration. It is administered by the Clifford B. Harmon Trust.

*Aviators*

1926. Lt. Col. George Pelletier-Doixy, France.

1927. Col. Charles A. Lindbergh, United States, for his New York to Paris flight.

1928. Col. Arturo Ferrarin, Italy, for his flight from Rome to Natal, Brazil.

1929. Maj. Dieudonne Costes, France.

1930. Maj. Dieudonne Costes, France.

1931. Air Marshall Italo Balbo, Italy, for leading a mass flight of 10 planes from Orteballo, Italy, to Rio de Janeiro, Brazil.

1932. Wolfgang von Gronau, Germany.

1933. Wiley Post, United States, for his world's flight.

1934. Charles W. A. Scott, Great Britain, for winning the MacRobertson race from London to Melbourne.

1935. Capt. Edwin Musick, United States, for his pioneer work in "China Clipper" flights.

1936. Howard Hughes, United States, for his fast coast-to-coast flight, Los Angeles to Newark in 7 hours 28 minutes and 25 seconds.

1937. Henry T. (Dick) Merrill, United States, for his successful return flight in the scheduled first commercial flight between New York and London, bringing to the United States the first motion pictures of the British coronation ceremonies.

1938. Howard Hughes, United States, for outstanding performance in his world flight—3 days 19 hours 8 minutes and 10 seconds.

1939. Alexander P. de Seversky, United States.

1940-49 (only one award made). Lt. Gen. James H. Doolittle, United States, for his outstanding personal leadership, skill, and courage, he is symbolic of all who flew in the Allied cause during the World War II.

1950. Col. David C. Schilling, USAF, for the first nonstop jet flight between Great Britain and the United States.

1951. Capt. Charles F. Blair, USAF, for being the first person to fly a single-engine fighter nonstop across the North Pole from Europe to North America.

1952. Col. Bernt Balchen, USAF, for Arctic operations including many flights.

1953. Maj. Charles E. Yaeger, USAF, for piloting the Bell X-1A rocket research plane at a speed of 1,650 mph.

1954. J. F. Coleman, Convair test pilot, for history's first transitional flight from vertical takeoff to level flight in the Navy's XFY-1 VTO experimental fighter.

1955. Group Capt. John Cunningham, RAF, for piloting the de Havilland Comet III on the first around-the-world commercial jet flight.

1956. Lt. Col. Frank Everest, USAF, for flights in the Bell X-2 in which he became the first man to fly near Mach 3.

1957. Gen. Curtis E. LeMay, USAF, for piloting a KC-135 on a nonrefueling record flight from Westover AFB, Mass., to Buenos Aires, Argentina.

1958. Maj. André Turcat of France, for the first flight at twice the speed of sound.

1959. Capt. Joe B. Jordan, USAF, for taking an F-104 to an altitude record of 103,395.5 feet; and Capt. Joseph W. Kittinger, Jr., USAF, for ballooning to a record 76,400 feet in an open gondola and then parachuting.

*Aviatrices*

1927. Lady Mary Bailey, Great Britain.

1928. Lady Mary Bailey, Great Britain.

1929. Winifred Spooner, Great Britain.

1930. Amy Johnson Mollison, Great Britain.

1931. Maryse Bastié, France, for Paris-Tokyo round-trip flight.

1932. Amelia Earhart Putnam, United States, for solo transatlantic flight.

1933. Maryse Helsz, France.

1934. Héléne Boucher, France.

——. Maryse Bastié, France (posthumously), for speed record.

1935. Amelia Earhart Putnam, United States.

——. Jean Batten, Great Britain.

1936. Jean Batten, for England-New Zealand flight.

1937. Jean Batten, for Darwin to England flight.

1938. Jacqueline Cochran, United States, winner of Bendix race.

1939. Jacqueline Cochran.

1940-49. Jacqueline Cochran.

1950. No award.

1951. Mme. Jacqueline Auriol, France, for jet speed record.

1952. Mme. Jacqueline Auriol, France.

1953. Jacqueline Cochran, United States, for being the first woman to break the sound barrier.

1954. No award.

1955. Mme. Jacqueline Auriol, France, for becoming world's fastest woman pilot.

1956. No award.

*Spherical Balloonist (Aeronauts)*
1926. Gen. Umberto Nobile, Italy.

1927. Comdr. Charles E. Rosendahl, USN.

1928. Dr. Hugo Eckener, Germany.

1929. Dr. Hugo Eckener, Germany.

1930. Dr. Hugo Eckener, Germany.

———. Ward T. Van Orman, United States.

1931. Dr. Hugo Eckener, Germany.

1932. Auguste Piccard, Switzerland.

———. Capt. Ernest A. Lehmann, Germany.

1933. Dr. Hugo Eckener, Germany.

———. Lt. Comdr. T. G. W. Settle, USN.

1934. Capt. Ernest A. Lehmann, Germany.

———. Jeannette Piccard, Switzerland.

1935. Capt. Orvil A. Anderson, United States.

———. Capt. H. von Schiller, Germany.

1936. Capt. Ernest A. Lehmann, Germany.

———. Ernest De Muyter, Belgium.

1940–49. Vice Adm. Charles E. Rosendahl, USN.

1950. No award.

1951. Lt. Carl J. Seiberlieh, USN.

1952. Walter L. Massie, United States.

1953. No award.

1954. Capt. Marion H. Eppes, USN.

1955. Lt. Comdr. Charles A. Mills, USN.

1956. Lt. Comdr. Malcolm D. Ross.

NATIONAL WINNERS UNITED STATES (DISCONTINUED)

*Aviators*
1926. Shirley J. Short.

1927. Charles A. Lindbergh.

1928. Capt. Carl Ben Eielson.

1929. Lt. James H. Doolittle.

1930. Lt. Comdr. Frank Hawks.

1931. Clyde Pangborn.

1932. Lt. Col. Roscoe Turner.

1933. Wiley Post.

1934. Dean Smith.

1935. Capt. Edwin C. Musick.

1936. Howard Hughes.

1937. Henry T. (Dick) Merrill.

1938. Howard Hughes.

1939. Col. Robert E. Olds.

*Aviatrices*
1932. Amelia Earhart Putnam.

1933. Anne Morrow Lindbergh.

1934. Laura Ingalls.

1935. Amelia Earhart Putnam.

1936. Louise Thaden.

1937. Jacqueline Cochran.

*Dirigible Pilots*
1932–33. Comdr. Charles E. Rosendahl.

1934. Herbert V. Wiley.

*Spherical Balloon Pilots*
1932–33. Lt. Comdr. T. G. W. Settle.

1934. Maj. William E. Kepner.

1935. Capt. Orvil A. Anderson.

## 18. LOUIS W. HILL SPACE TRANSPORTATION AWARD (IAS)

The Louis W. Hill Space Transportation Award is given for "significant contribution indicative of American enterprise and ingenuity in the art and science of space flight." Named after a transportation pioneer, this award is administered by the Institute of the Aerospace Sciences.

1958. Robert H. Goddard (posthumously), for significant contributions indicative of American ingenuity in the art and sciences of space flight.

1959. James A. Van Allen, State University of Iowa, by combining simple and direct techniques, he established beyond doubt the existence, intensity, and extent of the radiation belts above the earth's surface that now carry his name.

1960. S. K. Hoffman and Thomas E. Dixon, Rocketdyne Division of North American Aviation, for continuous leadership in the development, and for extraordinary improvements in the thrust efficiency and reliability of powerplants for major ballistic missiles and space vehicle boosters.

## 19. NELSON P. JACKSON AEROSPACE MEMORIAL AWARD (NRC)

Established in 1960 in honor of a founder of the National Rocket Club, the Nelson P. Jackson Aerospace Memorial Award is awarded annually to an aerospace industry "for an outstanding contribution to the missile, aircraft, and space field." It is administered by the National Rocket Club.

1961. The U.S. Air Force and the industry (Bell, Douglas, General Electric, Lockheed, and Rocketdyne of North American) responsible for the Discoverer satellite program, which made the first successful recovery of an object from space on August 11, 1960.

## 20. JOHN JEFFRIES AWARD (IAS)

Established in 1940 by the IAS for the purpose of honoring the memory of the American physician who made the first flight across the English Channel with Blanchard, the French balloonist, in 1785. The John Jeffries Award is given for "outstanding contributions to the advancement of aeronautics through medical research." It is administered by the Institute of the Aerospace Sciences.

1940. Dr. Louis H. Bauer, editor of *Journal of Aviation Medicine*.

1941. Maj. Harry G. Armstrong, USA MC, School of Aviation Medicine.

1942. Dr. Edward C. Schneider, Wesleyan University.

1943. Brig. Gen. Eugene G. Reinartz, AAF School of Aviation Medicine.

1944. Sir Harold E. Whittingham, Royal Air Force.

1945. Commodore John C. Adams, USN Aviation Medicine.

1946. Dr. Malcolm C. Grow, 8th Air Force.

1947. Dr. James W. Tice, RCAF.

1948. Dr. W. Randolph Lovelace II.

1949. Dr. Arnold D. Tuttle, United Air Lines.

1950. Brig. Gen. Otis O. Benson, Jr., School of Aviation Medicine.

1951. Capt. John R. Poppen, USN (MC).

1952. Lt. Col. John P. Stapp, USAF.

1953. Capt. Chas. F. Gell, USN.

1954. Dr. James P. Henry, Aeromedical Laboratory.

1955. Capt. Wilbur E. Kellum, Naval Medical Research Institute.

1956. Ross A. McFarland, Harvard University.

1957. Maj. David G. Simons, USAF (MC).

1958. Dr. Hubertus Strughold, USAF School of Aviation Medicine.

1959. Brig. Gen. Don Flickinger, USAF.

1960. Capt. Joseph W. Kittinger, Jr., USAF.

## 21. ALEXANDER KLEMIN AWARD (AHS)

Established in 1951 in honor of the memory of the late Dr. Alexander Klemin, eminent aeronautical engineer, educator, author, and pioneer of rotary-wing aeronautics, it is awarded each year for "engineering, design, and invention in the field of rotary-wing aircraft." It is administered by the American Helicopter Society.

1951. Igor I. Sikorsky.

1952. Louis Breguet, France.

1953. Raoul Hafner, England.

1954. Michael Glubareff, Sikorsky Aircraft.

1955. Bartram Kelley, Bell Aircraft.

1956. Charles H. Zimmerman, NACA Langley Aeronautical Laboratory.

1957. Capt. Wayne Eggert, USAF (posthumously).

## 22. THE LANGLEY MEDAL

Established in 1908 by the Regents of the Smithsonian Institution, the Langley Medal was initiated by Alexander Graham Bell for the purpose of presenting an American award to the Wright Brothers. Presented only upon occasion, the medal is awarded by motion of the Secretary of the Smithsonian and a designated committee "for specially meritorious investigations in connection with the science of aerodromics and its application to aviation."

1909. Wilbur and Orville Wright.

1913. Glenn H. Curtiss.

1913. Gustave Eiffel.

1927. Charles A. Lindbergh.

1929. Richard E. Byrd.

1929. Charles M. Manly.

1935. Dr. Joseph S. Ames.

1955. Dr. Jerome C. Hunsaker.

1960. Dr. Robert H. Goddard (posthumously).

## 23. ERIC LILJENCRANTZ AWARD (AMA)

Established in 1956, the Eric Liljencrantz Award is given annually for the best paper on basic research in the problems of acceleration and altitude. It is administered by the Aerospace Medical Association.

1957. Col. John P. Stapp, USAF (MC).

1958. Brig. Gen. Victor A. Byrnes, USAF (MC).

1959. Capt. Edward L. Beckman, MC USN.

1960. James D. Hardy, Ph. D, Bala-Cynwyd, Pa.

1961. Capt. Ashton Graybiel, MC USN.

## 24. RAYMOND F. LONGACRE AWARD (AMA)

Established in 1947, the Raymond F. Longacre Award is annually awarded for "outstanding achievement in the psychiatric and psychologic aspects of aviation medicine." It is administered by the Aerospace Medical Association.

1947. Ross A. McFarland (Boston, Mass.).

1948. Detlev Bronk (Baltimore, Md.).

1949. Sir Charles P. Symonds (London, England).

1950. Donald W. Hastings (Minneapolis, Minn.).

1951. Col. Neely C. Mashburn, USAF (San Antonio, Tex.).

1952. Sir Frederick Bartlett (Cambridge, England).

1953. Walter F. Grether (Wright-Patterson AFB, Ohio).

1954. John C. Flanagan (University of Pittsburgh).

1955. Roy R. Grinker (Chicago, Ill.).

1956. Saul B. Sells, SAM (San Antonio, Tex.).

1957. Brig. Gen. Eugene G. Runartz, USAF (MC).

1958. Col. Harry G. Moseley, USAF (MC).

1959. Capt. George E. Ruff, USAF (MC).

1960. Brant Clark (San Jose, Calif.).

## 25. ROBERT M. LOSEY AWARD (IAS)

Established by the Institute of the Aeronautical Sciences in 1940 in memory of the first U.S. officer in military service to die in World War II, the Robert M. Losey Award is annually given "in recognition of outstanding contributions to the science of meteorology as applied to aeronautics." It is administered by the Institute of the Aerospace Sciences.

1940. Henry G. Houghton, Jr., MIT.

1941. Horace R. Byers, University of Chicago.

1942. Francis W. Reichelderfer, U.S. Weather Bureau.

1943. Lt. Col. Joseph J. George, USAAF.

1944. John C. Bellamy, AAF Weather Service.

1945. Lt. Col. Harry Wexler, U.S. Weather Bureau.

1946. Carl G. Rossby, University of Chicago.

1947. Col. Benjamin G. Holzman, USAF.

1948. Paul A. Humphrey, U.S. Weather Bureau.

1949. William Lewis, U.S. Weather Bureau.

1950. Roscoe R. Braham, New Mexico School of Mines.

1951. Ivan R. Tannehill, U.S. Weather Bureau.

1952. Dr. Vincent J. Schaefer, General Electric.

1953. Henry T. Harrison, Jr., United Air Lines.

1954. Hermann B. Wobus, NAS Norfolk.

1955. Lt. Col. Robert C. Bundgaard, USAF Weather Service.

1956. Ross Gunn, U.S. Weather Bureau.

1957. Jule G. Charney, MIT.

1958. Patrick B. McTaggart-Cowan, Canadian Meteorology Office, Toronto, Canada.

1959. Herbert Riehl, University of Chicago.

1960. Thomas F. Malone, Travelers Insurance Co.

## 26. THEODORE C. LYSTER AWARD

Established in 1947 by the Aero Medical Association, this award is given "for outstanding achievement in the general field of aviation medicine." It is administered by the Aerospace Medical Association.

1947. Louis H. Bauer.

1948. Wilbur R. Franks (Toronto, Canada).

1949. Brig. Gen. Harry G. Armstrong, USAF (MC).

1950. Capt. Ashton Graybiel, MC USN.

1951. Rear Adm. Bertram Groesbeck, Jr., MC USN.

1952. Kenneth A. Evelyn (Montreal, Canada).

1953. Capt. Wilbur E. Kellum, MC USN.

1954. William R. Stovall, CAA.

1955. Brig. Gen. Otis O. Benson, Jr., USAF (MC).

1956. Brig. Gen. Don Flickinger, Jr., USAF (MC).

1957. Capt. Charles F. Gell, MC USN.

1958. Dr. Hubertus Strughold, USAF SAM.

1959. Capt. Clifford P. Phoebus, MC USN.

1960. Air Commodore A. A. G. Corbet, RCAF.

## 27. THE MACKAY TROPHY (USAF)

The Mackay Trophy is awarded "to the Air Force person or persons who made the most meritorious flight of the year." It is administered by the U.S. Air Force.

1912. Lt. Henry H. Arnold, for reconnaissance flight.

1913. Lt. J. E. Carberry and Lt. F. Snydel, for reconnaissance flight.

1914. Capt. T. F. Dodd and Lt. S. W. Fitzgerald, for reconnaissance flight.

1915. Lt. B. Q. Jones, for American duration record.

1916-17. No award due to war.

1918. Capt. E. V. Rickenbacker, for official record in combat.

1919. Lt. Belvin N. Maynard and others, for their flights between Atlantic and Pacific and return.

1920. Capt. St. Clair Streett, leader of Alaskan flight.

1921. Lt. J. A. Macready, for world's altitude record.

1922. Lts. J. A. Macready and C. G. Kelly, for world's duration record.

1923. Lts. J. A. Macready and C. G. Kelly, for nonstop transcontinental flight.

1924. Capt. Lowell H. Smith and others, for first round-the-world flight.

1925. Lt. James H. Doolittle and Lt. Cyrus K. Bettis, for Schneider Trophy and Pulitzer race awards, respectively.

1926. Maj. H. A. Dargue and others, for Pan American good-will flight.

1927. Capts. L. J. Maitland and A. F. Hegenberger, for nonstop flight to Hawaii.

1928. Lt. Harry A. Sutton, for spinning tests.

1929. Capt. Albert W. Stevens, for his long-distance and high-altitude photography.

1930. Maj. Ralph Royce, for "Arctic Patrol" flight.

1931. Brig. Gen. B. D. Foulois, for leadership in 1931 maneuvers.

1932. Lt. Chas. H. Howard and others, for Navajo relief flights.

1933. Capt. Westside T. Larson, for frontier defense flights.

1934. Brig. Gen. Henry H. Arnold, for Alaskan flight.

1935. Major A. W. Stevens and Capt. O. A. Anderson, for stratosphere flight in *Explorer II* balloon.

1936. Capt. Richard E. Nugent and others, for successful flight under adverse conditions from Langley Field to Allegan, Mich.

1937. Capt. Carl J. Crane and George V. Holloman, for developing and demonstrating automatic landing system.

1938. Lt. Col. Robert Olds and others, for B-17 flight to Argentina.

1939. Officers and men flying medical supplies from Langley Field to Santiago, Chile.

1940-45. No awards.

1947. Capt. Charles E. Yeager, for the first supersonic flight in XS-1.

1948. Lt. Col. Emil Beaudry, for rescue of 12 airmen from Greenland Icecap.

1949. Capt. James G. Gallaher and crew for *Lucky Lady II*, for first round-the-world nonstop flight.

1950. The 27th Fighter Wing, for its 1950 mass flight across the Atlantic.

1951. Col. Fred J. Ascani, for new world speed record at National Air Races.

1952. Maj. Louis H. Carrington, Jr., Maj. Fred W. Shook, and Capt. Wallace D. Yancy, for the first nonstop transpacific flight of an RB-45 jet flight, 3,460 nautical miles with two air refuelings.

1953. SAC's 40th Air Division, for longest nonstop flight over water ever flown by single-engine jet aircraft.

1954. SAC's 38th Air Division, for intercontinental maneuver with B-47 and determining fatigue limits of combat crews.

1955. Col. Horace A. Hanes, for establishing world's speed record.

1956. Capt. Iven C. Kincheloe, Jr., for flight of X-2 to world altitude record.

1957. SAC's 93d Bombardment Wing, for Operation Powerflight.

1958. TAC's Composite Air Strike Force, for deployment to the Far East.

1959. USAF Thunderbird Aerial Team, for good-will tour of Far East.

1960. USAF 6593d Test Squadron (Special), Hickam AFB, Hawaii, which made the first aerial recovery of an object from space orbit.

## 28. CHARLES MATTHEWS MANLY MEMORIAL AWARD (SAE)

Awarded "to the author of the best paper relating to theory or practice in the design or construction of, or research on, aerospace engines, their parts, components, or accessories which shall have been presented at a meeting of the society (SAE) or any of its sections during the calendar year." It is administered by the Society of Automotive Engineers.

1928. Sam D. Heron, "The In-Line Air-Cooled Engine."

1929. No. award.

1930. Dr. Oscar C. Bridgeman, "The Effect of Airplane Fuel-Line Design on Vapor Lock."

1931. No award.

1932. Ford L. Prescott, "Indicators as a Means of Improving Aircraft Engine Performance."

1933. A. H. R. Fedden, "Possible Future Developments in Air-Cooled Aero Engines."

1934. Rex B. Beisel, A. L. MacClain, and F. M. Thomas, "Cowling and Cooling of Radial Air-Cooled Aircraft Engines."

1935. Guy E. Beardsley, Jr., "Automatic Power and Mixture Control for Aircraft Engines."

1936. Raymond W. Young, "Air-Cooled Radial Aircraft Engine Performance."

1937. Richard S. Buck, "Flight Testing With an Engine Torque Indicator."

1938. A. L. Berger and Opie Chenoweth, "Supercharger Installation Problems."

1939. E. W. Hives and F. L. Smith, "High Output Aircraft Engines."

1940. No award.

1941. No award.

1942. John Dolza and H. C. Karcher, "Correlation of Ground and Altitude Performance of Oil Systems."

1943. J. O. Almen, "Shot Blasting To Increase Fatigue Resistance."

1944. Kenneth Campbell, "Engine Cooling Fan Theory and Practice."

1945. No award.

1946. Cearcy D. Miller, "Roles of Detonation Waves and Auto-Ignition in Spark Ignition Engine Knock as Shown by Photographs Taken at 40,000 and 200,000 Frames per Second."

1947. Erold F. Pierce and Harvey W. Welsh, "Engine Compounding for Power and Efficiency."

1948. Andrew Kalitinsky, "Atomic Power and Aircraft Propulsion."

1949. No award.

1950. No award.

1951. J. M. Mergen and J. H. Kasley, "Operating Characteristics of Propellers for Turboprop Airplanes."

1952. R. E. Gorton and B. E. Miller, "Instrumentation for Aircraft Gas Turbine Development."

1953. No award.

1954. John M. Tyler and E. C. Perry, Jr., "Jet Noise."

1955. George F. Wislicenus, "Principles and Applications of By-Pass Turbojet Engines."

1956. William A. Benser and Harold B. Finger, "Compressor Stall Problems in Gas Turbine Type Aircraft Engines."

1957. R. E. Matzdorff and C. F. Newberry, "Requirements, Parameters, and Design Considerations for Pneumatic and Inlet Control Systems."

1958. Robert H. Boden, "Ion Rocket Engine."

1959. No award.

1960. Donald B. Mackay, "Secondary Power Systems for Space Vehicles."

## 29. CLARENCE L. MEISINGER AWARD (AMS)

The Clarence Leroy Meisinger Award is given from time to time for outstanding research contributions by meteorologists under 35 years of age. It is administered by the American Meteorological Society.

1938. Jerome Namias, for application of thermodynamic tools to weather forecasting.

1941. Joseph Johnson George, for fog forecasting studies.

1946. Morris Nerburger, for studies of stratus clouds.

1947. Herbert Riehl, for aerological studies in tropical and subtropical meteorology.

1948. James Edward Miller, for studies in vertical motion.

1949. Jule Gregory Charney and Arnt Eliassen, for applying numerical methods to weather prediction.

1950. John C. Freeman, Jr., and Morris Tepper, for contributions in the application of hydraulic analogies.

1951. Dave Fultz, for application of experimental fluid mechanics to problems of atmospheric circulation.

1953-55. No award.

1956. Ernest J. Fawbush and Robert C. Miller, for pioneering studies on tornadoes and other local storms.

1957. Dave Atlas, for pioneering contributions in radar meterology.

1958. No award.

1959. Robert G. Fleagle, for research contributions on the dynamics of large-scale motions of the atmosphere.

1960. Philip D. Thompson and Norman A. Phillips, for theoretical and applied research in the field of numerical weather prediction.

## 30. HARRY G. MOSELEY AWARD (AMA)

Established in memory of Colonel Moseley, this award is given for an outstanding contribution to flight safety. First award will be given in 1961. Administered by the Aerospace Medical Association.

## 31. NACA/NASA DISTINGUISHED SERVICE MEDAL

The NASA Distinguished Service Medal is awarded to NASA personnel for "outstanding scientific achievement, outstanding leadership in aeronautical and space science, an outstanding contribution to public administration, or outstanding bravery." Recommendations for this award are reviewed by the NASA Incentive Awards Board. It is NASA's highest award.

NACA

January 19, 1956. Richard T. Whitcomb, for discovery and experimental verification of the area rule making possible supersonic flight by military aircraft.

April 19, 1956. Charles W. Littleton and John W. Moise, for outstanding bravery beyond the call of duty following explosion of X-1A airplane on August 8, 1955, risking life to save the life of a fellow employee.

March 21, 1957. Jerome C. Hunsaker, for fundamental contributions to aeronautical science since 1913.

April 18, 1957. H. Julian Allen, for discovery that blunt shapes for warheads of long-range missiles greatly reduced heat input and impart inherent stability during reentry into the atmosphere.

April 18, 1957. I. Irving Pinkel, for scientific research on causes of fire and of impact hazards in aircraft crashes leading to successful demonstrations and preventive design.

August 21, 1958. John Francis Victory, for vision and courage in strengthening NACA throughout 43 years and in promoting aeronautical research.

NASA

June 30, 1959. John W. Crowley, Jr., for accelerating progress in aircraft, missiles, and spacecraft by scientific investigations and promoting teamwork and cooperation with the military and industry.

May 5, 1961. Alan B. Shepard, Jr., for his flight as the first U.S. astronaut was an outstanding contribution to the advancement of human knowledge of space technology and a demonstration of man's capabilities in suborbital space flight.

## 32. NACA/NASA EXCEPTIONAL SERVICE MEDAL

The NASA Exceptional Service Medal is the second highest award in the NASA Incentive Awards Program. It is given to NASA personnel for "significant scientific achievement, significant leadership in aeronautical and space science, a significant contribution to public administration, or unusual courage or competence in an emergency." Recommendations are reviewed by the NASA Incentive Awards Board.

April 5, 1956. Stanley P. Butchart, Joseph A. Walker, and Richard G. Payne, for extreme courage and competence in an emergency during flight test of X–1A rocket research aircraft.

August 20, 1957. Seymour Lieblein, for significant scientific results in research on axial-flow compressors.

August 20, 1957. Robert G. Deissler, for significant scientific results in solution of fluid flow and heat transfer problems of great value in solution of problems associated with aircraft nuclear propulsion.

October 4, 1957. John B. Parkinson, for his analysis and experimental verification of the principle that high-beam ratios improve hydrodynamic and aerodynamic characteristics of seaplane hulls.

October 4, 1957. Anshal I. Neihouse, for leadership in the development of spin research techniques and the art of predicting spin-and-recovery characteristics.

## 33. NASA INVENTIONS AND CONTRIBUTIONS AWARD

Section 306 of the National Aeronautics and Space Act of 1958 (42 U.S.C. 2458) authorized the Administrator of NASA, upon recommendation of the NASA Inventions and Contributions Board, to make monetary awards not exceeding $100,000 for any scientific or technical contribution which has significant value in the conduct of aeronautical and space activities. Awards exceeding $100,000 must be reported to the appropriate committees of the Congress and if the Congress takes no action or does not veto the proposed award, it may be made.

January 17, 1961. Dr. Frank T. McClure in amount of $3,000 for the Satellite Doppler Navigation System Concept (which became the basis for the Navy's Project Transit). This was the first award.

## 34. NATIONAL AIR COUNCIL RESEARCH AWARD

Until the operation of the National Air Council was suspended in 1951, this award was given "to one member of the Air Force and one member of the Navy, who, in the opinion of the respective departments, contributed most in the previous fiscal year in the field of aviation research and experiment."

1948. Rear Adm. Theodore C. Lonnquest, USN.
Col. James M. Gillespie, USAF.

1949. Capt. William V. Davis, Jr., USN.
Capt. Vincent Mazza, USAF.

1950. Rear Adm. Calvin M. Bolster, USN.
Capt. James L. Hight, USAF.

1951. Capt. Walter S. Diehl, USN (Ret.).
Maj. John P. Stapp, USAF.

## 35. NATIONAL ROCKET CLUB AWARD

Established in 1960, the National Rocket Club Award is presented annually "in recognition of an individual or group of individuals who have added to the public understanding of the impact of the space age upon our Nation and all mankind." It is administered by the National Rocket Club.

1961. Edward R. Murrow, CBS.

## 36. G. EDWARD PENDRAY AWARD (ARS)

The G. Edward Pendray Award is granted annually for the outstanding contribution to astronautical literature. It is administered by the American Rocket Society.

1951. George P. Sutton.

1952. M. J. Zucrow.

1953. H. S. Tsien.

1954. Martin Summerfield.

1955. Walter Dornberger.

1956. Hermann Oberth.

1957. Grayson Merrill.

1958. Homer E. Newell, Jr.

1959. Ali Bulent Cambel.

1960. Luigi Crocco.

## 37. PROPULSION AWARD (ARS)

Established in 1948, the C. N. Hickman Award was dedicated to the memory of the American rocket pioneer and renamed the Propulsion Award in 1952. It is awarded annually for outstanding work in solid propellant rockets. It is administered by the American Rocket Society.

1948. Frank Malina.

1949. James A. Van Allen.

1950. Leslie Skinner.

1951. William Avery.

1952. A. L. Antonio.

1953. Charles E. Bartley.

1954. Harold W. Ritchey.

1955. F. S. Miller.

1956. Bruce H. Sage.

1957. Levering Smith.

1958. Barnet R. Adelman.

1959. Ernest Roberts.

1960. Ernest Stuhlinger.

## 38. SYLVANUS ALBERT REED AWARD (IAS)

Established in 1933 in memory of a founder-member of IAS, the Sylvanus Albert Reed Award is awarded annually "for a notable contribution to the aeronautical sciences resulting from experimental or theoretical investigations, the beneficial influence of which on the development of practical aeronautics is apparent." It is administered by the Institute of the Aerospace Sciences.

1934. C. G. Rossby and H. G. Willett, MIT, for practical application of the polar front theory.

1935. Frank W. Caldwell, Hamilton Standard Propeller, United Aircraft, for increasing aircraft effectiveness through development and improvement of controllable and constant speed propellers.

1936. Edward Story Taylor, for the development and practical application of the dynamic vibration absorber for aircraft engines.

1937. Eastman Nixon Jacobs, NACA, for his contribution to the aerodynamic improvement of airfoils.

1938. Alfred Victor de Forest, MIT, for development of method for testing metals magnetically.

1939. George J. Mead, NACA, for the design and development of high-output aircraft engines.

1940. Hugh Latimer Dryden, National Bureau of Standards, for his contributions to the mechanics of boundary layer flow and to the interpretation of wind tunnel experiments.

1941. Theodore von Kármán, California Institute of Technology, for development of theory of the influence of curvature on buckling characteristics of aircraft structures.

1942. Igor Ivan Sikorsky, for the creation and reduction to successful practice of a helicopter of superior controllability.

1943. Sanford A. Moss, General Electric, for the development of the turbosupercharger.

1944. Fred E. Weick, Engineering and Research Corp., for his contribution to the development of tricycle landing gear and to the two control nonspinning airplane.

1945. Charles S. Draper, MIT, for application of the gyroscope of computing sights for gunnery and to other computing devices.

1946. Robert T. Jones, NACA, for his contribution to the understanding of flow phenomena around wings and bodies at speeds below and above the speed of sound.

1947. Galen B. Schubauer and Harold K. Skramstad, National Bureau of Standards, for their contributions to the understanding of the mechanism of transition from laminar to turbulent flow.

1948. George W. Brady, Curtiss-Wright Corp., for development of the reversing propeller resulting in shorter landing runs for large aircraft.

1949. George S. Schairer, Boeing Airplane, for contributions to design and development of large sweptwing high-speed aircraft.

1950. Robert R. Gilruth, NACA, for the conception and development of new techniques for obtaining transonic and supersonic data using freely flying models.

1951. E. H. Heinemann, Douglas Aircraft, for design and development of experimental aircraft for investigating transonic and supersonic flight phenomena.

1952. John Stack, NACA, for his leadership in the design, development, and practical operation of transonic wind tunnels.

1953. Earnest G. Stout, Consolidated Vultee, for contributions to the design and development of high-speed, water-based aircraft.

1954. Clark B. Millikan, California Institute of Technology, for contributions to fluid mechanics, airplane aerodynamics, and wind tunnel technology.

1955. H. Julian Allen, NACA, for contributions and leadership in solving problems in the design of supersonic airplanes and missiles, especially the thermal problems at hypersonic speeds.

1956. Clarence L. Johnson, Lockheed Aircraft, for contributions in solving problems in design of supersonic airplanes and missiles, especially the thermal problems at hypersonic speeds.

1957. Raymond L. Bisplinghoff, MIT, for developing ways to calculate aircraft loads and stresses.

1958. Victor Carbonara, Kollsman Instrument, for basic engineering concept and leadership in the development of automatic celestial navigation systems for aircraft, and their integration in current weapon systems.

1959. Karel J. Bossart, Convair, for significant contributions to the design and development of the Atlas ICBM.

1960. John V. Becker, NASA, for contributions to the advancement of aeronautical science at hypersonic speeds and leadership in translating these advances into the designs of manned, winged, hypersonic vehicles.

## 39. CARL-GUSTAF ROSSBY AWARD (AMS)

Until 1958 the Award for Extraordinary Scientific Achievement, the Carl-Gustaf Rossby Award, is given for scientific achievement in the field of meteorology. It is administered by the American Meteorological Society.

1951. Hurd Curtis Willett, for contributions to synoptic meteorology.

1953. Carl-Gustaf Arvid Rossby, for contributions to dynamic meteorology leading to a better understanding of atmospheric motions and thermodynamics.

1955. Jerome Namias, for contributions to research in extended and long-range forecasting techniques.

1956. John von Neumann, for farsighted contributions to the science of meteorology and the national interests in developing the modern, high-speed electronic computer with meteorological application as its ultimate aim, and for support in organizing the world's first research group in numerical weather prediction.

1960. J. Bjerknes and Erik Palmen, for distinguished research contributions in atmospheric dynamics and synoptic aerology.

## 40. THE DAVID C. SCHILLING AWARD (AFA)

Awarded for distinguished services in the field of flight. Founded in 1948 as the Flight Trophy, award was renamed in 1957 in honor of the late Colonel Schilling. Administered by the Air Force Association.

1948. Herbert H. Hoover, NACA test pilot.

1949. Bill Odom, private pilot.

1950. Capt. James Jabara, world's first jet ace.

1951. Brig. Gen. Albert Boyd, Commanding General, Edwards Air Force Base.

1952. Col. David C. Schilling, USAF.

1953. Third Air Rescue Group, MATS.

1954. Charles Yeager, USAF.

1955. Maj. Stuart Childs, USAF.
George Welch (posthumously).

1956. Lt. Col. Frank K. Everest, USAF.

1957. Col. Patrick D. Fleming, USAF (posthumously).

1958. Capt. Iven C. Kincheloe, USAF (posthumously).

1959. Tactical Air Command, USAF.

## 41. SCIENCE TROPHY (AFA)

The Air Force Association Science Trophy is awarded for distinguished service in the field of aerospace science. It is administered by the Air Force Association.

1950. Dr. Theodore von Kármán, Chairman, USAF Scientific Advisory Board.

1951. Dr. George E. Valley, MIT.

1952. Dr. Edward Teller, University of California.

1953. Dr. Mervin J. Kelly, Bell Telephone.

1954. Lt. Col. John P. Stapp.

1955. Dr. John F. von Neumann, Atomic Energy Commission.

1956. Dr. Chalmers W. Sherwin, University of Illinois.

1957. Dr. Charles S. Draper, Massachusetts Institute of Technology.

1958. Mr. H. Julian Allen, Ames Aeronautical Laboratory.

1959. Dr. W. Randolph Lovelace and Brig. Gen. Don Flickinger.

1960. Dr. Louis N. Ridenour (posthumously).

## 42. SPACE FLIGHT AWARD (AAS)

Established in 1955 to acknowledge special efforts and achievements, this award is given annually "to the person who has contributed most to the advancement of the Astronautical Sciences." It is the premier award administered by the American Astronautical Society.

1955. Hermann Oberth.

1956. Comdr. George W. Hoover.

1957. Wernher von Braun.

1958. James A. Van Allen.

1959. No award.

1960. Homer E. Newell.

## 43. LAWRENCE B. SPERRY AWARD (IAS)

Endowed in 1936 in the memory of Lawrence B. Sperry, pioneer aviator and inventor, this award is given annually "for a notable contribution made by a young man to the advancement of aeronautics." It is administered by the Institute of the Aerospace Sciences.

1936. William Curtiss Rockefeller, California Institute of Technology, for the application of aerodynamics and meteorology in optimum air transport operation.

1937. Clarence L. Johnson, Lockheed Aircraft, for improvement of aeronautical design of high-speed commercial aircraft.

1938. Rossell Conwell Newhouse, Bell Telephone, for development and first practical application of the terrain clearance indicator.

1939. Charles M. Kearns, Jr., United Aircraft, for successful application of methods of measuring propeller vibration stresses in flight.

1940. William B. Oswald, Douglas Aircraft, for analytical studies in aerodynamics which greatly facilitated the design and economical operation of airplanes.

1941. Earnest Gordon Stout, Consolidated Aircraft, for contribution to the experimental determination of hydrodynamic stability of model flying boats and seaplanes.

1942. Edward C. Wells, Boeing Aircraft, for contributions to the art of airplane design with special reference to four-engined aircraft.

1943. William B. Bergen, Glenn L. Martin Co., for theoretical and experimental studies of dynamic loads on airplanes.

1944. William H. Phillips, NACA, for outstanding contribution in the field of stability and control of aircraft.

1945. Richard Hutton, Grumman Aircraft, for contribution to the development of carrier-based aircraft.

1946. Peter R. Murray, USAF, for radio control systems for guided missiles and pilotless aircraft.

1947. Comdr. Noel A. M. Gayler, USN, for outstanding contribution to synthetic training methods and devices in the field of aviation.

1948. Allen E. Puckett, Jet Propulsion Laboratory, Cal Tech, for contributions to the design and development of supersonic wind tunnels.

1949. Alexander H. Flax, Cornell Aeronautical Laboratory, for significant additions to the methods available for determining dynamic behavior of airplanes, helicopters, and missiles.

1950. Frank N. Piasecki, for his contributions to the design and development of helicopters.

1951. Robert C. Seamans, Jr., MIT, for system of determining dynamical characteristics of an airplane and matching them to automatic controls by measuring the pulse excitement response in flight.

1952. Dean R. Chapman, NACA, for contributions to the basic knowledge of skin friction, base pressure, and heat transfer at supersonic speeds.

1953. Donald Coles, California Institute of Technology, for fundamental contributions to the understanding of supersonic skin friction.

1954. A. Scott Crossfield, NACA, for important contributions in aeronautical flight research, especially at transonic and supersonic speeds up to Mach 2.

1955. Giles J. Strickroth, Martin Co., for contributions to the development of guidance system for the Matador.

1956. George F. Jude, Sperry Gyroscope, for contribution to the advancement of precision automatic flight control and safe all-weather flight.

1957. Clarence A. Syvertson, NACA Ames Laboratory, for solving problems of flight at speeds 30 times that of sound.

1958. Robert G. Loewy, Vertol, for work on rotary-wing aircraft.

1959. James E. McCune, Aeronautical Research Association of Princeton, for research contributions.

1960. Robert B. Howell, Lockheed Aircraft, for contributions to development of advanced theoretical guidance and control techniques.

## 44. SPIRIT OF ST. LOUIS MEDAL

Established in 1929 by citizens of St. Louis in honor of Charles A. Lindbergh, this award is given at approximately three-year periods "for meritorious service in the advancement of aeronautics." It is administered by the American Society of Mechanical Engineers.

1929. Daniel Guggenheim.

1932. P. W. Litchfield.

1935. Will Rogers.

1938. Maj. James H. Doolittle.

1940. John Elliott Younger, for contribution to airplane design.

1944. George W. Lewis, for encouragement of aeronautical research.

1947. John Knudsen Northrop, for development of flying wing.

1950. Reinout Pieter Kroon, Westinghouse Electric, for leadership in development of aviation gas turbines.

1953. No award.

1954. Arthur E. Raymond, Douglas Aircraft.

1955. Ralph S. Damon, Trans World Airlines.

## 45. THOMAS L. THURLOW NAVIGATION AWARD

Established in 1945 in honor of Colonel Thurlow to stimulate the development of the science of navigation in the United States. It is awarded "for the outstanding contribution to the science of navigation in the year." It is administered by the Institute of Navigation, University of California.

1945. Wing Commander K. C. Maclure, RCAF.

1946. Dr. John C. Bellamy, USAF.

1947. J. A. Pierce, Harvard University.

1948. Dr. D. H. Sadler, H.M. Nautical Almanac Office, England.

1949. Comdr. T. D. Davies, USN.

1950. Dr. E. G. Bowen, Institute of Navigation, Australia.

1951. Squadron Leader K. R. Greenaway, RCAF.

1952. Alton B. Moody, USN.

1953. Capt. F. J. Wylie, R.N., England.

1954. Paul D. Schrock, USN.

1955. Capt. Philip Van Dorn Weems, USN (Ret.).

1956. Dr. Charles S. Draper, Massachusetts Institute of Technology.

1957. Dr. Samuel M. Burka.

## 46. ARNOLD D. TUTTLE MEMORIAL AWARD

Established in 1952 in memory of Colonel Tuttle, USA MC, pioneer airline medical director, this award is given "for outstanding achievement in aviation medical research which was reported and published in the *Journal of Aviation Medicine* during the previous 2 years." It is administered by the Aerospace Medical Association.

1952. Edward H. Lambert, Rochester, Minn.

1953. James P. Henry, Wright-Patterson AFB, Ohio.

1954. John P. Marbarger, University of Illinois.

1955. Fred A. Hitchcock, Ohio State University.

1956. W. H. Johnson, Toronto, Canada.

1957. Maj. David S. Simons, USAF.

1958. Siegfried J. Gerathewohl, USAF School of Aviation Medicine.

1959. Lawrence E. Lamb, USAF School of Aviation Medicine.

1960. Hermann J. Schaefer, USN School of Aviation Medicine.

## 47. WRIGHT BROTHERS MEDAL (SAE)

Established in 1924, award is presented annually "for meritorious contribution to aeronautic engineering; that is, authorship of the best paper on aerodynamics or structural theory or research, or airplane design or construction, which shall have been presented at a meeting of the SAE, or any of its sections, during the calendar year." It is administered by the Society of Automotive Engineers.

1928. Lt. Comdr. Clinton H. Havill, USN, "Aircraft Propellers."

1929. Ralph Hazlett Upson, "Wings—A coordinated System of Basic Design."

1930. Theodore P. Wright, "Development of a Safe Airplane—the Curtiss Tanager."

1931. Stephen J. Zand, "A Study of Airplane and Instrument Board Vibration."

1932. Edward P. Warner, "The Rational Specifications of Airplane Load Factor."

1933. Eastman N. Jacobs, "The Aerodynamics of Wing Sections for Airplanes."

1934. Rex B. Beisel, A. L. MacClain, and F. M. Thomas, "Cowling and Cooling of Radial Air-Cooled Aircraft Engines."

1935. William Littlewood, "Operating Requirements for Transport Airplanes."

1936. R. J. Minshall, J. K. Ball, and F. P. Laudan, "Problems in the Design and Construction of Large Aircraft."

1937. Richard V. Rhode, "Gust-Loads on Airplanes."

1938. No award.

1939. Kenneth Alan Browne, "Dynamic Suspension—A Method of Aircraft Engine Mounting."

1940. Clarence L. Johnson, "Rudder Control Problems on Four-Engined Airplanes."

1941. Samuel J. Loring, "General Approach to the Flutter Problem."

1942. Charles Romain Strang, "Progress in Structural Design Through Strain-Gage Technique."

1943. Costas Ernest Pappas, "The Determination of Fuselage Moments."

1944. Kenneth Campbell, "Engine Cooling Fan Theory and Practice."

1945. Myron Tribus, "Report on Development and Application of Heated Wings."

1946. Frederick Van Horne Judd, "A Systematic Approach to the Aerodynamic Design of Radial Engine Installations."

1947. Henry B. Gibbons, "Experience of an Aircraft Manufacturer with Sandwich Material."

1948. Kermit Van Every, "The Aerodynamics of High Speed Airplanes."

1949. Homer J. Wood and Frederick Dallenback, "Auxiliary Gas Turbines for Pneumatic Power in Aircraft Applications."

1950. J. C. Floyd, "The Avro C 102 Jetliner."

1951. Orville A. Wheelon, "Design Methods and Manufacturing Techniques With Titanium."

1952. W. J. Kunz, Jr., "A New Technique for Investigating Jet Engine Compressor Stall and Other Transient Characteristics."

1953. Donald N. Meyers and Z. M. Ciolkosz, "Matching the Characteristics of Helicopters and Shaft-Turbines."

1954. John M. Tyler and E. C. Perry, Jr., "Jet Noise."

1955. Wendell E. Reed, "A New Approach to Turbojet and Ramjet Engine Controls."

1956. C. H. Zimmerman, "Some General Considerations Concerning VTOL Aircraft."

1957. Alf Fridtjof Ensrud, "Problems in the Application of High Strength Steel Alloys in the Design of Supersonic Aircraft."

1958. Kermit E. Van Every, "Design Problems of Very High Speed Flight."

1959. Milford Guy Childers, "Preliminary Design Considerations for the Structure of a Trisonic Transport."

## 48. WRIGHT BROTHERS LECTURERS (IAS)

Delivered annually in Washington, D.C., in commemoration of the first powered flights, on December 17, the Wright Brothers Lecturer is one of the highest honors of the Institute of the Aerospace Sciences. Selection of the lecturer is by a special committee, who alternate selection between the United States and abroad, of a person of "great distinction in the aerospace sciences." It is administered by the Institute of the Aerospace Sciences.

1937. B. Melvill Jones.

1938. Hugh L. Dryden.

1939. Clark B. Millikan.

1940. Sverre Pettersen.

1941. Richard V. Southwell.

1942. Edmund T. Allen.

1943. W. S. Farren.

1944. John Stack.

1945. H. Roxbee Cox.

1946. Theodore von Kármán.

1947. Sydney S. Goldstein.

1948. Abe Silverstein.

1949. A. E. Russell.

1950. William Bollay.

1951. P. B. Walker.

1952. William Littlewood.

1953. Glenn L. Martin.
1954. B. K. O. Lundberg.
1955. R. L. Bisplinghoff.
1956. Arnold Hall.
1957. H. Julian Allen.
1958. Maurice Roy.
1959. Alexander H. Flax.
1960. A. W. Quick.

## 49. THE WRIGHT BROTHERS MEMORIAL TROPHY (NAA)

The Wright Brothers Memorial Trophy is awarded each year to a living individual for "significant public service as a civilian of enduring value to aviation in the United States." It is administered by the National Aeronautic Association.

1948. Dr. William F. Durand.
1949. Charles A. Lindbergh.
1950. Grover Loening.
1951. Dr. Jerome Hunsaker.
1952. Lt. Gen. James H. Doolittle.
1953. Hon. Carl Hinshaw.
1954. Dr. Theodore Von Kármán.
1955. Dr. Hugh L. Dryden.
1956. Dr. Edward P. Warner.
1957. Senator Stuart Symington.
1958. Dr. John Francis Victory.
1959. William P. MacCracken, Jr.
1960. Frederick C. Crawford.

## 50. JAMES H. WYLD MEMORIAL AWARD (ARS)

The James H. Wyld Memorial Award is awarded annually to a person making an outstanding application of rocket propulsion. It is administered by the American Rocket Society.

1954. Milton W. Rosen.
1955. John P. Stapp.
1956. Louis G. Dunn.
1957. William H. Pickering.
1958. Holger W. Toftoy.
1959. Karel J. Bossart.
1960. Robert L. Johnson.

## 51. MISCELLANEOUS NASA AWARD WINNERS

The following well-known general awards have been accorded as indicated to NACA or NASA personnel. Such a listing properly belongs in a chronology.

ARTHUR S. FLEMMING AWARD:

1954. George E. Cooper (Ames).

1955. William M. Kauffman (Ames).

1956. Richard T. Whitcomb (Langley).

1957. Dr. Alfred J. Eggers, Jr. (Ames).

1959. Maxime A. Faget (Langley).

1960. Wolfgang E. Moekel (Lewis).
Joseph W. Siry (Goddard).

ROCKEFELLER PUBLIC SERVICE AWARD:

1953. Mrs. Dorothy Martin Simon (Lewis).

1954. Clinton E. Brown (Langley).

1955. Walter G. Vincentin (Ames).

1956. John C. Houbolt (Langley).

1957. Gerald Morrell (Lewis).

1958. William R. Mickelsen (Lewis).

1959. Dean R. Chapman (Ames).

NATIONAL CIVIL SERVICE LEAGUE AWARD:

1958. Dr. Hugh L. Dryden.

1959. Eugene S. Love (Langley).

APPENDIX E

# The Members of the National Advisory Committee for Aeronautics

## 1915–1958

THE LIST on the following pages has never been presented in the available historical record. As such it provides a useful index of persons prominent in the technical advancement of American and world aviation.

## CHAIRMEN OF THE NACA

Scriven, Brig. Gen. George P., USA, Chief Signal Officer . 1915–16
Durand, Dr. William F., Stanford University . . . . . 1916–18
Freeman, John R., Providence, Rhode Island . . . . . 1918–19
Walcott, Dr. Charles D., Secretary, Smithsonian Institution . . . . . . . . . . . . . . . . . . . . . 1919–27
Ames, Dr. Joseph S., Johns Hopkins University . . . . 1927–39
Bush, Dr. Vannevar, President, Carnegie Institution of Washington . . . . . . . . . . . . . . . . . . 1939–41
Hunsaker, Dr. Jerome C., Massachusetts Institute of Technology . . . . . . . . . . . . . . . . . . . . 1941–56
Doolittle, Dr. James H., Shell Oil Co . . . . . . . . 1956–58

## MEMBERS OF THE NACA

Abbott, Dr. Charles G., Secretary Smithsonian Institution . 1928–45
Adams, Hon. Joseph P., Civil Aeronautics Board . . . . 1952–56
Alison, John R., Assistant Secretary of Commerce . . . . 1947–49
Ames, Dr. Joseph S., Johns Hopkins University . . . . 1915–39
Arnold, Henry H., General of the Air Force . . . . . . 1938–46
Astin, Dr. Allen V., Director, National Bureau of Standards . . . . . . . . . . . . . . . . . . . . . 1952–58

Bane, Col. Thurman H., USA . . . . . . . . . . . 1919–22
Bassett, Preston, R., Sperry Gyroscope Co., Inc. . . . . 1953–58
Brett, Lt. Gen. George H., USAF . . . . . . . . . 1939–42
Briggs, Dr. Lyman J., Director, Bureau of Standards . . 1933–45
Bristol, Capt. Mark L., USN, Director Naval Aeronautics . 1915–16
Bronk, Dr. Detlev W., Rockefeller Foundation for Medical Research . . . . . . . . . . . . . . . . . . 1948–58
Burden, Dr. William A. M., Assistant Secretary of Commerce . . . . . . . . . . . . . . . . . . . . 1942–47
Burgess, George K., Director, Bureau of Standards . . . 1923–32
Bush, Dr. Vannevar, President, Carnegie Institution of Washington . . . . . . . . . . . . . . . . . . 1938–48

Carmichael, Dr. Leonard, Secretary, Smithsonian Institution . . . . . . . . . . . . . . . . . . . . . 1953–58
Cassady, Vice Adm. John H., USN, Deputy Chief of Naval Operations (Air) . . . . . . . . . . . . . . . . 1950–52
Clark, Col. Virginius, USA . . . . . . . . . . . . 1917–18
Combs, Vice Adm. Thomas S., USN, Deputy Chief of Naval Operations (Air) . . . . . . . . . 1952–53; 1955–56

Compton, Dr. Karl T., Research and Development Board . 1948–49
Condon, Dr. Edward U., Director, Bureau of Standards . . 1945–51
Connolly, Donald H. (retired Maj. Gen., USA), Administrator of Civil Aeronautics . . . . . . . . . . . 1940–42
Cook, Rear Adm. Arthur B., USN, Chief, BuAer . 1931–34; 1936–39
Craigie, Lt. Gen. Lawrence C., USAF . . . . . . . . 1951–54
Craven, Capt. Thomas, USN, Director of Naval Aviation . 1919–21
Crawford, Dr. Frederick C., Thompson Products, Inc. . . 1954–58
Curry, Maj. Gen. John F., USAF . . . . . . . . . 1924–26

Damon, Ralph S., Trans World Airlines, Inc. . . . . . 1953–56
Davis, Thomas W. S., Assistant Secretary of Commerce . . 1950–53
Davis, Vice Adm. William V., Jr., USN, Deputy Chief of Naval Operations (Air) . . . . . . . . . . . 1956–58
Doherty, Robert E., Carnegie Institute of Technology . . 1940–41
Doolittle, Dr. James H., Shell Oil Co. . . . . . . . . 1948–58
Duncan, Vice Adm. Donald B., USN, DCNO (Air) . . . 1947–48
Durand, Dr. William F., Stanford University . . 1915–33; 1941–45

Echols, Maj. Gen. Oliver P., USAF . . . . . . . . . 1942–45

Fagg, Dr. Fred D., Jr., Director, Bureau of Air Commerce . 1937–38
Fechet, Maj. Gen. James E., USA, Chief of Air Service . 1928–31
Fitch, Vice Adm. Aubrey W., USN, DCNO (Air) . . . 1944–45
Foote, Hon. Paul D., Assistant Secretary of Defense (Research and Engineering) . . . . . . . . . . . 1957–58
Foulois, Maj. Gen. Benjamin D., USAF, Chief of Air Corps . . . . . . . . . . . . 1929–30; 1932–36
Freeman, John R., Providence, R.I. . . . . . . . . . 1918–19
Furnas, Hon. Clifford C., Assistant Secretary of Defense (Research and Development) . . . . . . . . . . 1956–57

Gardner, Vice Adm. Matthias B., USN, DCNO (Air) . . 1952–53
Gilmore, Brig. Gen. William E., USA . . . . . . . . 1926–29
Gregg, Willis R., Chief, Weather Bureau . . . . . . . 1934–38
Guggenheim, Harry F., Long Island, N.Y. . . . . . . 1929–38

Harrison, Rear Adm. Lloyd, USN, Deputy and Assistant Chief of Bureau of Aeronautics . . . . . . . . . 1953–55
Hayford, John F., Northwestern University . . . . . . 1915–23
Hazen, Ronald M., Allison Division, General Motors . . . 1946–54
Hester, Clinton M., Administrator, Civil Aeronautics Authority . . . . . . . . . . . . . . . . 1938–40
Hinckley, Robert H., Assistant Secretary of Commerce . . 1939–42
Hines, Rear Adm. Wellington T., USN, Assistant Chief for Procurement, Bureau of Aeronautics . . . . . . . . 1957–58

Hunsaker, Dr. Jerome C., Massachusetts Institute of Technology . . . . . . . . . . . . . . . . . 1922–23; 1938–58
Kenly, Maj. Gen. William L., USA, Director of Military Aeronautics . . . . . . . . . . . . . . . . . . 1918–19
King, Rear Adm. Ernest J., USN, Chief, BuAer . . . . 1933–36
Kinler, Walter G. (retired Brig. Gen., USA) . . . . . 1939–40
Kraus, Rear Adm. Sydney M., USN, BuAer . . . . . . 1936–43

Land, Capt. Emory S., USN, BuAer . . . . . . . . 1923–29
Lindbergh, Brig. Gen. Charles A., USAFR . . . . . . 1931–39
Littlewood, William, American Airlines . . . . . . . 1944–53
Lonnquest, Rear Adm. Theodore C., USN, BuAer . . . 1947–52

MacCracken, William P., Jr., Assistant Secretary of Commerce . . . . . . . . . . . . . . . . . . . . . 1929–38
Marvin, Charles F., Chief, Weather Bureau . . . . . . 1915–34
McCain, Vice Adm. John S., USN, DCNO (Air) . . . . 1942–44
McCarthy, Charles J., Chance Vought Aircraft, Inc . . 1957–58
McIntosh, Col. Lawrence W., USAF . . . . . . . . 1923–24
Mead, George J., Hartford, Conn . . . . . . . . . . 1939–43
Menoher, Maj. Gen. Charles T., USA, Chief of Air Service . 1919–21
Mitscher, Vice Adm Marc A., USN, DCNO (Air) . . . 1945–46
Moffett, Rear Adm. William A., USN, Chief, Bureau of Aeronautics . . . . . . . . . . . . . . . . . . . 1921–33
Mulligan, Denis, Director, Bureau of Air Commerce . . . 1938
Murray, Robert B., Jr., Under Secretary of Commerce . . 1953–54

Newton, Byron R., Assistant Secretary of the Treasury . . 1915–18
Noble, Edward J., Chairman, Civil Aeronautics Authority . 1938–39
Nyrop, Donald W., Chairman, Civil Aeronautics Board . . 1951–52

Ofstie, Vice Adm. Ralph A., USN, DCNO (Air) . . . . 1953–54

Pace, Rear Adm. Ernest M., Jr., USN, BuAer . . . . . 1943–44
Patrick, Maj. Gen. Mason M., USA, Chief of Air Service . 1921–27
Pfingstag, Rear Adm. Carl J., USN, Assistant, Chief for Field Activities, Bureau of Aeronautics . . . . . . . 1955–57
Powers, Maj. Gen. Edward M., USAF . . . . . . . . 1945–49
Pratt, Maj. Gen. Henry C., USAF . . . . . . . . . 1930–35
Price, Vice Adm. John D., USN, DCNO (Air) . . . . . 1948–50
Pupin, Michael I., Columbia University . . . . . . . 1915–22
Putt, Lt. Gen. Donald L., USAF, Deputy Chief of Staff, Development . . . . . . . . . . . . . . . . . . 1949–58
Pyle, Hon. James T., Administrator of Civil Aeronautics . 1957–58

Quarles, Hon. Donald A., Assistant Secretary of Defense . 1954–56

Radford, Vice Adm. Arthur W., USN, DCNO (Air) . . . 1946–47

Raymond, Dr. Arthur E., Douglas, Aircraft Co, Inc . . . 1946–56
Reber, Lt. Col. Samuel, USA, In Charge Aviation Section
  Signal Corps. . . . . . . . . . . . . . . . . . 1915–16
Reichelderfer, Dr. Francis W., Chief, U.S. Weather Bureau . 1939–58
Rentzel, Delos W., Administrator of Civil Aeronautics;
  Under Secretary of Commerce . . . . . . . . . . 1948–51
Richardson, Holden C., USN, Naval Constructor . . . . 1915–17
Richardson, Rear Adm. Lawrence B., USN, BuAer . . . 1944–46
Rickenbacker, Capt. Edward V., Eastern Air Lines, Inc . . 1956–58
Robins, Brig. Gen. Augustine W., USA . . . . . . . 1935–39
Rothschild, Hon. Louis S., Under Secretary of Commerce
  for Transportation . . . . . . . . . . . . . . 1955–58
Ryan, Oswald, Civil Aeronautics Board . . . . . . .   1954

Sabine, Wallace C., Bureau of Aircraft Production . . .   1918
Saville, Maj. Gen. Gordon P., USAF . . . . . . . . 1950–51
Scriven, Brig. Gen. George P., USA, Chief Signal Officer . 1915–17
Spaatz, Gen. Carl, Chief of Staff, USAF . . . . . . . 1946–48
Squier, Maj. Gen. George O., USA, Chief Signal Officer . . 1916–18
Stevens, Rear Adm. Leslie C., USN, Assistant BuAer . . . 1946–47
Stratton, Samuel W., Director, Bureau of Standards . . . 1915–31

Taylor, Rear Adm. D. W., USN, Chief Naval Construction
  (civilian member from 1922) . . . . . . . . . . 1917–38
Towers, Rear Adm. John H., USN, Assistant and Chief
  BuAer . . . . . . . . . . . . . 1917–19; 1929–31; 1939–42
Twining, Gen. Nathan F., Chief of Staff, USAF . . . . 1954–57

Vandenberg, Gen. Hoyt S., Chief of Staff, USAF . . . . 1948–50
Vidal, Eugene L., Director, Bureau of Air Commerce . . . 1933–37

Walcott, Dr. Charles D., Secretary, Smithsonian Institution . . . . . . . . . . . . . . . . . . . . . 1915–27
Warner, Dr. Edward, private life and later Civil Aeronautics Board . . . . . . . . . . . . . . . . . 1929–45
Webster, William, Chairman, Research and Development
  Board . . . . . . . . . . . . . . . . . . . . 1950–51
Westover, Maj. Gen. Oscar, USA, Chief of Air Corps . . 1936–38
Wetmore, Dr. Alexander, Secretary, Smithsonian Institution . . . . . . . . . . . . . . . . . . . . . 1945–52
Weyerbacher, Cdr. Ralph D., USN, BuAer . . . . . . 1934–36
White, Gen. Thomas D., Chief of Staff, USAF . . . . . 1957–58
Whitman, Walter G., Chairman, Research and Development
  Board . . . . . . . . . . . . . . . . . . . . 1951–53
Wright, Dr. Orville, Dayton, Ohio . . . . . . . . . 1920–48
Wright, Dr. Theodore P., Director of Aircraft Production;
  Administrator of Civil Aeronautics; Cornell University . 1942–53

# SELECT BIBLIOGRAPHY*

Adams, Carsbie C., *Space Flight*. New York: McGraw-Hill, 1958.

Adams, Frank D., *N.A.S.A.: Aeronautical Dictionary*. Washington: GPO, 1959 (Bibliography, pp. 186–199).

Aerospace [Aviation] Industries Association, *The Aerospace [Aviation] Year Book*. Washington: American Aviation Publications [sic], 1920–date.

*Aircraft Year Book, 1919–1959*. New York: Manufacturers Aircraft Association [sic], 1919–1959.

Air Force, The Editors of, *Space Weapons*. New York: F. A. Praeger, 1959.

Akens, David S., *Historical Origins of the Marshall Space Flight Center*. Huntsville, Ala.: NASA MSFC, 1960.

Ambrose, Mary S., *The National Space Program*. Washington, D.C.: American University M.A. Thesis, 1961.

American Astronautical Society, *Advances in Astronautical Sciences*, 4 vols. of proceedings for 1956. New York: Plenum Press, 1958.

Angell, Joseph W., "Guided Missiles Could Have Won," *Atlantic Monthly*, December 1951, pp. 29–34; January 1952, pp. 57–63.

Arnold, Henry H., *Global Mission*. New York: Harper & Bros., 1949.

Arnold, Henry H., "Science and Air Power," *Air Affairs* (December 1946), pp. 184–85.

Baxter, James P., III, *Scientists Against Time* (OSRD). Boston: Little, Brown, 1946.

Benson, O. O., and Hubertus Strughold (eds.), *Physics and Medicine of the Atmosphere and Space*. New York: Wiley, 1960.

Bergaust, Erik, *Reaching for the Stars*. New York: Doubleday, 1961.

Berger, Carl, *History of the 1st Missile Division*. Vandenberg AFB, California: Strategic Air Command, 1960.

*This list of works examined includes those mainly of historical value for the general reader rather than the specialist. The full history of science and technology has not received its share of serious attention in general, and the fields of aeronautics and astronautics in particular.

Bergman, Jules, *Ninety Seconds to Space: The Story of the X–15*. New York: Hanover House, 1960.

Berkner, Lloyd V., and Hugh Odishaw (eds.), *Science and Space*. New York: McGraw-Hill, 1961.

Bernardo, James V., *Aviation in the Modern World*. New York: E. P. Dutton, 1960.

Blumenstock, David I., *The Ocean of Air*. New Brunswick: Rutgers University Press, 1959.

Bonney, Walter T., "The Research Airplane," *Pegasus* (June 1952), pp. 1–16.

Bonney, Walter T., "Prelude to Kitty Hawk," *Pegasus* (September 1953), pp. 3–32.

Bonney, Walter T., "The Heritage of Kitty Hawk," *Pegasus* (April 1955), pp. 7–11; (July 1955), pp. 6–11; (August 1955), pp. 5–13; (December 1955), pp. 10–15; (March 1956), pp. 7–15; (May 1956), pp. 10–15; (August 1956), pp. 10–15; (September 1956), pp. 10–15; (November 1956), pp. 10–15; (February 1957), pp. 1–10.

Boyce, Joseph C. (ed.), *New Weapons for Air Warfare* (OSRD). Boston: Little, Brown, 1947.

Boyd, R. L. F., and M. E. Seaton (eds.), *Rocket Exploration of the Upper Atmosphere*. London: Pergamon Press, 1954.

Brown, Willis C., *Selected Bibliography of Books About Jets, Rockets, and Space Exploration, 1958–60*. Washington: DHEW, 1960.

Burchard, John E. (ed.), *Rockets, Guns, and Targets* (OSRD). Boston: Little, Brown, 1949.

Bush, Vannevar, *Modern Arms and Free Men*. New York: Simon & Schuster, 1949.

Bushnell, David, *Contributions of Balloon Operations to Research and Development, 1947–1958*. Holloman AFB, New Mexico: AFMDC, 1959.

Bushnell, David, *History of Tracks and Track Testing at the AFMDC, 1949–1960*, 2 volumes. Holloman AFB, New Mexico: AFMDC, 1959, 1960.

Bushnell, David, *History of Research in Subgravity and Zero-G at AFMDC, 1948-1958*. Holloman AFB, New Mexico: AFMDC, 1958.

Caidin, Martin, *Vanguard*. New York: E. P. Dutton, 1957.

Caidin, Martin, *The Astronauts: The Story of Project Mercury*. New York: Dutton, 1961.

Chapman, John L., *Atlas: The Story of a Missile*. New York: Harper & Bros., 1960.

Clarke, Arthur C., *The Exploration of Space*. New York: Harper, 1959.

Cooke, David C., *Flights That Made History*. New York: Putnam's, 1960.

Crew, Henry, *Biographical Memoir of Joseph Sweetman Ames*. Washington: National Academy of Sciences, 1944.

Crossfield, A. Scott, and Clay Blair, *Always Another Dawn*. Cleveland: World, 1961.

Davy, M. J. B., *Interpretive History of Flight*. London: H.M.S.O., 1948.

Department of Defense, *Fact Sheet: Guided Missiles and Rockets*. Washington: Office of Public Affairs, 1960.

Doolittle, James H., "N.A.C.A.—The Following Years, 1955-1958," in *Annual Report of the NACA—1958*. Washington: GPO, 1959, pp. 29-31.

Doolittle, James H., "Science: Key to Supremacy," *Air Force* (June 1953), pp. 39-40.

Dornberger, Walter, *V-2*. New York: Viking Press, 1954.

Dryden, Hugh L., "The Aeronautical Research Scene—Goals, Methods and Accomplishments," *Journal of the Royal Aeronautical Society* (July 1949), pp. 623-666.

Dryden, Hugh L., "Fact Finding for Tomorrow's Planes," *National Geographic* (December 1953), pp. 757-780.

Dryden, Hugh L., "The Next Fifty Years," *Aero Digest* (July 1953), pp. 132 f.

Dryden, Hugh L., "A Half Century of Aeronautical Research," *American Physical Science Proceedings* (April 15, 1954), pp. 115-120.

Dryden, Hugh L., "IGY—Man's Most Ambitious Study of His Environment," *National Geographic* (February 1956), pp. 285-298.

Dryden, Hugh L., "Contribution of William F. Durand to Aeronautics," address at Stanford University, Stanford, California, Aug. 5, 1959.

Dryden, Hugh L., "Prospects for Space Travel," Penrose Lecture before the American Philosophical Society, Philadelphia, Pa., Apr. 21, 1960.

Dryden, Hugh L., "Future Exploration and Utilization of Outer Space," *Technology and Culture* (Spring 1961), pp. 112-126.

DuBridge, Lee A., "Sense and Nonsense About Space," *Harpers* (August 1959), pp. 21-28.

DuBridge, Lee A., *Introduction to Space*. New York: Columbia Univ. Press, 1960.

Dupree, A. Hunter, *Science in the Federal Government*. Cambridge: Harvard University Press, 1957.

Ehricke, Krafft, *Space Flight*. Princeton: Van Nostrand, 1960.

Emme, E. M., "Scientific Research and and Development," in *National Air Power and International Politics—A Select Bibliography*. Montgomery: Air University Documentary Research Study, 1950, pp. 87-96.

Emme, E. M., *Hitler's Blitzbomber: Historical Notes on High Command Decisions Influencing the Tardy Operational Use of the Me-262 in German Air Defense*. Montgomery: Air University, 1951.

Emme, E. M., *The Impact of Air Power: National Security and World Politics*. Princeton: D. Van Nostrand, 1959.

Emme, E. M., *Robert H. Goddard: World Rocket Pioneer*. Washington: NASA Historical Report, July 1960, reprinted in *The Airpower Historian* (October 1960), pp. 216-221 (Bibliography, p. 221).

Everest, Frank K., *Fastest Man Alive*. New York: Dutton, 1958.

Fisher, Allan C., "Exploring Tomorrow With the Space Agency," *National Geographic* (July 1960), pp. 48-89.

"For the Record," *Astronautics* (ARS), 1955–date.

Fortune, The Editors of, *The Mighty Force of Research*. New York: McGraw-Hill, 1956.

Franklin Institute, *Earth Satellites and Research Vehicles*, Monograph No. 2, Philadelphia: Franklin Institute, 1956.

Galland, Adolf, "The Development of Jet and Rocket Airplanes in Germany, 1938-1945," in *Development and Planning in the German Air Force*, Trans. A. Herling. Maxwell AFB, Alabama: Air University Library MS, 1951.

Gantz, Kenneth F. (ed.), *USAF Report on the Ballistic Missile*, New York: Doubleday, 1958.

Gantz, Kenneth F. (ed.), *Man In Space: the USAF Program for Developing the Spacecraft Crews.* New York: Duell, Sloan & Pearce, 1959.

Gantz, Kenneth F. (ed.), *Nuclear Flight.* New York: Duell, Sloan & Pearce, 1960.

Gartmann, Heinz, *The Men Behind the Space Rockets.* New York: D. McKay Co., 1956.

Gartmann, Heinz, *Man Unlimited: Technology's Challenge to Human Endurance.* New York: Pantheon, 1956.

Gatland, Kenneth W. (ed.), *Project Satellite.* New York: British Book Centre, 1958. Ch. I, "From Small Beginnings," by Wernher von Braun, pp. 19–49.

Gavin, James M., *War and Peace in the Space Age.* New York: Harpers, 1958.

Gibbs-Smith, Charles H., *The Aeroplane: An Historical Survey of Its Origins and Development.* London: H.M.S.O., 1960.

Glennan, T. Keith, "What's Next in Space," *U.S. News & World Report* (Feb. 6, 1959), pp. 102–103.

Glennan, T. Keith, "Our National Space Program," address before the Institute of World Affairs, Pasadena, Calif., Dec. 7, 1959.

Glennan, T. Keith, "A New Order of Technological Challenge in the Nation's Space Program" address at A.A.A.S., Chicago, Ill., Dec. 27, 1959.

Glennan, T. Keith, "Opportunities for International Cooperation in Space Exploration," *U.S. Department of State Bulletin* (Jan. 11, 1960), pp. 58–63.

Glennan, T. Keith, "Space Exploration," *Vital Speeches* (Apr. 16, 1960), pp. 401–404.

Goldberg, Alfred (ed.), *A History of the U.S. Air Force, 1907–1957.* Princeton: Van Nostrand, 1959.

Goddard, Robert H., *Rockets,* New York: American Rocket Society, 1946. Facsimile reprint of his 1919 and 1935 Smithsonian Reports.

Goddard, Robert H., *Rocket Development: Liquid Fuel Rocket Research, 1921–41,* ed. E. C. Goddard and G. Edward Pendray, New York: Prentice-Hall, 1948.

Gray, Charles G., *The History of Combat Airplanes.* Northfield: Norwich University, 1941.

Gray, George W., *Frontiers of Flight: The Story of NACA Research.* New York: A. Knopf, 1948.

Green, Murray, "Fifty Years of Airpower, 1907–1957," *Air Force* (August 1958), pp. 193–213.

Griffith, Elisabeth A., *The Genesis of the National Aeronautics and Space Act of 1958—A Legislative History.* New York: Columbia University M.A. Thesis, 1960.

Haber, Heinz, *Man in Space.* New York: Bobbs-Merrill, 1953.

Haggerty, James L., *First of the Spacemen: Iven C. Kincheloe.* New York: Duell, Sloan & Pearce, 1960.

Haley, Andrew G., *Rocketry and Space Exploration.* Princeton: D. Van Nostrand, 1959.

Hanrahan, James S., and David Bushnell, *Space Biology: The Human Factors in Space Flight.* New York: Basic Books, 1960.

Heinkel, Ernst, *Stormy Life,* trans. from German. New York: Dutton, 1956.

Holland, Maurice, and Thomas M. Smith, *Architects of Aviation.* New York: Duell, Sloan & Pearce, 1951.

Holley, I. B., Jr., *Ideas and Weapons: Exploitation of the Aerial Weapon by the U.S. During World War I.* New Haven: Yale Univ. Press, 1953.

Hubler, Richard G., *Big Eight: A Biography of an Airplane* (DC–8). New York: Duell, Sloan & Pearce, 1961.

Hunsaker, Jerome C., *Aeronautics at Mid-Century,* New Haven: Yale Univ. Press, 1952.

Hunsaker, Jerome C., "Forty Years of Aeronautical Research, 1915–1955," in *Smithsonian Report for 1955.* Washington: Smithsonian Institution, 1956, pp. 241–271.

Hyde, Margaret O., *Exploring Earth and Sky: The Story of the I.G.Y.* New York: Whittlesey House, 1958.

*Jane's All the World's Aircraft.* London: Sampson, Low, Marston, Macmillan; McGraw-Hill; 1909–date.

Jastrow, Robert (ed.), *The Exploration of Space.* New York, Macmillan, 1960.

Kármán, Theodore von, *Aerodynamics.* Ithaca: Cornell Univ., 1954.

Kármán, Theodore von, "Some Significant Developments in Aerodynamics since 1946," Daniel and Florence Guggenheim Memorial Lecture, Madrid, September 1958.

Killian, James R., Jr., and others, "Toward More Effective Research and Development," *General Electric Defense Quarterly* (October–December 1958), pp. 5–25.

Kirschner, Edwin J., *The Zeppelin in the Atomic Age: The Past, Present, and Future of the Rigid Lighter-than-Air Aircraft*. Urbana: Univ. of Illinois Press, 1959 (Bibliography, pp. 77–80).

Kolcum, Harry, "History of N.A.C.A.," *The Times-Herald* (Newport News, Va.), series of 12 articles, April–May 1958.

Kosmodemyansky, Prof., *Konstantin Tsiolkovsky: His Life and Work*, trans. X. Danko. Moscow: Foreign Languages Publishing House, 1956.

Krause, Ernest H., "High Altitude Research with V-2 Rockets," *Proceedings of the American Philosophical Society*, December 1947.

Krieger, F. J., *A Casebook on Soviet Astronautics*. Santa Monica: RAND Corp., 1957, R.M. 1760 and R.M. 1922.

Lear, John, "The Moon that Refused to be Eclipsed (Vanguard)," *Saturday Review of Literature*, Mar. 5, 1960.

Ley, Willy, *Rockets, Missiles and Space Travel*. New York: Viking Press, 1957. British Edition: London: Chapman, 1957 (Bibliography, pp. 489–520).

Loosbrock, John F., and Richard M. Skinner (eds.), *The Wild Blue*. New York: G. P. Putnam's, 1961.

Marberger, John F. (ed.), *Space Medicine*. Urbana: Univ. of Illinois, 1951.

McFarland, Marvin W. (ed.), *The Papers of Wilbur and Orville Wright*, 2 vols. New York: McGraw-Hill, 1953.

Medaris, John B., with A. Gordon, *Countdown for Decision*. New York: Putnam's, 1960.

Milbank, J., *The First Century of Flight in America*. Princeton: Princeton University, 1943.

Miller, J. A., "A History of the Development of Flight Refuelling," *Journal of the Royal Aeronautical Society* (November 1960), pp. 687–91.

Murray, Don, "O'Sullivan's Wonderful Lead Balloon," *Readers Digest* (February 1961), pp. 45–50.

Myrus, Don, *The Astronauts*. New York: Grosset & Dunlap, 1960.

National Academy of Sciences, *IGY Bulletin* (monthly), July 1957–date.

National Academy of Sciences, National Research Council, *Flight Summaries for the U.S. Rocketry Program for the IGY, 5 July 1956–31 December 1958*, two volumes. Washington: March, September 1959.

National Advisory Committee for Aeronautics, *Annual Reports, 1915–1958*. Washington: GPO, 1916–1959.

National Aeronautics and Space Administration, *Press Releases*, issued by the Office of Public Information, Washington: NASA, October 1958–date.

National Aeronautics and Space Administration, *Semiannual Reports to Congress, I, October 1, 1958–March 31, 1959; II, April 1, 1959–September 30, 1959; III, October 1, 1959–April 1, 1960; IV, April 1, 1960–September 30, 1960*. Washington: NASA, 1959–60.

National Aeronautics and Space Administration, *Space Activities Summary*, prepared by the Office of Public Information. Washington: N.A.S.A., 1958–date.

Newell, Homer E., *High Altitude Rocket Research*. New York: Academic Press, 1953 (Bibliography, pp. 278–279).

Newell, Homer E., Jr. (ed.), *Sounding Rockets*. New York: McGraw-Hill, 1959.

Northrop, John K., *Aviation History, 1903–1960*. Washington: Library of Congress, 1948.

Oberth, Hermann, *Man Into Space*. New York: Whittlesey House, 1953.

Ogburn, William F., *The Social Effects of Aviation*. New York: Houghton Mifflin, 1946 (Bibliography, pp. 725–737).

Ogburn, William F., (ed.), *Technology and International Relations*. Chicago: Univ. of Chicago Press, 1949.

Ordway, F. I., III, and R. C. Wakeford, *International Missile and Spacecraft Guide*. New York: McGraw-Hill, 1960.

Ostrander, Don R., "The U.S. Space Exploration Program," address at Western Michigan University, June 23, 1960.

Parry, Albert, *Russia's Rockets and Missiles*. New York: Doubleday, 1960.

Payne, L. G. S., *Air Dates (1783–1956)*. London: Heinemann, 1957.

Pendray, G. Edward, *The Coming Age of Rocket Power*, New York: Harper, 1945.

Piccard, Auguste, *Between Earth and Sky*. London: Falcon Press, 1950.

Pickering, William H., "Public Understanding in the Space Age," address to California Wing of AFA, Apr. 9, 1960.

Price, Don K., *Government and Science.* New York: New York Univ. Press, 1954.

Rees, Ed, *The Manned Missile (B-70).* New York: Duell, Sloan & Pearce, 1961.

Robertson, Bruce, *Spitfire: The Story of a Famous Fighter.* Letchworth, England: Harleyford, 1960.

Rosen, Milton W., *The Viking Rocket Story,* New York: Harper & Bros., 1955.

Royal Air Force, *A Historical Summary of the Royal Aircraft Factory and its Antecedents, 1878–1918,* prepared by S. Child and C. F. Caunter. Farnborough: Royal Aircraft Establishment, 1947.

Schlaifer, Robert, and S. D. Heron, *Development of Aircraft Engines and Aviation Fuels.* Boston: Harvard Business School, 1950.

Seifert, Howard (ed.), *Space Technology.* New York: Wiley, 1959.

Shelton, W. R., *Countdown: The Story of Cape Canaveral.* Boston: Little, Brown, 1960.

Shrader, Welman A., *Fifty Years of Flight: A Chronicle of the Aviation Industry in America, 1903–1953.* Cleveland: Eaton Manufacturing Co., 1953.

Simon, Leslie: *German Research in World War II.* New York: Wiley, 1947.

Simons, David G., and Don A. Schanche, *Man High.* Garden City, N.Y.; Doubleday, 1960.

Stehling, Kurt, *Project Vanguard.* New York: Doubleday, 1961.

Stemmer, Josef, *Raketenantriebe: Ihre Entwicklung, Anwendung und Zukunft.* Zurich: Schweizer, Druck-und-Velagshaus, 1952.

Sullivan, Walter, "The International Geophysical Year," *International Conciliation* No. 521 (January 1959), pp. 301–317.

Sullivan, Walter, *Assault on the Unknown: The International Geophysical Year.* New York: McGraw-Hill, 1961.

Tacker, L. J., *Flying Saucers and the U.S. Air Force.* Princeton: Van Nostrand, 1960.

Thetford, Owen, *Aircraft of the Royal Air Force, 1918–1957.* London: Putnam, 1957.

Thomas, Shirley, *Men of Space,* Vol. I. Philadelphia: Chilton, 1960. First of 12 volumes, each of which will contain 10 biographical profiles.

Turnbull, A. D., and C. L. Lord, *History of U.S. Naval Aviation.* New Haven: Yale University Press, 1949.

U.S. *Aeronautics and Space Activities, January 1 to December 31, 1958,* with transmittal message from the President to the Congress. Washington: GPO (H. Doc. 71), 1959.

U.S. *Aeronautics and Space Activities, January 1 to December 31, 1959,* with transmittal message from the President to the Congress. Washington, GPO (H. Doc. 349), 1960.

U.S. Air Force, *A Chronology of American Aerospace Events, 1903–1959* (AFP 190-2-2). Washington: USAF, 1960.

U.S. Air Force, Air Research and Development Command, *Experimental Research Aircraft.* Edwards AFB, Calif., AFFTC, 1960.

U.S. Army, Ordnance, *Hermes Guided Missile Research and Development Project, 1944–1954.* Washington: USA, 1959.

U.S. Army Air Forces, *Toward New Horizons,* series of 14 reports prepared by the AAF Scientific Advisory Board, Theodore von Kármán, Chairman. Washington: 1945.

U.S. Congress, House, Committee on Appropriations, Subcommittee on Independent Offices, *Hearings . . . National Aeronautics and Space Administration Appropriations.* Washington: GPO, 1959.

U.S. Congress, House, Committee on Appropriations, Subcommittee on Independent Offices, *Hearings . . . Independent Offices Appropriation for 1961,* Part 3, *NASA, National Science Foundation.* Washington: GPO, 1960.

U.S. Congress, House, Committee on Science and Astronautics, *Hearings . . . Part 1, No. 3, Review of the Space Program; No. 4, 1961 NASA Authorization; No. 5, To amend the National Aeronautics and Space Act of 1958.* Washington: GPO, 1960.

U.S. Congress, House, Committee on Science and Astronautics, *The Practical Values of Space Exploration* (H. Rept. 2091). Washington: GPO, 1960.

U.S. Congress, House, Committee on Science and Astronautics, *Hearings . . . Basic Scientific and Astronautic Research in the Department of Defense.* Washington: GPO, 1959.

U.S. Congress, House, Committee on Science and Astronautics, *Hearings . . . 1960 NASA Authorization*. Washington: GPO, 1959.

U.S. Congress, House, Committee on Science and Astronautics, *Hearings . . . Supersonic Air Transports*. Washington: GPO, 1960.

U.S. Congress, House, Committee on Science and Astronautics, *Panel on Science and Technology—First Meeting* (H. Doc. 1587). Washington: GPO, 1960.

U.S. Congress, House, Committee on Science and Astronautics, *A Chronology of Missile and Astronautic Events*, prepared by Dr. Charles S. Sheldon, II. Washington: GPO, 1961.

U.S. Congress, House, Select Committee on Astronautics and Space Exploration, *The National Space Program* (H. Rept. 1758). Washington: GPO, 1958.

U.S. Congress, House, Select Committee on Astronautics and Space Exploration, *The Next Ten Years in Space, 1959–1960* (H. Doc. 115). Washington: GPO, 1959.

U.S. Congress, Senate, Committee on Aeronautical and Space Sciences, Subcommittee of Governmental Organization for Space Activities, *Hearings . . . Investigation of Government Organization for Space Activities*. Washington: GPO, 1959.

U.S. Congress, Senate, Committee on Aeronautical and Space Sciences, *Project Mercury: Man-In-Space Program of N.A.S.A.* (Senate Report 1014). Washington: GPO, 1959.

U.S. Congress, Senate, Committee on Aeronautical and Space Sciences, *Hearings . . . NASA Authorization for Fiscal Year 1960*. Washington: GPO, 1959.

U.S. Congress, Senate, Committee on Aeronautical and Space Sciences, *Hearings . . . NASA Authorization for Fiscal Year 1961*, Parts I and II: Washington: GPO, 1960.

U.S. Congress, Senate, Committee on Armed Services, Preparedness Investigating Subcommittee, *Hearings . . . Inquiring Into Satellite and Missile Programs*, Parts I, II, III. Washington: GPO, 1958.

U.S. Congress, Senate, Committee on Armed Services, Preparedness Investigating Subcommittee, *Hearings . . . Missiles, Space, and Other Major Defense Matters*. Washington: GPO, 1960.

Van Allen, James A., *Scientific Uses of Earth Satellites*. Ann Arbor: University of Michigan Press, 1956.

Van Wyen, Adrian O., and Lee M. Pearson, *United States Naval Aviation, 1910–1960*. Washington: GPO, 1960.

Wagner, Ray, *American Combat Planes*. New York: Hanover House, 1961.

Warner, Edward P., *The Early History of Air Transportation*. Northfield, Vt.: Norwich University, 1937.

White, C. S., and O. O. Benson, *Physics and Medicine of the Upper Atmosphere*. Albuquerque: Univ. New Mexico Press, 1952.

Whitnah, Donald R., *A History of the U.S. Weather Bureau*. Urbana: University of Illinois Press, 1961.

Wilkinson, Paul H., *Aircraft Engines of the World*, issued annually, 1941–60. New York, Washington: P. Wilkinson, 1941–60.

Williams, Beryl, and Samuel Epstein, *The Rocket Pioneers on the Road to Space*. New York: Messner, 1955.

Williams, K. G., *The New Frontier: Man's Survival in the Sky*. Thomas, 1960.

Wright, Theodore P., *Aviation's Place in Civilization*. Washington: GPO, 1945.

Zaehringer, Alfred J., *Soviet Space Technology*. New York: Harpers, 1961.

# SUBJECT AND NAME INDEX

A3D, 62
A3J, 93, 133
A-4 (rocket), 43, 44
A-5, 38
A-9, 49
A-10, 49
A-20 (aircraft), 39, 44, 48, 50
AB-2, 3
Abbott, Charles G., 202
Abbott, Ira H., 50
Abdominal corset, 30
Aberdeen Proving Ground (See Army Ordnance)
Ablation, 81, 87, 130
"Able" (monkey), 109, 110
ABMA (See Army)
Acceleration, 29, 30
Accident prevention, 24, 27
   (See Safety)
Ackerman, John, 162
Acosta, B., 23
Acoustic detection, 27
AD-17, 39
Adams, C., 207
Adams, Elsie W., 172
Adams, Frank D., 207
Adams, Joseph P., 202
ADC (See U.S. Air Force)
Adelman, Barnet R., 190
*Adriatic,* 10
Advanced Research Projects Agency (DOD), 93–97, 102, 103, 113–16, 131
Advisory Committee for Aeronautics (See National Advisory Committee for Aeronautics)
Advisory Committee on Uranium, 38
A.E.F., 7, 9
Aerial mines, 41
Aerial Navigation, Commission of, 11
Aerial observation of fish, 10
Aerial photographs, 4, 18, 24
  at night, 8, 20
  of eclipse of sun, 29
Aerial reconnaissance, 36
Aerial spraying, 14
Aerial survey,
  of Alaska, 21, 25, 31
  of Antarctic, 56
  of Dominican Republic, 13
  of Florida, 22
  of Haiti, 13
Aerial targets, 34
Aerial torpedoes, 6
Aeroballistic research, 62
Aerobee rocket, 53, 54, 58, 59, 63, 67, 69, 70, 91; type 150–A, 121
Aerobee-Hi rocket, 77, 79, 82, 83, 86, 122, 123, 130, 131
Aero Commander, 128
Aerodynamic efficiency, 26
Aerodynamic research, 51
  wrong-flow method, 47
  in World War II, 52
  (See Wind tunnels)
Aerojet General Corp., 46, 51, 54, 82, 99, 121, 129
Aeromarine flying boat, 10, 12, 18

Aeromarine West Indies Airways, 12
Aero Medical Association, 25
Aeromedical Laboratory, 31, 43, 44, 45, 61, 62, 71
  Field Laboratory, 117, 134
  (See U.S. Air Force)
Aeromedicine, 7, 15
  (See SAM, Life Sciences, Aerospace Medical Assn.)
Aeronautic education, 22
Aeronautical awards, 167f.
Aeronautical Board (Army-Navy), 12
Aeronautical Engineering Society, 11
Aeronautical Laboratory, 181
Aeronautical patents (See Patents)
Aeronautical Research, 34
  Director of (NACA), 37, 58
Aeronautical Safety Conference, 24
Aeronautical sciences, 32, 49, 52
Aeronautics and Astronautics Coordinating Board, 127, 128, 130, 132
Aeronautics Patents and Design Board, 21
Aeronutronic, 122
Aerophysics Lab., 64
Aerospace Corporation, 124, 125
Aerospace Industries Association, 207
Aerospace Medical Association, 167, 182
Aerospace Medical Center, 91, 114
Aerospace Medical Laboratory, 122
Aerospace plane, 129
Aerospace technology, 133
AF (See U.S. Air Force)
AFBMD (See U.S. Air Force)
AFMTC (See U.S. Air Force)
Africano, Alfred, 41
Afterburner, development, 46, 52, 53
AGA beacon, 15
Agena, 120, 130, 133
Agriculture, Department of, 21
AH-10, 3
AH-14, 4
AI radar, 41
Ailerons, 18, 31
Air,
  viscosity of, 12
  Ocean of, 207
Air-to-air missile, 45
  (See Anti-aircraft rocket)
Airacomet (See Bell Aircraft)
Airborne television, 39
Air commerce,
  Act of 1926, 20, 21
  regulation of, 10
Air cooled aircraft engines
  with fins, 5
  Lawrence engine, 8, 11, 12, 14, 173
Air Coordination Committee, 24
Aircraft Industries Assn., 171
Aircraft,
  all-metal, 14, 17
  all-wood, 36, 40
  nuclear propulsion of, 54, 78, 92
  research and design for, 11, 34
  turbojet, propelled, 39
Aircraft accidents, 24
Aircraft Board, 7, 10, 20

213

Aircraft carrier, 25
  first, 14
  first landing on, 16
  first takeoff, 15
  first jet landing, 51, 52
  first turbojet operation, 54
  (See Navy, *Langley, Saratoga, Ranger*)
Aircraft Engine Research Lab. (See NACA, Aircraft Engine Research Lab.)
Aircraft engines
  air-cooled, 5, 8, 11, 12, 14, 173
  Allison, 27, 39
  Hispano-Suiza, 5
  nuclear, 54, 78, 92
  of the world, 212
  (See engine and propulsion types of engines)
Aircraft industry, 5,
  NACA recommendation to, 27
  (See companies by name)
Aircraft inspection,
  regulation of, 10
Aircraft Manufacturers Association, 6
Aircraft production
  British Ministry, 40
  Roosevelt's 50,000 airplanes, 46
  3-year program, 6
  U.S. in World War I, 5, 6, 7
Aircraft Production Board, 6
Aircraft records,
  altitude, 159–160
  distance, 156–58
  speed, 154–156
Aircraft Year Book, 207
Air defense, 33
Air Defense Research Committee, 33
Air Development Center, 66
Air Engineering Development Center, 54
Airflow Research Staff, 37
Airfoils,
  data on, 19, 50
  first NACA family of, 19, 30
  sections, 50
*Air Force*, Editors of, 207
Air Force Association, 184, 192, 193, 207
Air Forces (See Army Air Forces, German Air Force, RAF, RCAF)
Airframe, 41
Air-freight, 19
Airlift
  Berlin, 59
  Mecca, 70
Airliner (See aircraft type)
  Boeing 247, 30
Airlines
  increase in paid passengers, 27
  passenger carried, 134
  (See companies by name)
Air Mail, 10, 20, 31, 173
  AAC flying of, 31
  contract flying of, 20
  by rocket, 28, 33, 131
Air Mail Act of 1934, 31
Air Materiel Command, 57
Air Medical Association, 24
Air navigation, 3, 10, 14, 195
Air passenger traffic, 27, 34, 38, 105, 134
  first transatlantic, 38
  first transpacific, 34
  first jet transcontinental, 105
Air Policy Committee, 57
Air power, 208
Air Safety, 14, 27, 85, 120, 170
Air Service (See Army)
Airships, 10, 12, 14, 29, 37, 47, 69
Airship subcloud car, 27
Air surgeon, 42, 55

Air transportation, 68
  (See Airlines)
  early history, 212
  first domestic jet, 104
  first jet transcontinental, 105
  passengers, 134
Air Turbine Test Station, 67
Air-to-air radio, 7, 20
Air-to-air refueling, 14, 17, 63, 67, 72, 210
Air-to-air rockets, 42, 49
Airworthiness requirements, 27
Akens, David S., xi, 207
Akers, Frank, 31
*Akron* (ZR), 30
*Alabama*, 14
Alaska, 12, 21, 25, 28, 31, 63
Albert series, 62, 65
Alcock, John, 10
Algae, 101
Algeria, 56
Alison, John R., 202
All-inertial guidance, 93, 96
All-metal aircraft, 14, 15, 16, 17
All-wood aircraft, 36, 40
Allen, D. C., 21
Allen, Edmund T., 172, 177, 197
Allen, George V., 118
Allen, H. Julian, 69, 70, 81, 86, 93, 187, 191, 193, 198
Allen, James A., 62
Allen, Wm. M., 174
Allied Council of Ambassadors, 15
Allison engines, 27
  V-1710-33, 39
Almen, J. O., 185
Alsos Mission, 48
Altimeter calibration, 19
Altitude chamber, 7, 9, 44, 45, 84, 98, 102
Altitude flight record, 21, 25, 97, 115
  aircraft, 159
  balloon, 33, 163–65
  helicopter, 52
Alvarez, Luis W., 174
Aluminum Company of America, 4
Ambrose, Mary S., 207
American Airlines, 204
American Astronautical Society, 172, 176, 207
Amer. Assn. for Advancement of Science, 114
American Astronautical Society, 73, 74
American Bar Assn., 132
  report on the necessity of air law, 13
  recommendation for Federal legislation, 13
American Board of Preventive Medicine, 71
American Helicopter Society, 171, 181
American Institute of Biological Sciences, 98
American Interplanetary Society, 26, 27, 29, 30
  founded, 26
  renamed, 26
  experimental program of, 28
American Machine and Foundry, 99
American Medical Assn., 71
American Meteorological Society, 11, 167, 168, 169, 186
American Physical Society, 108
American Propeller Co., 15
American Rocket Society (ARS), 31, 37, 42, 91, 92, 94, 115, 167, 169, 176, 188, 198;
  Rocket tests, 30, 31
American Society of Engineers, 176
Americans, "inventive genius of," 3
Ames, Joseph S., 11, 22, 38, 182, 202, 208
Ames Aeronautical Laboratory (NACA), 37, 38, 40, 43, 59, 62, 70, 91
Ames Research Center (NASA), 102, 113, 120, 199
AMOL (See Navy)
Amphibian, 34, 37

Amundsen, Roald, 21
Anderson, Clinton P., 95
Anderson, Orvil A., 31, 33, 83, 161, 162, 179
Andrews, Frank M., 37
Angell, Joseph W., xi, 207
Annapolis (Md.), 35, 41, 42, 44, 45
Antarctic, Air survey of, 56
　expedition, 39, 87
　Treaty, 115
　(See South Pole)
Antennas, 11
Antiaircraft exercises, 36
Anti-aircraft rockets, 40, 41, 42, 49;
　British, 32, 47, 49
　Solid-propellant, 43
　(See Falcon, Genie, Sidewinder, etc.)
Antiaircraft projectile, 46
*Antietam*, 165
Antisubmarine missile, 43
Antonio, A. L., 190
Aphrodite missions, 47
Apollo program, (NASA) 93
Applied Physics Lab., 50, 52, 58, 86
Appropriations, 3, 5, 11, 27
Apt, Milburn G., 83, 125
Aptitude tests, 20
Arado 234–B, 46
Arc jets, 84, 130
Arctic expeditions, 18
　patrol, 185
ARDC (See U.S. Air Force)
Area rules, 68, 71, 174
Arends, Leslie C., 97
Argentina, 132
Argo D4, 110;
　D8, 131
Armed Forces Policy Council, 72, 83
Arms race, 128
Armstrong, H. G., 32, 37, 55, 180, 183
Armstrong, Neil, 133
Army (USA), 1, 5, 8, 43, 45, 46, 52
Army Air Corps, 21, 34–40, 46, 48, 51, 80;
　balloonists, 162
　Chief of, 136, 203–205
　Ferry Command, 41
　flying airmail, 31
　GHQ Air Force, 32, 37, 41
　Weather Service, 35
　(See School of Aviation Medicine, etc.)
Army Air Corps Act, 21
Army Air Forces, 38, 41, 42, 46;
　ATC, 41, 173
　AMC
　Chief of, 41, 46, 202, 205
　First Missile Group, 53
　and NACA, 49
　R&D policy, 51
　Scientific Advisory Board, 41, 48, 50, 51, 54, 173, 211
　(See School of Aviation Medicine, etc.)
Army Air Service, 8, 9, 10, 12, 20
　balloonist, 5, 162
　Chief of, 203, 204
　first bomber, 8–9
　Medical Research Laboratory, 9, 16
　Technical School, 20
　School of Aviation Medicine, 20, 22, 39
　pilots of, 12
Army Aviation, 171
Army Ballistic Missile Agency (ABMA), 81, 95–100, 102, 104, 105, 109–113, 115, 124
　(See Redstone Arsenal)
Army Bureau of Aeronautics, 8
Army Corps of Engineers, 72
Army Court Martial, 20

Army Ordnance, 9, 35, 46, 47, 48, 51, 55, 80, 100;
　Aberdeen Proving Ground of, 9, 49, 50, 59
　(See Corporal, Hermes, Nike, Redstone, Jupiter Loki, Wac, etc.)
Army Signal Corps, 1, 6, 8, 9, 35, 55, 106, 107, 132, 164;
　Laboratory of, 106, 128, 132, 164
　Weather Service of, 35
Arnold, Henry H., 17, 31, 36, 38, 41, 46, 48, 64, 174, 184, 202, 207
Arnold, Weld, 34
Arnold Engineering Development Center, 63, 66, 67, 96, 99, 110, 119
ARPA (See Defense Dept.)
Arrachart, Ludovic, 154
Artificial horizon, 3, 4, 10
Artillery projectile shapes, 9
Artillery proving ground, 27
　(See Aberdeen)
Ascension Island, 112
Ascani, Fred J., 184
Asp rocket, 80, 82, 102, 111, 112, 115
Assault drones, 42
Astin, Allen V., 202
Astronautics:
　ARS Magazine, 208
　Awards, 167f.
　early mathematical work on, 27
　French committee on, 23
　science of, 207
　USSR, 210
Astronauts, 208, 210
　equipment, 115
　names of, 108, 111, 117, 122
　selection of, 106, 108
　training, 109
　(See Project Mercury)
Astronomical navigation, 10
Astronomical Observatory Satellite, 117
Astronomical Society of the Pacific, 86
Astrophysics, 69
Astrophysics Observatory (See Smithsonian)
AT–1 (airship), 8
AT&T, 129
ATC (See Army)
Atlantic City (N.J.), 27
Atlantic Missile Range (AMR), 66, 67, 74, 87, 92, 98–135
　(See Cape Canaveral)
Atlas-Able rocket, 113, 115, 128, 133
Atlas-Agena, 120, 123
Atlas ICBM, 57, 60, 68, 74–76, 86, 87, 93, 97, 101, 104, 105, 111, 113, 114, 116, 123, 127, 128, 174, 208.
Atlas, of moon, 130, 132
Atlas, Robert, 186
Atmospheric drag, 134
Atmospheric, research, 85;
　turbulence, 110
　(See U–2)
Atomic Energy Commission, 56, 70, 72, 76, 80, 82–84, 92, 95, 106, 110, 116, 119, 121, 124, 129, 133, 193
　Chairman of, 116
Atomic test device, 50
Auburn (Mass.), 21, 25
Augsburg (Ger.), 37, 42, 45
Aurorae, 93, 101
Auroral zone, 93
Auriol, Jacqueline, 178
Australia, 128, 131, 134
　Woomera
Australian Radio and Physics Division, 73
Austria, 28
Auto, radio-controlled, 20

215

Autogiro
  development of, 29
  first U.S. flight, 24
  woman's altitude record, 27
Automatic homing device, 42
Automatic pilot, 11, 33
  landings, 35
  Macy, 3
  radio navigation, 33
  Sperry, 5
  Sperry Gyroscope system, 11
Avco Mfg., 122
Avery, William, 190
Aviation fuel, 13, 34, 42, 211
Aviation Health and Safety Center, 85
Aviation industry, 207, 211
  (See companies by name)
Aviation medicine, 71
  early manual on, 9
  JOURNAL of, 25, 180
  (See School of)
Aviation Medical Research Board, 7
Aviation Medical Research Unit, 38
Aviation, potentialities of, 22
Aviatrices, 178, 179
Avro Co., 197
Awards, aeronautics and astronautics, 167
Axial-flow turbojet powerplants, 44
Axis nations, 48
Azon (bomb), 42, 49
Azores, 47, 61

B-10, 10, 17
B-15, 35
B-17 (Fortress), 33, 36, 45, 46, 54, 72
B-24 (Liberator), 39, 42, 46, 48
B-25 (Mitchell), 39, 43
B-29 (Superfortress), 38, 39, 44, 57, 68, 72, 75
B-36 (Peacemaker), 50, 54, 61, 65, 68
B-45, 78
B-47 (Stratojet), 58, 72, 81
B-50, 61, 72, 66
B-52, 62, 69, 85, 102, 110, 133
B-57 (Canberra), 68
B-58 (Hustler), 66, 70, 80, 83, 86, 93
B-61 (Matador), 54, 67
B-70 (Valkyrie), 80, 93, 111, 119, 129, 211
BG-2 drone, 43
Baby Wac rocket, 50
Bahamas, 67
Baird, J. L., 42
Baka bomb, 50
"Baker" (monkey), 109–10
Baker Board, 31
Baker-Nunn Optical Network, 129
Baker-Nunn tracking, 92, 109
Balbo, Italo, 30, 175, 177
Balchen, Bernt, 23, 26, 178
Ball, Albert, 4
Ball, J. K., 196
Ballinger, E. R., 71
Ballute, 131
Ballistic missiles
  early U.S. development, 48, 75
  defense from, 53
  (See Atlas, Redstone, Thor, Jupiter, Polaris)
Ballistic Missile Committee, 81
Ballistic research, 9, 49, 62
  of projecticles, 96
Balloon, 4, 5, 7, 8, 29, 30, 36, 54, 56, 207;
  as tool of scientific research, X, 36, 57, 207
  chronology, 161–66
  NASA, 125
  plastic, 56, 63, 65
  record flights, 22, 23, 27, 29, 30, 33, 59, 82, 83

Balloon—Continued
  school, 5
  (See Rockoon, *Man High, Excelsior, Skyhook, Stratolab, Stratoscope*)
Baltic Sea, 35
Bancroft, W. D., 16
Bane, F. H., 16
Bane, Thurman H., 170, 202
  Award, 57, 170
Banshee, 62
Barbour, Laura Taber,
  Award, 170
Barcelona (Spain), 36
Barksdale, E. H., 18
Barling Bomber, 17
Barnaby, R. S., 17
Barnes, Wm. F., 158
Barnometer, rocket flight of, 25
Bartley, C. E., 190
Barwise, B., 119
Bassett, Preston R., 202
Battle of Britain, 33, 40
Bauer, Louis H., 25, 180, 183 ;
  Award, 167, 171
Baxter, James P., III, 207
Bay of Biscay, 45
Beacon satellite, 103, 112, 132
Beckman, E. L., 182
Becker, John V., 93, 191
Beechcraft AD-17, 39
Beisel, Rex B., 185, 196
Belgium, 132
Bell, Alexander Graham, 3, 181
Bell, Grover E., 167, 171
Bell, Laurence D., 171, 177
Bell Aircraft, 47, 49, 55, 70, 82, 99, 178, 207, 211
  helicopter development, 47
  P-59, 44, 52
  Rascal, 70
  (See X-1, X-2, and X-5)
Bell Telephone, 71, 193
  Laboratory, 60, 74
  System, 71, 79, 125
Bellamy, John C., 195
Bellanca, *Columbia*, 22
  aircraft, 28
Bellinger, P. N., 3
Bellonte, Maurice, 154
Bendix, Aviation, 46, 80, 99
Bendix Trophy Race, 34
Bennett, Floyd, 21
Bennett Field, 34
Benson, Otis O., 181, 183, 207, 212
Benser, Wm. A., 186
Bergaust, Erik, 207
Bergen, Wm. R., 194
Berger, A. L., 57, 170
Berger, Carl, 207
Bergman, Jules, 207
Berkner, Lloyd L., 99, 207
Berlin (Germany), 9, 24, 40 ;
  airlift, 59
Berliner, Henry, 15
Bernardo, James V., 207
Berstand, Floyd, 14
Besson, F. S., 133
Betts, Austin W., 116
"Big Joe" rocket, 112
Bikini, 82
Bickle, Paul E., 87
Bioclimatology Award, 171
Biology
  space research in, 55, 127, 209
Biomedical research, 71, 125
  on acceleration, 29, 30
  on deceleration, 29, 30

Biomedical research—Continued
   on multiple g-loads, 37
   on weightlessness, 40, 64, 67, 68, 70, 78, 79, 101, 119
Birnbaum, E., 30
Bisplinghoff, R. L., 191, 198
Bissell, C. L., 15
Bjerknes, J., 192
Black Knight missile, 101, 107
Black Widow (P-61), 54, 55
Blair, Charles F., 178
Blair, Clay, 208
Blair, Wm. R., 7
Bleakley, Wm. H., 24
Bleriot, Louis, 156
Blind flying, 26, 31
   first solo, 29
   physiological principles of, 22
Blitzbomber, 208
Blitzkrieg, 38
Blizna (Poland), 45
Blossom Project, 56
Blaw-Knox Equipment, 134
Blue Scout, rocket, 128
Blue Scout Junior, 130
"Blue skies," 48
Blumenstock, David I., 207
Blunt shape, 69
BMEWS, 128
BMW 185, 14
Boardman, R. N., 154
Boden, R. H., 186
Boeing Aircraft Corp., 38, 63, 72, 75, 99, 114, 174, 176, 191, 194;
   Bomarc, 86
   B-15, 35
   B-17, 33
   B-29, 44
   B-47, 58
   Gapa, 50
   247, 30
   Model 299, 30, 33
   Stratoliner, 307B, 40
   707 jet, 62, 75, 85
Boeing, Wm. E., 176
Bogart, G., xi
Bold Orion, 105, 108, 114
Bollay, Wm., 197
Bolling Field (D.C.), 15, 19
Bolster, Calvin M., 176
Bomarc, 64, 86
Bomber Command (See RAF)
Bombing, 44, 45
   bombsight (Norden), 13
   first electrical releases, 9
   "raid" on New York City, 13
   tests, 13, 14, 17
Bombing "raid" on New York City, 13
Bombsights, 13, 58
Bonney, Walter T., xi, 207
Booker, Henry G., 122
Boom refueling, 63
Boost-glide vehicle, 83, 91, 98, 99, 114;
   (See Dyna-Soar I)
Borden, W. A., 48
Boron fuels, 93
Bossart, Karel J., 191, 198
Boston, 79
Boushey, Homer A., 42, 93
Boundary layers
   control, 24, 26, 80
   demonstration flight, 47
   NACA Report on, 28
   theory, 14, 37
Bowen, E. G., 73, 195
Bowman, J. E., 93
Boyce, Joseph C., 207

Boyd, Albert, 57, 158, 173, 192
Boyd, R. L. F., 207
Brady, George W., 191
Braham, Rescoe B., 183
Brango, N., 99
Bread mold, 131
Breeding, E. L., 98
Bremen (Ger.), 35, 38, 45
Brequet, Louis, 181
Breslau (Ger.), 22
Brett, George H., 202
Brewer, Frank G., 172
   Award, 172
Brewster, F2A-3, 43
Bridgeman, Oscar C., 185
Bridgeman, Wm., 67, 70, 173
Bridges, Styles, 97
Briggs, Lyman J., 12, 19, 38, 202
Bristol, Mark L., 202
Britain, 36, 112, 120, 133;
   Air Ministry, 38
   Black Knight, 101, 107
   battle of, 33, 40
   first bombing of, 3
   joint satellite, 133
   Ministry of Aircraft Production, 40
   National Committee on Space Research, 107
   Purchasing Commission, 36
   rocket air defense, 3, 32
   Scientific cooperation with U.S., 40, 133
   Skylark rocket, 98
   space plans, 107, 109, 112
   War Office, 32
   (See Royal Aeronautical Society, Royal Aircraft Factor, Royal Air Force, Woomera, etc.)
British Interplanetary Society, 31, 62
Bronk, Detlev W., 101, 183, 202
Brooks AFB (Tex.), 126
Brooks Field, 28
Brossy, F. A., 28
Brown, A. W., 10
Brown, Clinton E., 199
Brown, J. R., 66
Brown, Willis C., 172, 207
Browne, K. A., 196
Bryan, Leslie A., 172
Bryant, Robert, 134
Bubble levels, 10
Buck, Richard S., 185
Buckingham, Edgar, 14, 15
Buenos Aires (Argen.), 36
"Bug," 9
Building-block rocket, 134
Bulganin, Nikolai A., 94-96
Bull, Harry W., 31
Bumble Bee missile, 53
Bumper project, 55, 65
Bumper Wac, 59, 61
Bundgaard, R. C., 183
Burchard, John E., 207
Burden, Wm. A. M., 101, 102
Bureau of Aircraft Production, 8
Bureau of Fisheries, 10
Burgess, George K., 202
Burgess, Neil, 174
Burgess, W. Starling, 173
Burka, S. M., 195
Burma, 49
Burton, John H., 172
Bush, Vannevar, 38, 58, 202, 208
Bushnell, David, 208
Butchart, Stanley P., 188
Byers, Horace R., 183
Byrd, Richard E., 10, 20, 23, 25, 26, 182
Byrnes, Victor A., 182

217

C-54, 58
C-69, 44
C-82, 61
C-119, 127, 130, 133
C-131, 101
C-133, 104
Cabot, Godfrey L., 9
Cahill, L., 87
Caidin, Martin, 208
Cajun rocket, 82, 84, 98, 125
Caldwell, Frank, 30, 174, 190
Caldwell, T., 57
Caldwell, T. F., 158
California Coast Range, 13
California Cooperative Tunnel, 56
California Institute of Technology, 22, 29, 34, 37, 42, 43, 46, 47, 48, 61, 190, 193, 194
    Guggenheim Aeronautical Laboratory, 52
    Rocket Research Project, 34, 38
    (See Jet Propulsion Laboratory)
California Radiation Laboratory, 85
California Rocket Society, 45
California, University of, 80, 85, 109; at Livermore, 101
Calvin, Melvin, 116, 127
Cambel, A. B., 189
Camber, 70
Cambridge Research Center, 109
    (Also see USAF)
Cambridge Research Division, 66
Campbell, Kenneth, 185, 197
Camp Cooke, 83
Camp Irwin, 48, 50
Canada, 110, 111, 125, 164;
    Canadian Meteorological Office, 183
    Defense Research Telecommunications, 108
Canberra, 62
Candle power, 15
Canright, Richard B., 169, 176
Cantilever wing, 4, 11
Cape Canaveral, 62, 65, 83, 124, 211 (See AMR)
Cape Hatteras, 17
Cape May (N.J.), 10, 35, 49
Caproni Aircraft, 31
    ducted jet, 42
Carbonara, Victor, 191
Carburetor, 7, 18, 19, 24, 191
Cargo Master (See C-133)
Carl, Marion E., 57, 72, 158
Carlson, Floyd, 47
Carmichael, Leonard, 202
Carolin, Norbert, 8
*Caroline Mars*, 60, 61
Carpenter, Malcolm S., 108
Carrier landing, first, 16
Carrier take-off, first, 15
Carrington, L. H., 185
Carroll, F. O., 50
Carver, N., 33
Casaccio, A., 78
Case bonding, 55
Cassady, John H., 202
Castable propellants, 55
Castle AFB, 85
Catalytic-cracking, 34
Catapult launching, 4, 16
Cause, accidents, 24
Cagley, Sir George, 1
CBS, 94, 189
Centaur rocket, 93, 103, 132
Central Powers, 6
Centrifuge, 61, 69
*Century of Progress*, 31
Century fighter series, 58

Ceramic-lining, 41
Cessna aircraft, 93
"Chaff," 45, 84
Chairman, of NACA, 3, 5, 11, 38
Chamberlain, Clarence D., 22, 154
Champion, C. C., 23, 24, 160
Chance Vought, 40, 71, 77, 86, 99, 174, 204
Chanute, Octave, 1, 172
Chanute Field (Ill.), 13
Chapman, Dean R., 194, 199
Chapman, John L., 208
Chemisch-Technische Reichsanstalt, 27
Charney, J. G., 183, 186
Chenoweth, Opie, 185
Chesapeake Bay, 14, 43
Chevalier, G., 16
Chicago, University of, 44, 134, 164, 183
Childers, Milford G., 197
Childs, Stuart, 192
*China Clipper*, 33, 34
China Lake (Calif.), 100, 111
Chincoteague (Va.), 53, 111
Chinese Nationalists, 102
Christmas aileron, 18
Christofilos, N. C., 101
Chrysler Corp., 65, 76, 106
    (See Redstone, Jupiter)
Churchill, Winston S., 45
Cierva, Juan de la, 17, 29, 175, 176
Cislunar spacecraft, 133
Civil Air Authority Act, 36
Civil Aeronautics Authority (CAA), 19, 36, 37, 172, 183, 203
Civil Aeronautics Board, 203–205
Civilian-Military Liaison Committee (CMLC), 103, 118, 121
Civil Service Commission, 132
Civil War, Spanish, 34
Clark University, 4, 30
Clark, Brandt, 182
Clark, Virginius, 202
Clarke, Arthur C., 208
Clarke, J. H., 78
Cleveland (Ohio), 38, 40, 43, 44, 45
    (See Lewis Research Center)
*Clipper* (PAA), 32, 33
Cloud, formation, 99
    measurement, 41
    seeding, 16
Clousing, Laurence A., 173
Clustered engine (See Saturn)
CMR, 75, 118
Coast and Geodetic Survey, 22
Coast Artillery Board, 46
Cobb, Jerrie, 128
Cochran, Jacqueline, 65, 72, 175, 178, 179
"Coke bottle" shape, 68
"Cold" rocket, 42
Cold testing, 43
Coleman, J. F., 75, 76, 178
Coles, Donald, 194
College Park (Md.), 15, 31
Collier (Robert J.) Trophy, 30, 173–5
Cologne, 43
Columb Bechar, 56, 107
Colonial Air Transport, 22
*Columbia*, 22
Columbia University, 37, 73, 204
Combined-loads testing, 39, 63
Comet (de Havilland), 69, 178
Combs, Thomas S., 202
Combustion, 37
Commerce Department, 9, 11, 13, 19, 20, 22, 24, 29, 31, 173, 203–5
    Bureau of Aeronautics, 11
    Bureau of Air Commerce, 22, 31
    Bureau of Aviation, 13

Commercial
   air transport, 10, 22, 27, 40 (See Airliners, Passengers)
   space operations, 129
Committee on Space Research (COSPAR), 102, 103, 107, 114, 115, 118, 123, 127, 128, 134
Communication
   air-to-air, 7
   air-to-ground, 7
   moon relay, 75, 118
   satellites use in, 129
   transcontinental, 3
Communications satellite, 129
   (See Project Score, Echo, Courier)
Compressed-air wind tunnel, 13
Compressor
   stall, 186, 197
   transonic, 68
Compton, Karl T., 203
Condon, Edward U., 203
Condor bomber, 20, 29
Coney, W. D. (USA), 13
Congress (See U.S. Congress)
Connecticut Aircraft, 3
Connolly, Donald H., 203
Conrad, C. E., 18
Conrad, Max, 124
Consolidated-Vultee, 66, 191
   MX–774, 54
Consolidated Aircraft, 23, 29, 31, 34, 39, 50, 66, 194
   B–24, 39, 42, 46, 48
   B–36, 50, 54, 61, 65, 68
   P2Y–1, 29, 31
   XC–99, 50
   (See Convair)
Constellation (C–69), 44
Controlled-trajectory, 42
Convair, 57, 67, 70, 75, 82, 109, 129, 175, 191
Cook, Arthur B., 202
Cooke, David C., 208
Cooke AFB, 87
Coolidge, Pres. Calvin, 18, 19, 21
Cooper, George E., 173, 199
Cooper, Leroy G., 108
Cooper, R., 111, 164
Coordination of research, 49
Coppens, Willy, 4
Corbet, A. A. G., 183
Cornell University, 16, 122, 127, 205;
   Aero Laboratory of, 194
Corona scope, 128
Corporal, 47, 68
Corporal E, 57
Corsair (F4U), 40, 71
Cosmic dust, 131
Cosmic rays, 89, 93;
   study of, 29, 70, 118
   rocket research on, 56
   research, 70
COSPAR (See Committee on Space Research)
Costes, Dieudonne, 154, 155, 177
Cosyns, Max, 29, 162
Cotton dusting, 21
Council of National Defense, 6, 40
*Courier* satellite, 126, 128, 129, 150
Cowling, 24, 173, 185
Cox, H. Roxbee, 197
Cox-Klemin, 21
Craigie, Lawrence C., 203
Crane, Carl J., 35
Cranwell (Eng.), 41
Craven, Thomas T., 203
Crawford, Frederick C., 198, 203
Crew, Henry, 208
Crocco, Luigi, 189

Craft, H. O., 39
Crop-dusting, 21
Crosby, Harry, 47
Crossfield, A. Scott, 68, 73, 107, 110, 131, 133, 169, 173, 176, 194, 208
Cross-licensing, 5
Crowley, John W., 187
CSAGI, 88
Culbert, F. P., 8
Cunningham, A. A., 8
Cunningham, John, 178
Civil Air Patrol, 172
Curry, John F., 203
Curtis, Edward P., 174
Curtiss, Glenn H., 3, 156, 173, 181
Curtiss-Wright Corp.,
   18–T (Kirkham), 10, 11
   HS–1, 8
   JN4 (Jenny), 7, 20
   P–40, 36, 39, 43
   P–36, 37
   PW–8, 18
   R2C, 17, 20
   SB2C, 38
   Tanager, 26
Curtiss Wright Aero., 38, 191
Curvature of earth, 60
Cushing, W. H., 13
Cusk, 56, 64
Cutrell, E. A., 173

D–558, 49, 56, 59, 63, 67, 72, 73
Dahlgren (Va.), 40
Dallenback, Frederick, 197
Damon, Ralph S., 195, 203
Daniels, Josephus, 3
Dart, 75
Dauntless (SBD), 32
da Vinci, Leonardo, 1
Davies, T. D., 154
Davis, J. F., 128
Davis, Leighton I, 170
Davis, P., xi
Davis, Thomas W. S., 203
Davis, William V., Jr., 189, 203
Davy, M. J. B., 208
Dayton (Ohio), 10, 29
Dayton Wright Co., 8
DC–1, 30
DC–2, 174
DC–3, 30, 32, 33
DC–4, 43, 56, 58, 60
DC–8, 209
Deacon rocket, 57, 72, 74, 78, 83
Deadstick landing, 22
de Barnardini, Mario, 42
DeBothezat, George, 12
DeBothezat helicopter, 16
Debus, Kurt, 124
Deceleration, 29, 30
Decompression sickness, 32
Defense Department (DOD), 72, 81, 93–97, 102, 113–116, 129, 131, 211–212
   Arpa, 93–97, 102, 103, 116, 131
   Ballistic Missile Committee, 81
   JCS, 54, 64
   Mission in Space, x, 211–212
   and NASA, 127
   Reorganization, 72, 211
   (See CMLC, etc.)
de Florez, Luis W., 174
De France, Smith J, 28, 40
de Forest, A. B., 191
de Havilland, Geoffrey, 20, 177
de Havilland, aircraft,
   Comet, 69, 178
   DH–4, 6–10, 13, 15, 17, 18, 19

219

de Havilland—Continued
  Goblin, 45
  Mosquito, 36, 41
  Moth, 20
de Seversky, Alexander P., 34, 36
Delta (Thor-Vanguard) rocket, 108
Delta-wing, 28, 67, 72, 73, 128
Denmark, 132
Density, of atmosphere, 19, 134
  (See Standard atmosphere)
Department of Aviation Medicine and Physiological Research, 46
Derwent turbojet, 57
Design competition, 11
Design of landing gears, 26
"Design Requirement for Airplanes," 5
Designation, of missiles, 56, 57
Dessau (Ger.), 25, 27
Detroit, Michigan, 7
*Deutsche Luftreederei* (Ger.), 9
DEW Line, 87
Dewey Report, 52
Dexter (Kans.), 36
D-GA aircraft, 13
DH-4, 6-10, 13, 15, 17-19
Diehl, Walter S., 19, 189
Diesel engine, 24, 28
Dietz, E. L., 53
Dirigibles
  *Akron,* 30
  *Los Angeles,* 18, 29
  *Macon,* 30, 32
  *Navy,* C-7, 14
  (See Zeppelin)
Disasters, 12, 30
*Discoverer* I, 107, 143
*Discoverer* II, 108, 143
*Discoverer* III, 110
*Discoverer* IV, 110
*Discoverer* V, 112, 119, 144
*Discoverer* VI, 112, 144
*Discoverer* VII, 14, 146
*Discoverer* VII, 114, 146
*Discoverer* VIII, 115, 146
*Discoverer* X,
*Discoverer* XI, 119, 122
*Discoverer* XII, 124
*Discoverer* XIII, 126-127, 148
*Discoverer* XIV, 126-127, 149
*Discoverer* XV, 127, 149
*Discoverer* XVI, 129
*Discoverer* XVII, 130-133, 150
*Discoverer* XVIII, 133, 151
*Discoverer* XIX, 134, 151
Disney, Walt, 76
Disinsectation, 44
Dive bombing, 26, 29, 32, 38
Dive-recovery flap, 43
Dixon, T. F., 176
DN-1 airship, 3
Doenhoff, A. E. von, 50
Doherty, Robert E., 203
Doblhoff No. 1, 46
Dollfus, A., 86, 164
Dominican Republic, 13
Donati, Renato, 31
Doolittle, J. H., 15, 20, 22, 24, 25, 28, 29, 39, 43, 65, 94, 96, 101, 108, 158, 175, 177, 178, 179, 184, 195, 198, 202, 203, 208
Dopler radar, 53
Dopler navigation, 188
Dordilly, Francis, 154
Dornberger, Walter, 29, 35, 45, 169, 189, 208
Dornier DO-X, 25, 28
Douglas, Donald W., 174, 177
Douglas, William H., 93

Douglas Aircraft Co., 51, 65, 94, 128, 180, 191, 194, 205
  A-20, 43, 48
  DC-1, 30
  DC-2, 174
  DC-3, 30, 32, 33
  DC-4, 43, 56, 58, 60
  DC-8, 209
  D-558, 49, 56, 59, 63, 67, 72, 73
  SBD, 32
  XB-43, 54
  Project RAND, 53
  World Cruiser, 18
  (See Thor)
Dover AFB, 104
Drag, atmospheric, 134
  balloon, 131
  research, 51
"Dragon", 41
Draper, Charles S., 191, 193, 195
Drew, Adrian E., 158
Drift indicator, 4
Drinker, C. K., 29
Drone aircraft, 34, 36, 46, 54
  (See Pilotless)
Drussler, Robert G., 188
Dryden, H. L., Foreword by, iii, 19, 37, 44, 58, 66, 100, 108, 127, 129, 177, 190, 197-199, 208
Dry-ice, 55
DT-2, 19
DT-6, 20
DuBridge, Lee A., 208
DuBusk, A. Gib, 131
Duke, N. F., 158
Dulles, John Foster, 95
Duncan, Donald B., 203
Dunkirk fighter (HA), 8
Dunn, Louis G., 198
Duralumin, 17
Dupree, A. Hunter, 207
Durand, William F., 5, 8, 12, 19, 41, 51, 176, 198, 202, 203
Dyna-Soar I, 85, 91, 98, 114, 122

Eaker, Ira C., 25, 34
Earhart, Amelia, 27, 35, 178, 179
Earth satellites projects, 51, 52, 53, 58, 60
  (See *Discoverer, Echo, Explorer, Vanguard, Sputnik,* etc.)
Earthquakes, 13
Eastern Air Lines, 205
Eastman Kodak, 4
*Eastwood,* 70
Eaton Manufacturing Co., 21, 211
E-boat pens, 48
*Echo I,* 114, 118, 121, 126-128, 130, 134, 148, 165
Echols, Oliver P., 203
Eckener, Hugo, 18, 175, 176, 177, 179
Eclipse, of sun, 19, 102
  aerial photograph of, 29
Ecology, 58, 117
Edgar, L., 69
Edinburgh (Scot.), 36
Edison, Thomas A., 3
Edwards, George R., 177
Edwards (Calif.), 55
  AFB, 66, 67, 70, 72, 73, 98, 107, 113, 131, 211
EES 3401
  (See Project TED)
Eggers, Alfred J., Jr., 80, 82, 87, 93, 120, 199
Eggert, Wayne, 181
Eglin AFB, 44, 53, 67, 119
Ehricke, Krafft, 169, 208
Eighth Air Force, 46

Einstein, Albert, 38
Eisenhower, President Dwight D., 72, 74, 75, 79, 80, 85, 87, 91, 92, 95, 96, 99–102, 104, 107, 114, 118, 119, 126, 132
Ejection seat, 62
Electric arc engine, 126
Electronic intercept, 60
Elfrey, G. A., 10
Ellsworth, Lincoln, 21
Elmendorf, Armin, 12
Emergency Managements, Office of, 41
Emme, E. M., 208
Emulsion, 74
Encyclopedia on interplanetary travel, 25
Endurance records, 25, 28
Engines
    air-cooled, 25
    fairings, 27
    liquid-cooled, 26, 27
    location of nacelles, 27
    manufacturer's test of, 30
    performance, 9
    suitability tests, 15
    (See Propulsion, Rocket)
Engineering Division (McCook Field), 13
Engineering and Research Corp., 191
England (see Britain)
Englehardt, N. L., 172
Eniwetok Proving Ground, 70, 91, 124, 133
Ensrud, Alf F., 197
Entry simultator, 120
Environment
    of earth, 89
    of Mars, 93
    of Space, 89
Ephrata (Wash.), 45
Eppes, M. H., 74, 179
Epstein, Samuel, 212
Ercoupe airplane, 42
Esnault-Pelterie, Robert, 27, 28
Europe, 1, 10, 27
    (See Britain, Germany, etc.)
Evans, Evan, 172
Everest, Frank K., 62, 73, 75, 82, 158, 173, 178, 192, 208
*Excelsior* (Project), 114, 115, 164–165
Exhaust valves, 15
Exosphere, 122
Exos rocket, 102, 119
*Explorer I* (balloon), 81
*Explorer I* (satellite), 89, 95, 98, 124, 131, 140
*Explorer II* (balloon), 33
*Explorer II* (satellite), 96, 98
*Explorer III*, 96, 99, 141
*Explorer IV*, 100, 101, 141
*Explorer V*, 101
*Explorer VI*, 111–113, 114, 122, 144
*Explorer VII*, 113, 115, 117, 119, 122, 123, 128, 129, 145
*Explorer VIII*, 134
Extraterrestrial, iii, 89

F-1 engine, 77, 119, 123
F2 A-3, 43
F2H-1, 62
F2Y-1, 75
F4D, 62
F4F, 35
F4H, 115, 127
F4U, 40, 71
F5L, 10, 11, 12, 16
F8U, 82
F11F, 68, 97
F-94C, 69
F-80, 43, 60, 66
F-84, 77
F-84F, 66

F-84G, 72
F-86 (Saberjet), 63, 70, 72, 102
F-89 (Scorpion), 60
F-100, 73, 79
F-101, 62
F-101B, 85
F-102, 66, 67, 70 ;
    F-102A, 73
F-104, 80, 93, 98, 100, 116, 174, 178
F-105, 98
F-105A, 62
F-106, 66, 116
F-107, 83
F-108, 111
F-860, 81
FA-61, 35
FHF, 93
F J-2, 81
FR-1, 43, 45, 51
Faget, Maxime, 93, 199
Fagg, Fred D., Jr., 203
Fahrney, D. S., 34
FAI, 129
    FAI, gold medal, 175
Fairchild Aircraft, 50
Farman, Henri C., 154, 156
Farmingdale (N.Y.), 15
Farnborough (England), 5, 7
Farrell, Donald G., 95
Farren, W. S., 197
"Father of Space Travel," 33
Fatigue of metals, 19
Fawbush, E. J., 186
Fechet, James E., 203
Fedden, Alfred H. R., 36, 176, 185
Federal Airways system, 14
Federal Aviation Act of 1958, 100
Federal Aviation Agency, 100
Federal Aviation Commission, 32
Federal Communications Commission, 125, 129
Federal Council for Science and Technology, 105, 107
Fermi, Enrico, 38
Ferri, Antonio, 78
Ferry service, 42
Fighter pilots, 30
Finger, Harold B., 127, 186
Finletter, Thomas K., 57
Finletter Commission, 59
Fireball (FR-1), 43, 45, 51
Firsts,
    liquid fuel rocket flight, 21
    rocket-powered aircraft flight, 24
    transatlantic nonstop flight, 10
    transatlantic passenger service, 38
    transpacific passenger service, 34
    U.S. rocket program, 38
    U.S. rocket aircraft flight, 44
First Aero Squadron (USAS), 4
Fischer, C. Fink, 43
Fisher, Allen C., 208
Fitch, Aubrey W., 203
Fitzmaurice J., 24
Flax, Alexander H., 194, 198
Fleagle, R. G., 186
Fleet exercises, 25
Fleming, P. D., 192
Fleming, Wm. A., 53
Flemming Award, 199
Flickinger, Don, 172, 181, 183, 193
Flight Refuelling Ltd., 63
Flight Research Center (NASA), 55, 113
Flight surgeons, 13, 15, 32
Flint, L. E., 115, 160
Florida State University, 131
Floyd, J. C. 197
Fluorine, 68, 92, 104, 117

Flying boat, 9
  F5L (USA), 10, 11
  hull design of, 3
  PBY amphibian, 37
  prototype four-engined, 34
  XPB2Y-1, 34
"Flying Bomb," 9
  (See V-1)
Flying Flapjack, 39
Flying laboratory, 98
Flying saucer, 63, 71, 106, 211
Flying wind tunnel, 47, 60
Focke-Achgelis,
  FA-61, 35
Fog-dispersal, 19
Fokker, Anthony H. G., 3
Fokker aircraft
  *American,* 23
  C-2-3, 25
  Monoplane, 21
  T-2, 17
Foote, Paul D., 203
Ford, Henry, 19
Ford Motor Co., 122
Ford trimotor, 26
Fordney, Chester L., 162
Forest fires, 12, 14
Forest Service, 12
Forman, Edward, 34
Formation flying, 26
Forrestal, James, 61
Fort Bliss (Texas), 50, 51, 52, 54
Fort Churchill, 111, 125, 164
Fort Knox (Ky.), 110
Fort Monmouth (N.J.), 106, 128, 132, 164
Fort Omaha (Nebr.), 5
Fort Worth (Tex.), 9
"For the Record," *Astronautics* (ARS)
*Fortune,* Editors of, 1956, 208
Foulois, Benjamin D., 4, 6, 184, 203
Four-engine bomber, 33, 39
France, 8, 36, 62, 132
  Aircraft in WWI, 3
  plutonium bomb, 119
  rocket range, 56, 107
  use of aircraft rockets, 4
  and V-1's, 47
  (See Veronique)
Frankford Arsenal, 46
*Frankfurt,* 13
Franklin Institute, 208
Franks, Wilbur R., 183
Fuel pumps, for rockets, 17, 30
*Frau in Mond,* 25
F-region, 61
Free-flight tunnel, 37
Freeman, J. C., 186
Freeman, John R., 202, 203
Frelinghuysen, Peter, Jr., 97
Friedriehshafen, 15
Frigitorium, 43
Froelich, H., 83, 163
Fruin, J. L., 62
Fruit flies, 56
Fuel, aircraft, 211
  consumption, 15
  High-energy, 74
Fuel tanks,
  drop type, 16
  leakproof, 11
Fugo balloon, 49, 162
Fuller, Edgar, 172
Fulton, James G., 97
Fultz, Dave, 186
Furbay, John H., 172
Furnas, Clifford C., 203
Fuselage, 19
Fusion, 78

Gallagher, James G., 61, 184
Galland, Adolf, 208
Gallaudet plane, 17
Gallauder 59A, 4
Gallet, Roger M., 122
GAM-77, 108
*Ganet,* 21
Gantz, Kenneth F., 208, 209
Gapa missile, 50
Garber, Paul E., XI, 172
Gardner, George N., 172
Gardner, Lester D., 53
Gardner, Matthias B., 203
Gardner, Trevor, 73, 77
Gardner, Lester O., 30, 177
Gargoyle, 48, 53
Garros, Roland, 3, 159
Gartmann, Heinz, 209
Gas tank, jettisonable, 16
Gas turbine development, 35, 41, 44, 68
Gasoline, as aircraft fuel, 13, 34, 42, 211;
  as rocket fuel, 13, 28
Gas-pressurized fuel system, 46
Gatland, Kenneth W. (ed.), 209
Gatty, Harold, 28
Gaven, James M., 209
Gayler, Noel A. M., 194
GB-4 glide bomber, 48
GB-8, glide bomber, 42
Gee Bee Monoplane, 29
Gell, Charles F., 172, 181, 183
General Electric Co., 8, 11, 41, 44, 48, 52, 55, 57, 80, 99, 119, 122, 129, 180, 183
General Mills, 55, 63, 74, 83, 87, 162, 163
General Motors Corp., Allison Division, 27
General Vehicle Co., 4
Geni rocket, 87
Geodetic Satellite, 127, 130, 134
Geomagnetic pole, 70
Geophysical science, 54, 134
George, J. J., 169, 183, 186
Gerathewohl, S. J., 67, 78, 196
Germany
  Aircraft Industry, 209
  Air Force, 32, 39, 40, 49
  Army Ordnance, 27, 29, 32
  First bombing of Britain, 3
  Flying boat DO-X, 28
  Infantry, 9
  Pilots, 24
  Research in World War II, 211
  Rocket development, 10, 18, 26, 27, 29, 32, 33, 49, 51, 52
  (See Hitler, V-1, V-2, Zeppelin)
Getting, Ivan A., 125
GHQ Air Force (See Army)
Gibbons, Henry B., 197
Gibbs-Smith, Charles H., 209
Gillan, J. W., 36
Gillespie, James M., 170, 189
Gilmore, William E., 203
Gilruth, Robert R., 41, 47, 80, 191
Giraud, Peter F., 83
Girders, Zeppelin-type, 4
Givson, Professor, 5
Glacier, 87
Glenn, John H., 87, 108
Glennan, T. Keith, 100, 101, 115, 129, 131, 134, 209
Glide bombs, 41, 42, 43, 46, 48
Glider,
  Rocket-powered, 24, 25, 27
  Speed record, 87
Global cloud cover, 89
Glomb missile, 41, 53

Gloster aircraft,
  E28/39 jet, 41
  Meteor, 44, 51
Gloves (flight), 25
Glubareff, Michael, 181
Gnome engine, 4
Goddard, George W., 20, 170
Goddard Robert H., 9, 10, 25, 27, 29, 30, 32, 34, 40, 42, 44, 61, 109, 176, 208, 209 ;
  Award named for, 176
  first liquid-fuel rocket, iii, 4, 21
  military demonstration, 9, 34
  writings of, 34, 208, 209
Goddard Space Flight Center (NASA), 109, 122, 217, 134
Godunov, K., 30
Goering, Hermann, 32
Goett, Harry J., 109
Goggles (flight), 25
Goldberg, Alfred, 209
Goldstein, Sidney S., 197
Goldstone Lake (Cal.), 43
Goldstone Tracking Station (JPL), 113
Goodlin, Chalmers, 55
Goodrich, B. F., Co., 102, 115
Goodyear Corp., 17, 69, 116
Goose (See SM-73)
"Gordo" (Monkey), 104
Gorgon (air-ram), 45, 49, 53, 57
Gorky, Maxim, 33
Gottfield, Robert E., 122
Gottinger University (Ger.), 14
Gough, Melvin, 170, 172
*Graf Zeppelin*, 25, 28, 40
Grand Central Rocket, 86, 129
Granville Brothers, 29
Graveline, Duane E., 119
Gray, George W., 209
Gray, H. C., 22, 23, 162
Graybrel, Ashton, 182, 183
Great Britain (See Britain)
Green, Murray, xi, 29
Greenaway, K. R., 195
Green Bank (W. Va.), 109, 117
Greenbelt (Md.), 109
Greene, Ben. F., Jr., 170
Greenland, 20, 70, 78, 128
Gregg, Willis R., 203
Gregory, H. F., 170
Grether, W. F., 182
Grey, Charles G., 209
Griffin, Leigh M., 16
Griffin, V. C., 15
Griffith, E. A., 209
Grissom, Virgil I., 108
Gromov, Mikhael, 154
Grow, H. B., 10
Grow, Harold J., 157
Grow, M. C., 55, 181
Grubb, O. E., 25
Grumman, Leroy R., 177
Guderley, Gottfield, 170
Guggenheim, Daniel, 20, 195 ;
  Medal named for, 195
Guggenheim, Harry F., 40, 170, 203
Guggenheim, Aeronautical Lab., 34
Guggenheim Foundation, 61, 85, 123
Guggenheim Fund for the Promotion of Aeronautics, 20, 21, 26, 61, 85, 123
Guggenheim Institute, 73
Guggenheim Medal, 26, 29, 34, 36, 43, 55, 167, 176
Guggenheim Safe Aircraft Competition, 26
Guidance
  inertial, 93, 96, 120
  Polaris, 102
  radio-inertial, 101

Guided Missiles Committee (JNW), 49
Guided missiles (See type by name)
  first U.S., 9
  in Warfare, 46
Gunn, Ross, 183
Gusts in atmosphere, 4, 66
  (See U-2)
Guymon, V. M., 26
Gyrocompass, 10, 46
Gyro-controlled rocket flight, 29, 32, 34, 38
Gyro horizon, 27
Gyroscope, 4, 13
Gyrostabilized bombsight, 13

H-1 rocket engine, 109
H-13 airplane, 82
H-16, 8, 9
H-21 helicopter, 82
HA seaplane, 8
Haber, Fritz, 64, 67
Haber, Heinz, 64, 67, 209
Habluetzel, W. D., 127
Hagen, John P., 69, 119
Haggerty, James L., 209
Haiti, aerial survey of, 13
Haley, Andrew G., 209
Halford H1 turbojet engine, 45, 47
Hall, Arnold, 198
Hall, Charles F., 70
Hall, E. N., 176
Hall, E. J., 6
Hall-Scott Motor Car Co., 6
Halvorson, G., 62
Hamburg (Ger.), 27, 38, 45
Hamilton, A. G., 13
Hamilton Standard Propeller Co., 30, 173, 174, 190
Hampton (Va.), 5
Hampton Roads (Va.), 9, 10, 12, 13, 14, 15
Handley Page Aircraft, (See DH-4) ; wing Slots, 26
Hanes, Horace A., 71, 158, 185
Hanrahan, James S., 209
Harding, President Warren G., 13, 16
Hardy, James D., 182
Hargreaves, J. S., 127
Harmon, E. E., 10
Harmon Trophies, 65, 177
Harriman Mission, 42
Harper Dry Lake (Calif.), 47
Harris, H. R., 13
Harrison, Henry T., 169, 183
Harrison, Lloyd, 203
Hartz, R. L., 10
Harvard University, 29, 43, 85, 195
Harvard trainers, 36
Haskell, Harry G., 97
Hastings, Donald W., 182
Havana (Cuba), 12
Hawaii, 61, 108
Hawk missile, 79
Hawker Hurricane, 33, 36
Hawks, Frank, 25, 179
Hayden Planetarium, 74
Hayford, John F., 203
Hazelhurst Field, 7
Hazen, Ronald M., 203
Hean, J. H., 43
Heat balance, 128
Heath, Leroy, 133
Heat transfer, 62, 73 ; study of, 72
Hegenberger, A. F., 29, 174, 184
Heim, J. W., 37
Heinemann, E. H., 174, 191
Heinkel, Ernst, 209

Heinkel aircraft
  HE-111, 47
  HE-112, 33, 35
  HE-162, 52
  HE-178, 38
  HE-S-3B jet engine, 38
Heinrich, Helmut G., 170
Helicopter
  Berliner, 15
  DeBothezat, 16
  FA-61, 35, 41
  French, 18
  Jet rotor, 46
  ramjet, 58
  single rotor, 38
  two-man, 58
  (See Sikorsky, Bell, Focke)
Helios Balloon, 54, 55, 162
Helium, 36, 38
  U.S. monopoly, 14
  use in airships, 14, 17
  use in balloons, 15, 33
Helium gun catapult, 68
Helium plants, 36
*Helldiver*, 38
Henry, James P., 70, 181, 196
Hermes Project, 48, 56, 60, 64, 68, 76, 211
Herndon, Hugh, 28
Heron, S. D., 21, 24, 185, 211
Hester, Clinton M., 203
Hewitt, P. C., 5
Hewlett, R. G., xi
Hibbs, Albert, 131
Hickman, C. N., 40
High-altitude pressure flying suits, 43, 44
High-lift, 24
High-performance aircraft engines, 41
High-speed photography, 19
High-speed wind tunnel, 19
High-temperature alloys, 40
Hight, J. L., 189
Hill Space Transportation Award, 167, 180
Hiller Helicopters, 66
Hiller X-18, 115
Hinckley, Robert H., 203
Hindshaw, Carl, 198
Hines, Wellington T., 203
Hiroshima, 51
Hispano-Suiza engine, 5
Histories
  AFMDC, 208
  aeroplanes, 209
  air transport, 212
  balloons, 207
  chronology, 210, 212
  First Missile Division, 208
  IGY, 209, 211
  Naval Aviation, 211, 212
  of combat aircraft, 209
  of flight, 208, 209
  OSRD, 207
  Royal Aircraft Factory, 211
  USAF, 209
  Vanguard, 20
  Weather Bureau, 212
  (See bibliography, 207f.)
Hitchcock, Fred A., 196
Hitler, Adolf, 32, 45, 65, 208
Hives, E. W., 185
H.M. Nautical Almanac Office, 195
Hobbs, Leonard S., 174
Hoelker, R. F., 169
Hoffman, E. L., 173
Hoffman, Samuel K., 176
Hohenemser, Kurt H., 171
Holaday, William M., 86, 92, 103, 121
*Holiday* balloon, 125
Holland, Maurice, 209

Hollandia, 50
Holley, I. B., Jr., 209
Holloman, George V., 35, 184
Holloman AFB (N. Mex.), 55, 64, 71, 79, 96, 104, 107, 162-164
Holzman, Ben G., 183
Honest John rocket, 65
Hooten, E. A., 43
Hoover, George W., 193
Hoover, Herbert H., 59, 173, 192
Hoover Commission, 78
Hopkins, Philip S., 172
Horner, Richard E., 47, 108, 125
*Hornet* (carrier), 43
Horsepower, Progress in aircraft engines, 52
Houbolt, John C., 199
Houghton, Henry G., 183
Hound Dog (See GAM-77)
Howe, Clarence D., 177
Howell, Robert B., 194
Howland Island (Pacific O.), 35
HPAG, 74
HS-1, flying boat, 8
HU-1, helicopter, 125
Hubler, Richard G., 209
Hudsons (Lockheed), 36
Hueco Range, 50
Hughes, Howard, 158, 172, 174, 177, 179
Hughes Aircraft, 70, 123, 125, 134
Hull, G. F., 19
Hullsmeyer, Christian, 15
Human tissue, 132, 133
Human tolerance to multiple g-loads, 37
Humphrey, Paul A., 183
Hunsaker, Jerome C., 4, 5, 6, 7, 9, 12, 30, 42, 52, 108, 176, 182, 187, 198, 202, 204, 209
Hunt, J. R., 85
Huntsville (Ala.) 81, 120, 128
HS-293 (glide bomb), 46
Hutton, Richard, 194
Hurricane
  first photo, 76
  flight into, 48
Hurricane fighter, 33, 36
Hustler (See B-58)
Hybrid rocket design, 45
Hyde, Margaret O., 209
Hydroaeroplane, 4
Hydrocarbons, 12
Hydrogen device, 70
Hydrogen-fluorine engine, 92, 122
Hydrogen-oxygen engine, 93, 105, 117
Hydro-ski, 72
Hynek, J. Allen, 106
Hypersonic
  test vehicles, 55
  flow, 58
  research vehicle, 72, 83, 87
Hypoxia, 37, 40

I-16 engine, 44
Icarus, 1
ICBM, 53, 54, 74, 79, 94
  Soviet development, 58, 87
  (See Atlas, Titan, Minuteman)
Ice formation, on aircraft, 24
Ice jam, use of aircraft on, 18
Ickes, Harold L., 36
Ide, John J., 13, 27
IGY (See International Geophysical Year)
IGY/IGC-59, 117
IGY Warning Center, 130
I1-2 (Stormairk), 42
Illinois, Univ. of, 64
Impact
  of air power, 207
  of technology upon society, iii

Imperial Airways, 35
Incendiaries, 4, 49
India, 131
Indian Head (Md.), 40, 41
Inertial coupling, 59
Inertial guidance, 83, 120
Inflatible spheres, 99
In-flight refueling, 9
 (See refueling)
Infrared, 121
Infringement suits, 5
Ingalls, Laura, 179
Insecticide, aerial spray, 18
Institute of Technology, 54
Institute of the Aeronautical (Aerospace) Sciences, 29, 30, 32, 39, 53, 58, 62, 129, 167, 170–172, 180, 182, 190, 193, 197
Institute of World Affairs 115
Instrument flying, 18, 26, 29, 31
 first solo, 29
 landing, 31
Instrument standardization, 8
Intake valves, 15
Interference lift, 81
Interior, Department of, 21, 36
International Academy of Astronautics, 112, 123
International Air Congress, 17
International airmail service, first U.S., 12
International Air Navigation, 1919 Convention of, 12
International Aircraft Exposition, 10
International air transport, 33
International Astronautics Federation (IAF), 65, 67, 70, 72, 75, 78, 112, 123, 126
International Civil Aviation Conference (Chicago), 48
International Civil Aviation Organization, 35, 134
International Congress on Aerial Safety, 27
International Convention on Air Navigation (Paris), 10, 11
International Council of Scientific Union, 70, 102 (ICSU)
International Geophysical Year (IGY), 68, 70, 75, 76, 78, 86, 91, 92, 95, 100, 102, 105, 117, 130, 208, 209, 211
 IGY/IGC 59, 113, 117
 Soviet Committee, 96, 98, 100
 U.S. National Committee, 71, 75, 77, 78, 85, 91, 92, 102, 105, 113
International law, 10
International Mars Committee, 86
International Polar Year, 68, 70
International power politics, role of aviation in, 32
International Rocket and Satellite Conference, 88
International Space Science Symposium, 118
International Telecommunications Union, 111
Interplanetary travel, 26, 75
Inventions and Contributions Board (NASA), 101, 188
Inventions, NACA review of, 21
Inyokern, China Lake (Calif.), 46, 70
Ion engine research, 85, 111, 123
Ionosphere, 61, 89, 126
Iowa, State University of, 39, 70, 78, 93, 180
IRAC, 122
IRBM, 77, 85 (See Jupiter, Polaris, Thor)
Iris rocket, 125
Iroquois helicopter, 125
Irvine, C. S., 154
Irvine, Rutledge, 17
Irwin, Walter A., 98
Irwin, W. W., 158, 174

Italy, 36, 132
 Air force of, 30

J–1 engine, 4, 18
J30 engine, 45
J34 engine, 45, 63
J40 engine, 45
J46 engine, 45
J47 engine, 81
Ju–52 aircraft, 30
J54 engine, 45
J57, 62, 71
J65, 66, 68
J71, 78
Jabara, James, 192
Jacobs, James, 192
Jacobs, Eastman V., 37, 190, 196
Jackass Flats (Nev.), 110, 124, 128
Jackson, Nelson P., 180
 NRC award, 180
Jamaica, 32
Jane's All The World's Aircraft, 209
Japan, 49, 99, 110, 130
 air force, 35
 naval air units, 42
 Kappa rocket, 99
Jarvis Island, 64
Jastrow, Robert, 109, 134, 209
Javelin rocket, 110, 116
JB–2 (robot bomb), 48
Jeep (NACA jet engine), 44
Jefferies Award, 180
Jenny (JN4), 7, 20
Jet aircraft
 first flight, 38
 first four-engine, 52
 Italian ducted jet, 42
 rotor helicopter, 46
 and propeller fighter, 45
 U.S. design, 44
 (See Turbo-Jet, Rampet, etc.)
Jet-assisted takeoffs (JATO), 37, 40, 42, 50
 first in Germany, 25
 first use in USSR, 30
 liquid fuel, 41, 42, 43, 44, 45
Jet exhausts, 39
Jet propulsion
 Bureau of Standard's study, 28
 early NACA study, 14
 revaluation of, 37
 Laboratory, 45
 Special NACA Committee on, 41
Jet Propulsion Laboratory (JPL), 34, 47, 49, 50, 51, 55, 67, 68, 73, 80, 83, 95, 96, 100, 104, 105, 111, 113, 128, 132, 194
 ORDCIT, 47, 49
 (See Corporal, Hermes, etc.)
Jet transport, 75
Jet-type wind tunnel, 23
Jettisonable gas tank, 16
J–13 aircraft, 14
Jacksonville (Fla.), 9
Jet VTOL aircraft, 80
JH–1, 35, 36
JN4 (Jenny), 7, 9, 20, 22
Jodrell Bank (Eng.), 109, 124, 127
Johns Hopkins University, 38, 52
Johnson, Clarence L., 47, 174, 191, 193, 196
Johnson, Howard C., 98, 160, 174
Johnson, J. B., 170
Johnson, John A., 109
Johnson, Louis, 60
Johnson, Lyndon B., 94, 97, 134
Johnson, Richard L., 60
Johnson, Robert L., 158, 198
Johnson, Roy W., 95

225

Johnson, Thomas, 162
Johnson, W. H., 196
Johnson, Island, 100
Johnston, A. M., 173
Johnston, S. Paul, 53
Johnsville (Pa.), 45, 100, 112, 122
Joint Army-Navy Research and Development Board, 54
Joint Army and Navy Technical Aeronautical Board (JANTAB), 8, 12
Joint Chiefs of Staff, 54, 64
(See Defense Dept.)
Joint Committee on New Weapons and Equipment (JNW), 43, 52
Joint Meteorological Satellite Committee, 109
Joint Research Airplane Committee, 75
Joint Technical Board on Aircraft, 6
Jones, B. Melville, 197
Jones, B. Q., 3, 184
Jones, Gene, 133
Jones, B., 18
Jones, Robert T., 49, 191
Jongbloed, J. (Ger.), 26
Jordan, Joseph B., 116, 160
*Journal of Aviation Medicine,* 25
JRM-2, 60
Juarez (Mex.), 57
Judd, F. V. H., 197
Jude, Geo. F., 194
Junkers, Hugh, 4
Junkers aircraft, 11, 25, 33, 35, 45
  J-1, 4
  J-13, 14
  .005 Engines, 46
  JU-33, 24, 25
  JU-52, 30
  JU-287, 52
Juno II, 96, 104, 111, 113, 121, 130
Jupiter C, 83, 92, 96, 111
Jupiter IRBM, 80, 81, 82, 83, 86, 91, 92, 99, 104, 106, 109
Jupiter (Planet), 127
*Jupiter* (USS), 14

K-2 airship, 37
K-123, 47
Kalitinsky, Andrew, 186
Kaman helicopter, 68, 116
Kaplan, Joseph, 169
Kappa rocket, 99
Kauffman, William M., 199
Kazan (USSR), 35
KC-135, 62, 75, 85, 95, 97, 102, 178
Kearley, J. G., 44
Kearns, C. M., 193
Keating, Kenneth, 97
Keirn, Donald J., 41, 92, 170
Keller, K. T., 65
Kellett, R., 154
Kelley, Bartram, 181
Kellum, Wilbur E., 181, 183
Kelly, C. G., 184
Kelly, Mervin J., 193
Kelly, Oakley C., 155
Kelly, O. G., 17
Kelly, P. L., 170
Kelly Air Mail Act, 20
Kelly Field (Tex.), 18
Kelsey, Ben, 37, 173
Kennedy, President-elect John F., 134
Kenly, William L., 7, 204
Kepler, Johannes, 1
Keplerian trajectories, 68
Kepner, Wm. E., 31, 58, 162, 179
Kessler, F. W., 33
Kettering, Charles, 9
Kety, Seymour, 112

Key West, 12
Key West Agreement, 59
Khrushchev, Premier Nikita, 112, 113, 128
Killian, James R., Jr., 77, 95, 105, 109, 210
Killian Committee, 77
Kilner, W. C., 38
Kilner-Lindbergh Board, Establishment of, 38
Kincheloe, Iven C., 82, 100, 126, 169, 176, 185, 192
Kindelberger, James H., 174
King, Ernest J., 30, 204
Kingfisher Missile, 53
Kinler, Walter G., 204
Kipfer, P., 162
Kirkham racer, 10
Kirkham-Packard, 23
Kirschner, Edwin J., 210
Kistiakowsky, George B., 109
Kite Balloon, 8
Kittinger, Joseph W., 86, 114, 116, 126, 163, 164, 181
Kitt Peak (Ariz.)., 117
Kitty Hawk (N.C.), 24, 207
Kitty Hawk, of rocketry, 21
Kiwi-A, 110, 129,
  Prime, 124
Klemin, Alexander, award for, 181
Klieforth, H., 69
Knight, William, 10
Knipfer, Charles, 28
Knudson, John, 195
Köchl, Herr, 24
Kokkinaki, K. K., 128
Kolcum, Harry, 210
Kollsman Instrument Co., 191
Komet (Me-163), 48, Me-163B, 47
Korean War, 64, 67, 72
Kosmodemyansky, Prof., 210
Kotanchik, J. N., 39
Kraus, Sydney M., 204
Krebs, Geoirge, 56
Kroon, R. P., 195
Krause, Ernest H., 210
Krieger, F. J., 210
Kuiper, G. P., 130
Kummersdorf, 27, 29, 35
Kunz, W. J., 197
KVW-1 (Loon), 53
  (Also see JB-2, V-1)
Kwajalein, 115

Lacrosse missile, 65, 75
Lahaina (Hawaii), 15
"Laika" (dog), 98
Laird *Super-Solution*, 28
Lake Constance (Switz.), 25
Lakehurst (N.J.), 18, 35
Lamb, Lawrence E., 196
Lambert, E. H., 196
Lambert, Sgt., 54
Laminar flow, airfoil, 39
  wind tunnel, 35
Land, Emory S., 204
*Land of the Soviets,* 25
Landing gears
  first Navy aircraft with, 27
  oleo, 20
  retractable, 12, 26, 31
  track, 61
  tricycle, 31
Lange, Fritz, 25
Langley, Samuel P., 181
*Langley* (CV), 14, 15, 16
Langley Field (Va.), 5, 9, 24, 35, 36
Langley Medals, 26, 124

Langley Memorial Aeronautical Laboratory (NACA), 8, 12–16, 19, 21, 27, 34, 35, 37, 47, 48, 55, 58, 63, 65, 66, 92, 93
   Engineer-in-charge, 48, 84
   Flight Research, 47, 48
   PARD, 51, 54, 56, 60, 67, 68, 76, 82, 96
   Wind tunnels, 8, 26, 27
   (See Wind tunnels, Wallops Island)
Langley Research Center (NASA), 105, 107, 111, 122
   Structures Research Division, 130
Lark missile, 50, 63
*L'Astronatique*, 27
Lasser, David, 26
Lassiter, O. F., 57, 156
Latham, Hubert, 159
Laudan, F. P., 196
Launching rope, 24
Launch Operations Directorate (NASA), 124
Law, of outer space, 132
Lawrance, Charles L., 173
Lawrance aero engine, 8, 11, 12, 14
Lawrence, Lovell, 176
LBD–1 (Gargoyle), 48
Leahy, William D., 53
Lear, John, 210
Leduc, Rene, 62
*Lee, Robert E.*, 134
Leipzig (Ger.), 45
Leaflet bombing, 38
League of Nations, 11
Lees, W., 28
Legal doctrine, 48
Legislation, NACA analysis of, 20, 21
   (See U.S. Congress)
Lehmann, Milton, xi
Lehmann, Ernst A., 179
Leipzig (Ger.), 9
LeMaitre, Henri, 154
LeMay, Curtis E., 52, 92, 178
Leningrad (USSR), 35
LePere, 7, 11, 14
Le Prieur rockets, 4
Levine, Charles A., 22, 154
Lewey, F. H., 40
Lewis, George W., 18, 34, 37, 40, 58, 60, 176, 195
Lewis, M. Lee, 62, 82, 83, 91, 100, 110
Lewis, William, 183
Lewis Flight Propulsion Laboratory (NACA), 38, 46, 47, 49, 52, 60, 71, 74, 78, 79, 84, 92, 93, 99
Lewis Research Center (NASA), 99, 102, 103, 117, 120, 163
   (See Plum Brook)
*Lexington*, 25
Ley, Willy, 33, 210
Liberty engines, 6, 7, 8
Licensing of pilots, 10
Lieblein, Seymour, 188
Lick Observatory, 111
*Life* magazine, 50
Life sciences, 125
   discipline of, ix
   use of satellites in, 98
Lift, coefficient, 28
   interference, 81
Lift-drag ratio, 79
Light planes, 20
Lighter-than-aircraft,
   first USN nonrigid, 3
   first Atlantic crossing, 10
   (See Airships, Zeppelins)
Lilienthal, Otto, 1
Liljencrantz award, 182
Lilly, Howard C., 59

Lindbergh, Charles A., 22, 38, 154, 175, 177, 179, 181, 195, 198, 204
   and Mrs., 28, 30
   Atlantic survey flight, 30
   Pacific survey, 28
Link, Edwin A., 172
Lippisch, Alex., 24, 25
Liquid-fuel rocket
   first European flight, 27
   first Soviet flight, 30
   first world flight, 21
   of Esnault-Pelterie, 28
Liquid-fuel rocket motor, 34, 35, 41–44, 46
   (See types of motors)
Litchfield, P. W., 195
Little America, 26
Little, Arthur D., Co., 164
Little Joe booster, 53, 113, 114, 118, 130
Little Joe missile, 50
Littleton, Charles W., 187
Littlewood, Wm., 177, 196, 197, 204
Lockheed Aircraft Corp., 25, 28, 35, 36, 56, 82, 84, 130, 180, 191, 193, 194
   *Air Express*, 25
   C–69, 44
   Electra, 31
   F–94–C, 69
   P–38, 37
   PV–2, 55
   Sirius, 28
   U–2, 73
   Vega, 30
   *Winnie Mae*, 28
   XC–35, 35
   XP–80, 47, 80
Lodge, Henry C., 101, 108, 116
Loening, Grover C., 4, 173, 177, 198
Loening aircraft
   Amphibian, 20, 21
   M–3, 8
Loewy, Robert G., 194
Loki rocket, 67, 75, 78
London (Eng.), 3, 5, 6, 36, 47, 48
Long, K., 83, 163
Longacre Award, 182
Longham, Sgt., 15
Long-Range Missile Proving Ground, 54, 62
   (See AFMTC, AMR, Cape Canaveral)
Long-range rockets, 45, 46
Long-range studies, of NASA, 109
Lonnequest, T. C., 189, 204
Loon missile, 48, 53, 56, 64
Loosbrock, John F., 210
Lorin tube, 30
Loring, Samuel J., 196
Los Alamos Scientific Laboratory, 85, 108
Los Angeles (Cal.), 34
*Los Angeles* (ZR–3), 18, 29
Losey Award, 182
Love, E. S., 193, 199
Lovelace, W. R., 45, 181, 193
Lovelace Foundation, 66
Low-drag wing, 37, 39
Low-turbulence wind tunnel tests, 39
LSD Point Reference, 102
*Lucky Lady*, 63
*Lucky Lady II*, 61
*Luftwaffe* (See Germany, Air Force)
Lunar exploration, 109
Lunar maps, 126, 130, 132
Lunar programs, 90, 102
Lundberg, B. K. O., 198
Lundquist, E. E., 39
*Lunik I*, 106, 121, 143
*Lunik II*, 112–114, 144
*Lunik III*, 130, 145
L. W. F. Engineering Co., 6
Lyster Award, 183

M8, 43
MacClain, Albert L., 172, 196
MacCracken, William P., 198, 204
MacDairmid, D. B., 173
Mach 3, 81
Mace missile, 131
Machine-gun, aerial use of, 3
Mackay, Donald B., 186
Mackay Trophy, 23, 184–185
Macmillan, Prime Minister Harold, 109
Macmillan arctic expedition, 20
*Macon*, 30, 32
Macauley, T. C., 9
Macready, J. A., 14, 17, 18, 21, 155, 159, 184
Macy automatic Pilot, 3
Magnesium airframe, 78
Maitland, L. J., 184
Maitland, R. L., 16
Magnetic fields, 93, 122
    tape, 134
Majorca (Spain), 36
Malina, Frank J., 34, 43, 190
Malone, Thomas F., 183
Manchester, Univ. of, 109
    (See Jodrell Bank)
*Man-High balloon*, 86, 87, 103, 163
Manila (Phil. Is.), 34
Man in space, 71, 104, 208, 209, 210
Manly, Charles M., 26, 182;
    Award, 185
Manned reentry, 87
Manned space flight, 71
Manufacturers Aircraft Association, 5
Marberger, John F., 196, 210
March Field (Cal.), 29, 42
Marconi, Gughlielmo, 15
Margaret, Princess, 120
Marine Corps Aviation,
    aerial survey by, 13, 26
    aviation in 1917, 6
    first Aviation Force, 8
Mars (planet), environment of, 93, 127
    life on, 73
Mars chamber, 84
Marshall, George C., 124, 127
Marshall Islands, 74
Marshall Space Flight Center (NASA), 120, 131, 134, 207
    (See Saturn, Redstone, etc.)
Martin, Glenn L., 49, 52, 134, 177, 198
Martin, James, 170
Martin, John, 59
Martin Aircraft Co., 3, 12, 17, 26, 34, 35, 54, 67, 78, 86, 99, 114, 129, 134, 194
    (See *China Clipper*, Matador, Titan, Vanguard, etc.)
Marvin, Charles F., 204
Maryland, Univ. of, 49, 79
Massachusetts Institute of Technology (MIT), 3, 11, 12, 19, 21, 36, 63, 95, 107, 183, 190, 193–195
Mass production techniques, 6
Massey, H. S. W., 107
Massie, Walter L., 79
Matador missile, 54, 67
Materiel Division, 24 (See McCook, Wright Field)
MATS (See U.S. Air Force)
Matzdorff, R. E., 186
Maugham, Lt. R. L., 18, 157
May, Gene, 56
McAllister, Lt. C. D., 29
McAvoy, Wm. H., 172
McCain, John S., 204
McCarthy, Charles J., 174, 204
McClure, Clifton M., 103
McClure, D., 164

McClure, Frank T., 188
McCook Field (Dayton, Ohio), 6, 11, 12, 17, 18, 19, 20
McCormack, John W., 197
McCune, James E., 194
McDonnell Aircraft, 58, 85, 106, 121, 125, 127, 171
    LBD-1, 48
    Mercury capsule, 69
    XFD-1, 45, 54
McDonough, Gordon L., 97
McElroy, Neil H., 92, 95, 97, 113
McFarland, Marvin W., 210
McFarland, Ross A., 182
McGee, J. C., 52
McIntosh, Lawrence W., 204
McKinney, J. C., 8
McPherson-Smith, Ross, 11
McTaggart-Cowan, P. B., 183
McVeigh, D. R., xi
Mead, George J., 190, 204
*Mechta*, 143 (See *Lunik I*)
Medaris, John B., 210
Medical research, 55
Medical Research Laboratory
Mehrens, Harold E., 172
Meisinger, Clarence L., 18, 186
    Award, 186
Menoher, Charles T., 204
Mercedes aircraft motor, 4
Mercury capsule, 69
Mercury program, 100, 102, 106 (See Project Mercury)
Mercury-Redstone, 106, 131, 134
Mergen, J. M., 186
Merrell, A. E., 173
Merrill, Grayson, 189
Merrill, Henry T., 177, 179
Mertes, Ray C., 172
Messerschmitt Aircraft
    Me 109, 37
    Me 163A, 42
    Me 163B, 45 47, 48
    Me 262, 44, 45, 49, 208
Metals, NASA subcommittee on, 19
Meteor jet aircraft, 46, 47
Meteorgraphs, 33
Meteoric dust, 73
Meteorological research, use of satellites, 74, 109, 121, 129
    survey of U.S., 73
    (See *Vanguard II*, *Tiros I*, Radiosondes)
Meteorology
    and aeronautics, 7
    science of, 11
    study of, 24
Methane, 13, 27
Metz, Richard, 3, 45
Metzger, C. A., 122
Mexico, 4, 112
Meyers, Donald N., 197
Miami (Fla.), 9, 36
Mice, 65
Michigan, Univ. of, 22, 81, 82, 125, 127, 131
Michikawa rocket center, 99
Mickelsen, Wm. R., 199
Microbiology, 84
*Midas* satellite, 85, 120, 121, 123, 134, 147
*Midway* (CV), 58
MiG-15, 57, 66, 72
Mikesell, A., 98, 164
Milbank, J., 210
Miller, C. D., 37, 185
Miller, F. H., 190
Miller, J. A., 210
Miller, J. E., 186
Miller, T. H., 127
Millikan, Clark B., 191, 197

Millikan, Robert A., 29
Mills, Charles A., 179
Millstone Hill Radar Observatory, 110
Minitrack, 83, 88
Minneapolis-Honeywell, 99
Minnesota, Univ. of, 83, 109, 162, 164
Minshall, R. J., 196
Minuteman ICBM, 96, 112, 113, 130
*Mirak II*, 29
Missile ranges (See AMR, PMR, WSPG, etc.)
Mistran, 125
Mitchell, H. F., 126
Mitchell, R. J., 34
Mitchell, William, 13, 14, 20, 157
Mitchell Field (N.Y.), 12, 37
Mitscher, Marc A., 204
ML alloy, 52
Moby Dick balloon, 71, 73, 163
Model Test Basin, Taylor, 48
Moekel, Wolfgang E., 199
Moeller, C. A., 52
Moffett, William A., 14, 30, 204
Moffett Field (Calif.), 37, 38, 40, 52, 113
Moise, John W., 187
Mollison, Amy Johnson, 178
Mollusks, 15
Molybdenum casting, 106
Monkey (rocket payload), 62, 63, 109–110, 118
Monoethylanline, 46
Moody, A. B., 195
Moon, as radio reflector, 25, 55, 109
  eclipse of, 13
  impact of *Lunik II*, 113
  landing on, 122
  maps of, 126, 130, 132
"Moon rocket," 25
Moonwatch, 128
Moore, Charles B., 63, 115, 163
Moore, J. H., 116
Mortar-fired bomb, 42
Morrell, Gerald, 199
Morrow Board, 20
Morse, Al, 170
Moscow (USSR), 33, 35
Moscow Radio, 75, 132
Moscow-to-San Jacinto (Calif.) flight, 35
Mosely, C. C., 12
Mosely, Harry G., 182
  Award, 187
Moslem, 70
Mosquito bomber, 36, 40, 41
Moss, Sanford A., 11, 174, 191
Mount Clemens (Mich.), 14
M.O.U.S.E., 72
Mousetrap rocket, 42
Muchio's Field (N.J.), 20
Mulligan, Denis, 204
Multi-engine aircraft, 27
Multiple g-loads, human tolerance to, 37
Munich (Ger.), 36
Munitions Board, 72
Murock Dry Lake (Calif.), 44, 47, 57, 61
Murphree, E. V., 81
Murray, Arthur, 75, 76
Murray, Robert B., Jr., 204
Murry, Don, 210
Murry, Peter R., 194
Musick, E. C., 33, 177, 179
Mustin, H. C., 4
MX-324, 47
MX-776 (Rascal), 53, 70
Myers, D. A., 22
Myers, John, 46
Mylar plastic, 83, 126
Myrus, Don, 210

N2C-236
N-9, Navy "Flying Bomb," 6, 7, 9, 10
  Seaplane, 18
Nagasaki, 51
NATIV Missile, 59
Namias, Jerome, 186, 192
Nash, J. Slade, 70, 158
Natcher, Wm. H., 97
National Academy of Sciences, 34, 38, 40, 77, 78, 88, 94–96, 99, 101–103, 105, 108, 113, 117
  *IGY Bulletin of*, 210
  National Research Council of, 5
  Space Science, Board of, 99, 102, 108
National Advisory Committee for Aeronautics (NACA)
  aeronautical research, iii, 3, 12, 39, 41, 45
  airmail recommendation, 5
  Annual Reports, 187–88
  awards of, 210
  Chairman of, 3, 5, 11, 38, 42, 202
  committees of, 19
  cowling development, 24, 173
  early history, 3–7
  Executive Committee of, 11, 19, 42, 69
  facilities of, 37, 50, 62
  Federal legislation, 9, 12
  Final Report, 108, 208
  high-speed camera, 37
  history of, 209–10
  inspections and conferences of, 21, 24, 26, 28
  international U.S. policy, 10
  jet engine of, 44
  members of, 3, 25, 202–5
  Memoranda, 14
  Muroc Flight Test Unit, 55
  National Scientific Research Center, 54
  Office of Aero. Intelligence, 54
  Paris Office, 40
  personnel, 170, 172, 174, 181, 190
  Reports of, 12, 16, 19, 26, 28, 37, 41, 108
  rules and regulations, 62
  Secretary of, 23
  Space Technology committee, 92
  special committees, 7, 37, 41, 49
  tank, 28
  transfer to NASA, iii, 100–101
  wasp waist concept, 73, 83
  wind tunnels of, 8, 20, 21, 26
  and industry, 4
  and USAF, 98–101
  (See Ames, Langley and Lewis Laboratories, Wallops Island, etc.)
National Aeronautics and Space Administration (NASA)
  Administrator, 100, 101, 109, 116, 118, 131, 134
  advisory committees of, 104, 106
  Associate Administrator, 125
  awards, 187–88, 199
  Committee on Long-Range Studies, 109
  Deputy Administrator, 100
  Founding of, 89, 94, 95, 97, 99, 100–2, 207, 208, 212
  Inventions and Contributions Board, 104, 188
  Launch Operations Directorate, 124, 131
  Launch record, 134
  Office of Launch Vehicles, 116, 117
  Office of Life Sciences, 104, 112, 118, 120
  Office of Space Flight, 99
  Office of Technical Information and Educational Programs, 123
  Office of Public Information, 139, 210
  Office for the United Nations Conference, 118
  programs of, 102, 212

National Aeronautics and Space Administration—Continued
  Space Task Group, 103
  Ten Year Plan, 118, 212
  organization of, 99, 101, 102, 112, 112–7, 212
  and Congress, 212
  and Space Act of 1958, 100, 212
  and AEC, 127, 129, 133
  and ARPA, 102
  and DOD, 127, 130, 212
  and Space Council, 101, 212
  (See Ames, Langley, Lewis and Flight Research Centers, and Goddard and Marshall Space Flight Centers, JPL, WOO, and Wallops Island)
National Aeronautic Association, 16, 37, 153, 173
National Air Council, 189
  Award, 189
National Airlines, 104
National Air Museum, 54, 172
National Booster Program, 105
National Aviation Education Council, 172
National aviation policy, 11, 32;
  NACA recommendations on, 11
National Bureau of Standards, 7, 9, 19, 24, 33, 37, 41, 109, 112, 190, 203–205
  aircraft engine laboratory, 12
  Director of, 38, 203–205
  jet propulsion study, 14, 28
  Mech. and Sound Division, 44
  radio-echo altimeter, 26
  Altitude engine laboratory, 7, 9
National Center for Atmospheric Research, 131
National Civil Service League Award, 199
National Committee for IGY (U.S.), 71, 75, 77, 78, 85, 91, 92, 95, 98, 100, 102
National Conference on Aeronautical Education, 26
National Defense Act of 1940, 37
National Defense Research Committee, 40, 41, 42, 43, 46
  Division, 5, 43, 46
National Geographic, 31, 33, 162, 208
National goals, 132
National Institutes of Health, 5
National Parks, aerial observations of, 14
National Program, for guided missile, 53
National Radio Astronomy Observatory, 109, 117, 121
National Research Council, first meeting of, 5
National research effort, 52, 212
National Rocket Club, 91, 167, 169, 180, 189
National Science Foundation, 64, 74, 75, 77, 82, 94, 98, 107, 114, 117, 131, 133
National sovereignty, in air space, 11, 48
National space program and policy, 92, 97, 209, 212
  (See Congress, NASA, USAF, etc.)
NATO Advisory Group for Aeronautical Research, 69
*Nautilus*, 74
Navaho missile, 64, 70, 72, 73, 74, 76, 77, 86
Navy Department (USN)
  Aircraft Factory (NAF), 6, 8, 26, 38, 39, 41, 43, 45, 81
  Air Stations (NAS)
    Anacostia, 10, 20, 26, 36, 43
    Chincoteague, 111
    Glynco, 123
    Hampton Roads, 9
    Lakehurst, 17, 37, 69
    Miami, 8
    Mountain View, 75
    New York, 47

Navy Department—Continued
  Air Stations (NAS)—Continued
    Norfolk, 92, 102, 183
    Pensacola, 3, 7, 8, 15
    Rockaway, 13
    San Diego, 22
  AMAL, 100, 112
  AMOC, 110
  ARL, 15
  Bureau of Aeronautics, 14, 15, 34–35, 115, 202–5, 212
  Bureau of Construction, 34
  Bureau of Naval Weapons, 115
  Bureau of Ordnance, 10, 13, 40, 43, 46, 115
  Bureau of Steam Engineering, 8, 11
  CEFSR, 51, 53
  CNO, 34, 54
  Gun Factory, 46, 69
  Medical Research Institute, 181
  NAMC, 45, 46, 72, 181
  NAMTC, 55, 61, 80, 95
  NOL, 62, 46, 53
  NOTS, 66, 70, 100, 111, 121
  NPG, 18, 40
  NRL, 16, 34, 42, 52, 53, 69, 70, 75, 83, 103, 118
  Observatory, 4, 19, 98
  ONR, 54, 67, 70, 74, 76, 78, 79, 98, 101, 163–65
  Secretary of, 46
  School of Aviation Medicine, 15, 38, 86, 181, 183, 196
  (See aircraft carriers, programs and projects by name, and Loon, Polaris, Regulus, Skyhook, Sidewinder, Stratolab, Vanguard, Viking and Transit)
NB–1,
NB–36H, 78
NC (flying boat)
  NC–1, 6, 7, 9
  NCO4, 10
Nebraska, Univ. of, 172
Neely, James T., 18
Neihouse, Anshal I., 188
Neill, T. T., 41
Neiman, Jack, 98
Nene turbojet, 57
NEPA, 54, 78
Neptune (PV–2), 55
Neptune rocket, 53
Nerburger, Morris, 186
Netherlands, The, 132
Neuhardenberg, 35
Neumann, Gerhard, 174
Nevada Test Site, 128, 129
Newell, Homer E., Jr., 116, 189, 193, 210
Newhouse, R. C., 193
*New Jersey* (U.S.S.), 17
New Mexico School of Mines, 183
New Rochelle (N.Y.), 37
New York University, 22, 57, 64, 162, 163
Newfoundland, 47
Newton, Byron R., 204
Nieuport, Edouard, 156–157
Nieuport Fighter, 4
Night, flying at, 16
  wind studies at, 11
Nike I missile, 47, 49, 69, 70
Nike-Asp (sounding rocket), 102, 111, 112, 115
Nike-Cajun, 82, 98, 125
Nike-Deacon, 73, 78
Nike-Recruit, 84
Nike-Zeus (A–ICBM), 114, 132
Nitric acid, 46
Nitrogen, 40
Noble, Edward J., 204
Noise, jet, 186, 197
Noiseless airplane, 57

230

Nomenclature, for aeronautics, 11
NOMTE, 84
NORAD, 130
Norden, Carl L., 13, 58
Norden bombsight, 58
Nonrigid airships, 14, 47
Norfleet, Comdr. J. P., 15
*Norge*, 21
North American Aviation, 39, 59, 64, 70, 72, 76, 79, 83, 93, 119, 133
  B-25, 39, 43
  Harvard trainer, 36
  P-51 Mustang, 39
  XB-45, 56
  (See Rocketdyne)
*North Carolina* (USS), 4
North Korea, 65
North Island Flying School, 6
Northolt (Eng.), 36
North Pole, 21, 35, 70
Northrop, John K., 210
Northrop Aircraft Corp., 32, 41, 70, 81, 195
Northwestern Univ., 203
*Norton Sound*, 58, 61, 62, 63, 64, 70, 101
Norway, 132
Nose cone, 69, 86, 109
NOTS (See Navy)
Noville, 23
Noyes, W. Albert, 102
Nozzles, of rockets, 61
Nuclear chain reaction, 44
Nuclear emulsion, 127
Nuclear Engine Propulsion (aircraft) (NEPA), 54, 68, 78, 92
Nuclear-powered aircraft, 92, 186, 209
Nuclear power, for rockets, 80
Nuclear Propulsion Office (NASA-AEC), 127
Nuclear rocket program, 116
Nuclear weapons, 56
Nugent, Richard E., 184
Nyrop, Donald W., 204


O2U, 26
Oak Ridge National Laboratory, 76
Oberth, Hermann, 18, 26, 79, 189, 193, 210
*Observation Island*, 115, 121
Ocker, W. A., 22
Odishaw, Hugh, 207
OGO satellite, 134
Odom, Wm., 192
Oehmichen helicopter, 18
Office of Education, 167
Office of Emergency Management, 41
Office of Scientific Research and Development (OSRD), 41
Ofstie, Ralph A., 19, 204
Ogburn, William F., 210
Ohio State Univ., 196
OJ-2, 31
Old Ferris Point (N.Y.), 35
Olds, Robert, 36, 179, 184
Opel Sander Rak. 1, 25
Operation Far Side, 91
Operation Highjump, 56
Operation Paperclip, 50, 52
Operation Plumbob, 87
Optical measurements, 71
Orange, 109
Orbits polar (See *Discoverer*)
ORDCIT, 47, 49
Ordnance (See Army Ordnance)
Ordway, F. I., III, 210
Origin of universe, 90
*Osaviakhim* (USSR), 162
Oscillations, high frequency, 71
Ostrander, Donald R., 116, 210

O'Sullivan, Wm. J., Jr., 169, 210
Oswald, Wm. B., 194
Otis, R. M., 16
Outside loop, first flown, 22
Oxygen, 13, 24, 25, 26


P2Y, 29, 31
P5M-2, 134
P6M, 78
P-36, 37
P-38, 37
P-40 fighters, 36, 39, 40
P-41, 39
P-47, 43
P-51, 39, 47
P-59, 47, 52
P-61, 54, 55
P-80, 47
P-80R, 57
Pace, Ernest M., Jr., 204
Pacific Missile Range (PMR), 95, 99, 123, 124, 127, 131
Pacific survey flight, 28
Packard Motor Car Co., 6, 12, 23, 24, 173
Page, A. H., 26
PBM (Mariner), 35
PBY (Catalina), 23, 37, 45
PBY-5A, 43, 49
Paimboeuf (Fr.), 8
*Palaemon* barge, 133
Palmer, Eric, 192
Pan American Airways, 32, 33, 34, 35, 38, 42, 174
  Pan American World Airways, 74, 103
Pangborn, Clyde, 28, 179
Paperclip, (project) 50, 52
Pappas, Costas E., 196
Parabolas, aerodynamic, 68
Parachute,
  aircraft recovery, 24
  records, 12 13, 15, 18, 65, 114, 116, 126
  rocket recovery, 38
Parasite, fighter, 65
  wind tunnel, 56
Paris (Fr.), 10, 11, 48
Paris Office (NACA), 10
Parkinson, John B., 188
Parry, Albert, 210
Parsons, John W., 34
Parsons, Jack, 37
Pas de Calais, 46
Pasadena (Cal.), 34
Passengers,
  aircraft records, 25, 27, 28, 69
  air compared to Pullman, 68
  commercial flights, 27
  first jet, 69
  statistics, 35
Passenger service,
  first transpacific, 34
  first transatlantic, 38
  first jet transcontinental, 105
Patents, 125
  cross-licensing and, 5
  for jet engine, 27
  infrigement of, 5
  NACA and, 6
  on radar, 15
  rights, 18
Patent Office (U.S.), 3, 58
Patterson, Robert, 51
Patrick, Mason M., 65, 204
Patrick AFB, 65, 108
  (See AFMTC, Cape Canaveral)
*Patrick Henry*, 113, 129
Patuxent River (Md.), 60
Paul, Wilfred J., 27

Pavillac (Fr.), 8
Payload,
  of rocket, 25
  IGY, 99, 100
  (See Rockets and Satellites by name)
Payne, L.G.S., 210
Payne, Richard G., 188
Peaceful uses, 89, 94, 95, 96, 97, 128
Peacemaker (See B-36)
Pearl Harbor, 42
Pearson, Lee M., xi, 212
Peenemünde, 35, 43, 44, 50, 51;
  V-1's flown at, 47 (See V-2)
*Pegasus*, Magazine, 207
Peiping, 113
Pelican missile, 43, 47, 48
Pendray, G. Edward, 26, 27, 189, 209, 210
  Award, 189
Pensacola (Fla.), 39
Pentagon, 50
Pershing rocket, 120, 133
Petras, T. A., 39
Petterssen, Sverre, 197
Pfingstag, Carl J., 204
Phantom (See XFD-1)
Philadelphia (Pa.), 46
Philippines, 61
Phillips, William H., 59, 194
Phoebus, C. P., 183
Photoelectric detector, 41
Photographic measurements, 71
Photography,
  aerial, 4, 18, 24
  combustion analysis, 37
  at night, 8, 20
  of satellites, 109
  from satellites, 111, 113
  of moon, 130
  of eclipse of sun, 29
Physiological research, 29, 35
Piasecki, Frank N., 194
Piasecki helicopter, 63
Piccard, Auguste, 28, 29, 162, 179, 210
Piccard, Donald, 125, 163
Piccard, Jean, 31, 54, 162
Piccard, Jeanette, 31, 162, 179
Pickering, William H., 198, 210
Pierce, Erold F., 185
Pierce, H. F., 35
Pierce, J. A., 195
Pikes Peak (Colo.), 37
Pilotless aircraft, 5, 6, 9, 28, 34, 47, 48
Pine Castle (Fla.), 53
Pinkel, I. Irving, 39, 170, 187
Piper Comanche, 124
*Pioneer I* (space probe), 103, 142
*Pioneer II*, 103, 104, 142
*Pioneer III*, 89, 104, 105, 142
*Pioneer IV*, 107, 124, 143
*Pioneer V*, 89, 120-122, 146
Pioneers of rocketry, 212
  (See Goddard, Oberth, Ziolkovsky)
Pitcairn, Harold F., 24, 173
Pitcairn aircraft, 27
Pittsburgh, Univ. of, 182
Planets, 90, 93
  (See Mars, Venus, etc.)
Plasmadyne Corp., 126
Plasma propulsion, 105
Platte river, 18
*Pleiades* balloon, 162
Plug nozzle, 119
Plum Brook reactor (NASA), 80
Pluto, 80
Plywood, development of, 12
Point Arguello (Calif.), 111, 127

Point Mugu (Calif.), 53, 65, 80, 95, 99
Poland, 38
Polando, J., 154
Polaris, IRBM, 84, 92, 95, 96, 102, 106, 110, 115, 116, 118, 121, 131
  A-2, 132
Politburo, 58
Polyethylene balloons, 57
Pomerantz, Martin, 57
Poppen, John R., 29, 30, 32, 181
Port Darwin (Aust.), 11
Port Washington, 8
Porter, Richard W., 79, 176
Post Office Department (U.S.), 9, 130
  first airmail route, 8
  regular daily airmail route, 13
  contract of air mail, 20
Post, Wiley, 28, 175, 177, 179
Powers, Edward M., 204
Prandtl, Ludwig, 14, 26, 37, 176
Prather, Victor G., 165
Pratt, Fletcher, 26
Pratt, Henry C., 204
Pratt-Whitney, 21, 24, 29, 41, 62, 66, 103, 117
Pre-set guidance, 42
Pressed-powder propellant, 42
Pressure chamber, 98
Pressure suit, 62, 102
Pressurized-cabin airplane, 13, 35, 40
Price, Don K., 210
Price, G. E., 39
Price, John D., 204
Price, Ward, 32
Princeton University, 57, 61, 66, 111, 163
Private A rocket, 47, 48
Private F rocket, 50
Probe and drogue, 63
Project Advent, 132
Project Apollo, 125, 127, 129
Project Argus, 101, 107
Project Defender, 131
Project Bumper, 55, 65
Project Discoverer, 104, 180
  (See *Discoverer I*, etc.)
Project Excelsior, 114, 115, 164-165
Project Far Side, 86, 163
Project Fox, 43
Project Hardtack, 100
Project Helios, 54, 55, 162
Project Hermes, 48, 56, 60, 64, 68, 76, 211
Project ICEF, 164
Project Mercury, 100, 102, 106, 108-135, 212
  (See Astronauts, McDonnell, STG, etc., Big Joe, Little Joe, Redstone, Atlas)
Project Midas, 120, 121, 123, 134, 147
Project MOUSE, 72
Project NEPA, 54
Project, NERV, 127
Project Orbiter, 75, 78
Project OZMA, 121
Project Paperclip, 50, 52
Project Pilot, 105
Project Pluto, 80
Project RAND, 53, 58, 60
Project Rover, 82, 124, 129
Project Saturn (See Saturn)
Project Saucer, 63
Project Score, 105, 143
Project Skyhook, 58
Project Stratolab, 76, 82
Project Transit, 188
Project TED, 42
Project Vanguard, 79, 84, 88, 91, 92, 96, 98, 102, 108, 110, 143, 208, 210
Projectile ballistics, 96
Prokofiev, G., 30, 162
Propaganda, 71
Propane fuel, 13

Propeller
  and jet aircraft, 51
  early research and tests, 11, 12, 15, 19
  five-bladed, 51
  pusher-type, 31
  reversible pitch, 11, 15, 60
  variable pitch, 30
  (See Turboprop)
Propulsion (See subtopics)
Provisional International Civil Aviation Organization (PICAO), 48
Public Health Service, 13
Puckett, Allen E., 194
Puerto Rico, 128
Pulitzer Trophy, 12, 17
Pullman passengers, 68
Pulse-jet, 45, 67
Pupin, Michael I., 204
Purser, Paul, 105
Putnam, Amelia Earhart (See Earhart)
Putt, Donald L., 204
Pyle, James T., 204

Quarles, Donald A., 107, 204
Quesada, Elwood R., 25
*Question Mark*, 25

R-4 helicopter, 38
R4D aircraft, 56
R-5, 52
R-34 airship, 10
R-1830 engine, 39
R-2800 engine, 41
Race,
  aircraft, 12, 17, 19, 34
  arms, 128
  balloon, 15
Radar, 15, 41, 42, 45, 61, 74
Radford, Arthur W., 204
Radial engine development, 11, 25
Radiation belts, 113, 115, 133
  (See Van Allen Belts)
Radiation Lab., U. of C., 80
Radio
  air-to-air, 5, 20
  beacon, 20
  compass, 16, 30, 33
  direction finder, 9
  loop antennas, 11
  meteorgraphs, 36
  transmission in space, 55
Radio-biology Lab., 64
Radio controls, of aircraft, 4, 8, 9, 11, 34, 36, 39, 41, 42, 43, 44, 45, 47, 48, 53, 54
  of glide bomb, 41, 46
Radio Corporation of America (RCA), 41, 107
Radiosondes, 33, 36
Radio Technical Commission for Aeronautics, 174
Raff, G. E., 182
Rainfall, 73
*Raketen flugplätze Berlin*, 27
Ramjet aircraft, 60, 62
  fuels, 69
  helicopter, 66
  missile, 64
  propulsion, 30, 45, 57
Ramo-Wooldridge, 99
RAND Corp., 60, 65, 74
Randolph Field (Tex.), 28
Randt, Clark, 120
*Ranger* (CU-4), 30, 36
Range safety, 86
Rapid City (S. Dak.), 33, 83
Rascal missile, 53, 70
Raven Industries, 164

Raymond, Arthur E., 195, 205
RB-50A, 72
RDB, 58
Reaction Motors, Inc., 42, 55, 63, 70, 119
Read, Albert C., 10
Reber, Samuel, 205
Rechlin (Ger.), 45
Reckford, Frank, 48
Redstone Arsenal, 64, 68, 72, 75, 80, 81, 102, 103, 109, 124
Redstone Missile, 72, 75, 76, 80, 81, 83, 100, 101, 102, 107, 131, 134
Reed, Sylvanus Albert, Award, 32, 173
Reed, W. F., 8
Reed, Wendell E., 197
Reeder, John P., 173
Re-entry, into atmosphere, 69, 80, 84, 93
Rees, Ed, 211
Reforestry, by aircraft, 21
Refrigerated wind tunnel, 24
Refueling,
  air-to-air, 9, 14
  history of, 210
  hose, 97
  jet aircraft, 67, 72
  pipe, 17
  techniques of, 63
Regeneratively-cooled thrust chambers, 36
Regulation
  of international air commerce, 11
  of civil aviation, 21
Regulus missile, 66, 72, 92, 93, 110, 121
Reichelderfer, Frances W., 188, 205
Reid, Henry J. E., 21, 48, 58
Reinartz, E. G., 180
Reitsch, Hanna, 35
Remote-controlled flight, 38
Rentschler, Frederick B., 177
Rentzel, Delos W., 205
Republic Aviation, 66, 98 ;
  XP-47, 41
Research and development
  conduct of, 21
  policy on, 21
  in space technology, iii
Research and Development Board, 203–205
Retrorocket, 43
Reversible-pitch propeller, 11, 15, 60
RF-80, 67
Rhesus monkey, 109, 110
Rhine River, 15
Rhode, Richard V., 196
Rhön mountains, 24
Rich Tool Co., 21
Richardson, Holden C., 9, 205
Richardson, Lawrence B., 205
Richie, Wm., 65
Richter, J. P., 17
Rickenbacker, Edward V., 184, 205
Ridenour, Louis N., 63, 193
Riehl, Herbert, 183, 186
Ritchey, Harold W., 190
Rittenhouse, David, 7
Robbins, H. W., 170
Roberts, Ernest, 190
Roberts, Walter Orr, 131
Robertson, Bruce, 210
Robertson, W. A., 6
Robins, Augustine W., 205
Robinson, Russell G., 34
Robot, airplanes, 41
  bombs, 53
  (See Pilotless, Radio-control)
Rochester (N.Y.), 20
Rochester, Univ. of, 131
Rockair, 78
Rockaway Beach (N.Y.), 9
Rockefeller, Alfred, xi

233

Rockefeller, Wm. C., 193
Rockefeller Public Service Award, 71, 199
Rockets,
 air mail by, 28, 33, 131
 pioneers, 212
 potentialities of, 25
 (See types by name)
Rocket Development Branch (Army), 46
Rocket-powered,
 aircraft, 33, 46, 47, 48, 51, 52 (See X-1, X-2, X-15)
 auto, 24
 bomb, 40
 glider, 24, 25;
 ram, 46;
 sled, 65, 72, 76, 96, 98, 107
 torpedo, 41
Rocket propulsion, 45
 fuel pumps, 17, 30
 principle demonstrated in vacuum, 4
 (See types of fuels)
Rocket Research Project (Cal. Tech.), 34, 38
Rocketdyne (NAA), 71, 74, 77, 104, 123, 180
 (See North American)
Rocket-sonde research, 52
Rocket thrust chamber tests, 35, 37
 ceramic-lined, 41
Rockets and Satellites Research Panel, 86
Rockwell Field (Calif.), 15
Rockoon, 62, 67, 70, 87, 110, 163
Rodert, Lewis A., 174
Rogers, J. W., 116, 158
Rogers, Will, 195
Rohlfs, Roland, 8, 11
Rohrbach, Adolph (Ger.), 11
Roll coupling, 59
Roll damping, 60, 61
Rolls Royce, 10;
 Trent engine, 51
*Roma*, 14
Rome (It.), 42
Romorantin, France, 8
Roosevelt, President Franklin D., 6, 33, 37, 38, 40, 41, 42
*Roosevelt*, USS *Franklin*, 54
Roosevelt Field (N.Y.), 14
Roots-type supercharger, 19
Rosen, Milton W., 198, 211
Rosendahl, Charles E., 179
Ross, Chandler C., 176
Ross, Malcolm D., xi, 82, 83, 91, 98, 100, 111, 115, 161, 163, 179
Rossby, Carl-Gustav A., 168, 169, 183, 190, 192
Rossi, Maurice, 154
Rossow, Vernon J., 79
Roswell (N. Mex.), 27, 29, 32, 34
Rotary-wing aircraft
 first, 17
 first Navy, 27
Rothschild, Louis S., 205
Rotterdam (Neth.), 40
Round-the-World flights
 nonstop flights, 61
 nonstop jet, 85
 of *Winnie Mae*, 28
Round Three Committee (NACA), 85, 86, 87, 91
Rover Program, 80, 124, 129
Roy, Maurice, 198
Royal Aeronautical Society (Br.), 12, 58, 62
Royal Aircraft Factory, 5, 7, 211
Royal Air Force (RAF), 27, 32, 34, 36, 40, 42, 43, 47, 181, 211
 proposal for, 6
 Battle of Britain, 33, 40
 Bomber Command, 38, 43
 History of, 211
Royal Canadian Air Force (RCAF), 183, 195

Royal Flying Corps (RFC), 5
Royal Radar Establishment (Malvern), 112
Royce, Ralph, 184
Ruhr (Ger.), 38, 40
Runartz, E. G., 182
Russell, A. E., 197
Ryan, Oswald, 205
Ryan Aircraft, 45, 83
 FR-1 Fireball, 51
 X-13 of, 80
Rynin, Nikolai A., 25


Sabine, Wallace C., 205
Sabrejet (XF-86), 58
Sachs, Alexander, 38
Sadler, D. H., 195
Safety in flight, 14, 24, 85, 120, 170
Sage, Bruce H., 190
Sailplanes, 54
"SAM" (monkey), 115
Sam, "Miss" (monkey), 118
*Samos*, 129
San Andreas rift, 13
San Antonio (Tex.), 12
San Clemente Island, 96, 121
Sandacz, V. L., 102
San Diego (Calif.), 3, 6, 9, 11, 39
Sänger, Eugen, 31
Santos Dumont, A., 154, 156
*Saratoga*, 25
Sargent, Fred., II, 171
Satellite of earth
 application to communication (See *Echo*, *Courier*, *Score*)
 application to weather analysis, 74 (See *Tiros*)
 early U.S. development, 51, 52, 58, 59
 scientific use of, 212
 summary of successful launchings, 140-151
Satellite Research Panel, 94
Satre, Pierre, 175
Saturn launch vehicle, 102, 109, 110, 115, 116, 121, 124, 132-134
Saufley, R. C., 4
Saville, Gordon P., 205
Savoia-Marchetti, 30
Sayer, Fl. Lt., 41
SBD Dauntless, 32
Scapa Flow, 39
Schaefer, H. J., 196
Schaefer, Vincent J., 55, 169, 183
Schairer, G. S., 191
Schanche, Don A., 211
Schilling, David C., 65, 178, 192;
 Award named for, 192
Schilt, C. F., 22
Schlaifer, Robert, 211
Schmidt, Paul, 29
Schmiedel, Friedrich, 28
Schneider, Edward C., 8, 180
Schneider Cup Race, 17, 19, 34
Schock, G. J. D., 79
School of Aviation (Aerospace) Medicine, 7, 15, 16, 22, 28, 45, 55, 58, 60, 65, 73, 84, 95, 110, 117, 127 133 180 181 183 196
 (See Army Air Service AAC, AAF, Navy, USAF)
Schriever, Bernard A., 75, 108, 109
Schroeder, R. W., 9-11, 159
Schubauer, Galen B., 191
Schweinfurt-Regensburg raid, 45
Schweizer sailplanes, 87
Science, 51, 211
 and government, 210
Scientific Advisory Board (AAF), 41, 48, 50, 51, 54, 211

Scientific Research and Development, Office of, 41
Scorpion (F-89), 60
Scott Field (Ill.), 22
Scout launch vehicle, 87, 92, 99, 100, 107, 118, 122, 124, 128, 132, 133
 (See *Explorer IX*)
Scriven, George P., 3, 202, 205
Scylla, 121
S.E.-5 (Br. fighter), 7
Sea Dart, 72, 75
Seal, John C., 178
Seamans, Robert C., 125, 194
Seamaster (See P6M)
Sea sled, 9
Secretary
 of Air Force, 60
 of Defense, 57, 78, 101, 115
 of Navy, 3, 6, 12, 21
 of Smithsonian Institution, 3
 of Treasury, 10
 of War, 3, 6, 12, 49, 51
Sedov, L. I., 78
Seeler, Henry, 170
Seifert, F. W., 16
Seifert, Howard, 211
Seismological Society of America, 13
Selfridge Field (Mich.), 16
Sells, Saul B., 182
Sergeant rocket, 80, 134
 Delta, 118
Sessums, John, 34
Settle, T. G. W., 30, 162, 179
Sextant, 3, 10
Sharp, Edward R., 44
Sheldon, Chas. S., xi, 212
Shelton, W. R., 24
*Shenandoah* (ZR-1), 17, 18, 20
Shepard, Alan B., Jr., 108, 187
Sherwin, Chalmers W., 193
Shesta, John, 176
Shook, Fred W., 185
Shotput, 121
Shrader, Welman A., 210
Siberia, 28
Sibole, J. G., 124
Sidewinder missile, 70, 73, 82, 102
Signal Corps (See Army)
Sikorsky, Igor I., 38, 41, 171, 175, 177, 181, 190
Sikorsky Aircraft Co., 43, 63, 70, 181
Silo, 132
Silverstein, Abe, 49, 52, 99, 197
Simon, Dorothy M., 71, 199
Simon, Leslie, 211
Simons, David G., 87, 163, 172, 175, 181, 196, 211
Simulated,
 altitude, 98 102
 entry, 120
 space cabin, 81, 127
 space flight, 96
Singer, S. Fred, 62, 72
Single-rotor helicopter, 41
*Sirius*, 28, 30
Siry, Joseph W., 199
Skinner, Leslie, 190
Skinner, Richard M., 210
Ski-seaplane 67
Shramstad, S. K., 191
Skybolt missile, 118
*Skyhook* balloon, 119, 162, 163
Skylark rocket, 98
Slayton, Donald K., 108
Sled, rocket research, 65, 72, 76, 96, 98, 107
Sleeve valve, 36
Slotted-throat tunnel technique, 56, 63
Slow-burning rocket propellant, 37

SM-73, 86
*Small-World* Balloon, 104, 164
*SMART*, 78
Smith, A. M. O., 34, 176
Smith, D. F., 81
Smith, L. H., 17
Smith, Levering, 190
Smith, Thomas M., 209
Smith, Wilfred J., xi
Smithsonian Institution, 10, 11, 15, 21, 34, 54, 60, 92, 124, 203-205, 209
 Astrophysical Observatory of, 92, 106, 109, 127
 Secretary of, 11, 203-205
Smuts, Jan Christian, 6
SNAP, 84, 106, 114,
 SNAP-2, 114, 124
 SNAP-8, 106, 121
Snark (B-62), 62, 70, 77, 81, 83, 91, 99, 108, 132
Snow, artificial, 55
Social effects of aviation, 210
*Société Astronomique Francaise*, 23
Society of Automotive Engineers, 176
Society for Space Travel (*Vfr*), 22
Socony-Vacuum Oil Co., 34
Sodium-filled valves, 21
Solar eclipse, 102
Solar flares, 93, 109, 113, 130, 134
Solar-power, 79
Solar storms, 134
Solar telescope, 117
Solar protons, 109
Solid-fuel, rocket development, 43, 55, 112, 129
 (See Minuteman, Polaris, Scout)
Soncek, A., 25, 26, 160
Sorenson, Frank E., 172
Sounding balloon, 27
 (See Radiosondes)
Sounding rockets, 34, 62, 83, 102, 114, 117, 209-211
 USSR, 75
 (See Aerobee, Asp, Blue Scout, Nike, Rockoon, Scout, V-2)
Sound-location detection, 27
South Africa, 127
South Pole, 25, 26
Southwell, Richard V., 197
Sovereignty, in air space, 11, 48
Spaatz, Carl A., 25, 38, 174, 205
Space,
 age, 89, 210
 environment of, ix, 89, 93, 209
 next 10 years, 212
Space biology, 55, 127, 209
 (See Life Sciences, Biomedical)
Space cabin, 81, 126, 127, 132
Spacecraft,
 Apollo, 120
 interplanetary, 123
 Mercury capsules, 103f, 128
 USSR, 123, 126, 132, 147
*Spacecraft* I, 148
*Spacecraft* II, 126, 149
*Spacecraft* III, 151
Space Council, 101
Space exploration, practical values of, 211
Space flight, manned, 71, 104, 209
Space medicine, 61
Space probes, summary of US and USSR, 140-151
Space Research Pilot Course, 117
Spaceship, 75
Space Task Group (NASA), 103, 127
 (See Mercury)
Space technology, 211
Space Technology Laboratories (STL), 120, 125, 134, 175

235

SPACETRACK, 119, 130
Space weapons, 207
Spain, 121, 132
Spanish Civil War, 34, 36
Sparrow missile, 71
Sparrowbee missile, 125
Spaulding, Ralph, 48
Spark plugs, 41
SPASUR, 130
Special Committee, for the IGY, 76, 83
  of NACA, 37, 41
  (See IGY)
Spectrograph, 115
Spectrometers, 131
Sperry, Elmer A., 4, 173
Sperry, Lawrence B., 15 ;
  award named for, 193-194
Sperry Corp., 5, 9, 10, 11, 27, 33, 83, 134, 194
Spins
  study of, 23
  wind tunnel study of, 27
Spitfire fighter, 33, 34, 210
Spitsbergen, 21, 108
*Sputnik I,* 89, 91, 140
*Sputnik II,* 91, 92, 97, 98, 140
*Sputnik III,* 98, 100, 103, 121, 134, 141
Sportplanes, 20, 28
Springer, T. E., 8
Squier, George O., 205
Squirrel monkey, 104
St. Louis (Mo.), 11, 37
  Spirit of, 195
  Stack, John, 174, 191, 197
Stalin, Joseph, 58
Stamer, Friedrich, 24
Standard Aircraft Corp., 8
Standard atmosphere, 16, 19
Standardization, 17
  instrument, 8
Stanford University, 12, 22, 119
Stanley, Robert, 44
Stans, Maurice H., 97
STOL, 113
Stapp, John P., 58, 72, 76, 181, 182, 189, 193, 198
*Star of Poland,* 38
State Department, 12
Stehling, Kurt, 211
Stemmer, Josef, 211
Stevens, A. W., 15, 18, 24, 29, 31, 33, 38, 83, 162, 184
Stevens, Leslie C., 205
Stewart Homer J., 79
Stewart Committee, 79
Stever, H. Guyford, 92, 94
Stievers, Louis, 50
Stinson, Edward, 14
Stormovik (Il-2), 42
Stout, Earnest G., 191, 194
Stout, Raymond K., 35
Stout ST-1, 15
Stovall, Wm. R., 183
Strange, Charles R., 196
Strategic Missiles Evaluation Committee, 72, 74
Stratojet (B-47), 58
*Stratolab* balloon, 83, 163
*Stratolab High* balloon, 91, 100, 115, 163, 164, 165
*Stratoscope* balloon, 87, 111, 128, 163
Stratosphere, 86
  balloon flights, 28, 29, 31, 162-165
  observations in, 98
Stratotanker (see KC-35)
Stratovision, 52
Stratton, Samuel W., 205
Streetcar Satellite, 134
Streett, St. Clair, 12, 24, 184

Stressed-skin concept, 11
Strickroth, Giles J., 194
Strongarm rockets, 125
Strughold, Hubertus, 65, 67, 73, 181, 183, 207
Struve, Otto, 109
Stupalith, 63
Stuhlinger, Ernest, 190
Sturtevant Aeroplane Co., 4
Submarine
  aircraft launching from, 17
  first sinking by aircraft, 6
  rocket launch from, 53, 56
  S-1 21 ; Flights from, 21
  use of rocket against, 47
Subroc, 111
Suicide missions, 47
Suit, full-pressure, 102
Sullivan, Walter, 211
Sullivan, William N., 44
Summerfield, Martin, 189
Sun
  artificial satellites of, ix
  eclipse of, 19, 29
  sun-earth relationship, 89, 113
Sunflower I, 122
Suomi, Verner E., 128
Supercharger, 191
  NACA-type, 24
  Roots-type, 19
  Turbine drive, 18
Supermarine, 34 (See Spitfire)
Super Sabre (see F-100)
*Super Skyhook,* 74, 163
*Super-Solution,* 28
Supersonic, first flight, 58
  air transports, 212
  fighter, 72
  research, 19, 51
  (See X-1, X-2, X-15, B-70)
Surveillance, 92, 123
Sutton, George P., 189
Sutton, H. A., 23, 184
Sverdrup and Parcel, Inc., 54
Swan, William G., 27
Swarthmore College, 57
Sweden, 47, 132
Swept-back wings, 48, 49
  variable, 67
Swept-forward wings, 52
Switzerland, 132
SXB2C-1, 38
Symington, Stuart, 61, 198
Synchronizing gear, 3
Syverton, Clarence A., 81, 194
Szilard, Leo, 38


Tabor, R. H., 102
Tacker, L. J., 211
Tailless, airplane, 61
  glider, 24
Talos missile, 53, 66
Tank, of NACA, 28
Tannehill, I. R., 183
Target-viewing television, 41, 43
Tartar missile, 100
*Tass,* 62 87
Taylor, Albert H., 15
Taylor, David W., 6, 34, 205
  Model Basin, named for, 34, 63, 69
Taylor, Edward C., 190
Taylor, M. B., 43
Taylor Cub, 20, 28
TBF, 47
Teak, 109
Teapot Committee, 73
Technical change, iii, iv, 1, 89
*Technology and Culture,* 208

Telemetry, 48, 51, 56, 61
Television, 41, 46
Teller, Edward, 193
Tepper, Morris, 186
Terrapin, 83
Terrier missile, 69, 104
Test pilots, 55, 69, 72, 77, 108, 117, 124
Test Support Office (LOD), 131
Texas, Univ. of, 74, 76, 101
TG-2 engine, 39, 43
TG-100 engine, 41
Thaden, Louise, 179
Thatcher, H. B., 76
Thetford, Owen, 211
Thiokol Chemical Corp., 119, 129
Thomas, F. M., 196
Thomas-Morse (MB-3), 10
Thomas, Shirley, 24
Thompson, P. D., 186
Thompson, Shelby, 123
Thompson Products, 203
Thompson-Ramo-Wooldridge, 122
Thor-Able, 98, 99, 108, 111, 116, 120, 121
Thor-Able-Star, 121, 126, 128, 132
Thor-Agena B, 133
Thor-Delta, 123, 126, 131
Thor-Hustler, 107
Thor IRBM, 80, 82, 85, 86, 91, 92, 94, 95, 98, 99, 104, 108, 110, 113, 119, 128, 129
Thorner, N. W., 40
Thule (Greenland), 70, 128
Thunderbird Aerial Team, 185
Thunderchief (See F-105)
Thunderjets (See F-84G)
Thurlow Award, 195
Tiamat research missile, 50, 51
Tice, James W., 181
Tikhonravov, M. K., 33, 68
Tilt-wing aircraft, 115
Tilton, T. H., 13
*Tiros I*, 89, 109, 121, 123, 124, 128, 131, 146
*Tiros II*, 131, 151
Titan ICBM, 77, 106, 116, 119, 121, 124, 129, 132, 134
Tissandier, Paul, 156
Titanium, 197
Tizard, Sir Henry, 40
TM-61 (See Matador)
Toftoy, Holger W., 198
Tokyo, 43
Torpedo Airplanes, 39
Torpex, 48
Tory reactor, 133
Towers, John H., 37, 205
Tracking, antenna, 134
    network, 115 ; for IGY, 83
Transatlantic flight
    by NC-4, 10 ;
    first nonstop, 22 ;
    by DO-X, 28 ;
    first blind, 34 ;
    first passenger service, 38
    by nonrigid airship, 47
    by helicopter, 70
    non-stop round trip, 85
    first jet passenger service, 103
    passenger statistics, 105
Transcontinental flight
    first flights, 9
    first in 24 hours, 13
    flights, 25, 28, 36
    by helicopter, 82
Transistor, 60, 71, 74
*Transit* satellite, 121, 122, 124, 125, 148
Transonic
    aircraft development, 47
    bump technique, 55

Transpacific, air route 32
    air mail, 33
    first nonstop flight, 28
    first passenger service, 34
Trans World Airlines, 195
Trapnell, F. M., 173
Travelers Insurance, 183
Treasury, Secretary of, 10
Treaty of Versailles, 10
Tribus, Myron, 170, 197
Trimotor Aircraft (See Ford, Ju-52)
Trippe, Juan T., 177
Transonic, transport, 197
    (See B-70)
*Triton*, 123
Troop Carrier, 61
Troy (Ohio), 14
Traux, Robert C., 35, 37, 176
*Truculent Turtle*, 55
Truman, President Harry S., 52, 56, 57, 58, 62, 67
Truscott, Starr, 28
Tsien, H. S., 34, 189
Tu-114, 112
Tucker, Charles, 61
Tullahoma (Tenn.), 66, 96, 99
*Tunny*, 72
Tupelev, Audrey M., 175
Turbine wheels, 40
Turbo-compressor, 8
Turbojet
    engine development, 27, 39, 41, 44, 45, 46, 62, 105
    first U.S. flight, 47
    operation at altitude, 47
Turboprop aircraft, 51, 61, 186
    first flight, 51
    airliners, 61
Turbosupercharger, 191
Turbulence of atmosphere, 85
Turcat, Andre', 178
Turkey, 110, 122
Turnbull, A. D., 211
Turner, Roscoe, 179
Tuttle, Arnold D., 181
    award named for, 70, 196
Twining, Nathan F., 77, 174, 205
Twiss, L. Peter, 175
Tyler, John M., 186
Tymczyszym, Joseph J., 173

U-2 (aircraft), 73, 85, 110, 122
U-36 (sub.), 6
Ultraviolet radiation, 93
Underground sites, 96, 132
Unidentified Flying Objects (UFO's), 71, 106
Unification of armed services, 52, 57, 59
Unitary Wind Tunnel Act, 62
United Aircraft Corp., 174, 193
United Air Lines, 181, 183
United Kingdom (See Britain)
United Nations, 94, 104, 116, 128
    Committee on Peaceful Uses of Outer Space, 108, 109, 110, 111
*U.S. Aeronautical and Space Activities* reports, 212
U.S. Air Force
    ADC, 87
    AEDC, 63, 66, 67, 96, 99, 110, 119
    Aeromedical Lab., 31, 43, 45, 61, 62, 71, 117, 134
    Air Force Council, 68
    ARDC, 63, 66 67, 70, 74, 75, 83, 108, 109, 170, 211, 212
    BMD 75, 97, 102, 108, 115, 120
    Chief of Staff, 131, 203-205
    created, 57
    MATS, 70

U.S. Air Force—Continued
  missile responsibilities, 96
  MTC, 67, 72, 108, 134, 183 (See Cape Canaveral)
  Moon atlas, 130
  SAC, 74, 94, 112, 133, 185, 207
  SAM, 7, 15, 16, 22, 28, 65, 73, 84, 91, 95, 110, 117, 127, 133, 180, 181, 183, 196
  Scientific Advisory Board 41, 173, 193, 211
  TAC, 185, 192
  and NACA, 101
  and NASA, 75, 77
  (See Army Air Service, Army Air Forces, Atlas, *Discoverer, Midas, Samos,* Thor, Titan, etc.)
U.S. Army (See Army)
U.S. Bureau of Fisheries (See Bureau)
U.S. Bureau of Mines, 106
U.S. Congress, 38, 91–100, 118, 120, 123, 131, 132, 211–212
  legislation, 91–99
U.S. Congress, House
  Committee on Science and Astronautics, 96, 102, 123, 132;
  reports of, 211–212; Panel on Science and Technology, 118, 212
U.S. Congress, Senate
  Committee on Aeronautical and Space Sciences, 100, 131, 132
  Reports and Hearings, 211–212
U.S. Department of Defense (See Defense)
U.S. Forest Service (See Forest)
U.S. Information Agency, 118
U.S. Navy (See Navy, Naval)
U.S. Post Office Department (See Post Office)
U.S. Public Health Service (See Public)
U.S. successful earth satellites, 134
U.S.S.R., 18, 25, 33, 35, 39, 68, 77, 79, 88, 112, 113, 118, 119, 128, 134, 210
  Academy of Sciences, 79
  balloons, 30, 32, 33, 162
  Central Committee, 18
  Encyclopedia on space travel, 25
  H-bomb, 72
  ICBM development, 77
  IGY Committee, 96, 98, 100
  Premier, 58, 112, 113, 128
  record flights, 35, 128
  spacecraft, 132
  rocket test center, 35
  successful satellites and space probes, 134, 140–51
  in war, 39, 41
  Unmanned aircraft, 9, 10, 18, 28, 35, 54
Unwins, Cyril F., 29
Upper-air-soundings, 20
  (See Sounding rockets)
Upper atmosphere research panel, 53, 59, 63, 86, 210
Upson, Ralph H., 196
Uranium, potential military importance of, 38
*Utah*, 36

V-1, 30, 46
  first combat firing, 47
  use in World War II, 50, 207
V-2 rocket, iii, 43–45, 56, 62–68, 208, 210
  first launch, 44
  use in World War II, 47, 48, 50
  firing from U.S., 51, 52, 53
  firing from ship, 57
Valier, Max, 24
*Valley Forge*, 118, 119, 164
Valkyrie (See B-70)
Valley, George E., 193
Van Allen, James A., xi, 70, 87, 93, 95, 98, 163, 190, 193, 212

Van Allen radiation belts, 89, 93, 95, 98, 104, 119
Vandenberg, Hoyt S., 59, 205
Vandenberg AFB, 83, 102, 106, 108, 112, 119, 130
Van Every, Kermit, 197
*Vanguard I*, 96, 109, 119, 141, 210
  (See Project Vanguard)
*Vanguard II*, 106, 143
*Vanguard III*, 113, 116, 145
Van Orman, Ward T., 179
Van Straten, F., 99
Van Wyen, Adrian O., 212
Variable-density wind tunnel (NACA), 15, 23
Variable-pitch propellers, 11, 30
Variable-sweep aircraft, 67
VB-1 (Azon), 42, 49
V-E Day, 41
Vega booster, 109, 116
Venus (planet), 115, 116, 127
  radar returns from, 95, 107
Verden, J. B., 73, 158
*Verein fuer raunschiffahrt (Vjr)*, 22, 26, 29, 65
Vertijet (See X-13)
Veronique rocket, 107
Versailles, Treaty of, 10, 32
Vertical wind tunnel, 27
*Vertiplane*, 34
Vertol Aircraft Corp., 171, 194
Verville-Packard, aircraft, 12, 200
Vickers aircraft
  Vimy-2, 10
  Viscount, 61
  Wellington, 34
Victory, John F., xi, 23, 125, 187, 198
Vidal, Eugene L., 205
Viking rocket, 53, 55, 62, 64, 65, 67, 70, 74, 84, 86, 134, 211
Vincent, J. G., 6
Vincentin, Walter G., 199
Viner, D. D., 52
Virginia Capes, 13
*Virginia*, 17
*Volksjaeger*, 52
von Braun, Wernher, 29, 62, 64, 124, 169, 193, 209
von Diringshofen, Heinz, 37, 40
von Gronau, Wolfgang, 177
von Hindenburg, 35
von Karman, Theodore, 31, 46, 48, 51, 54, 69, 76, 112, 169, 176, 177, 190, 193, 198, 210, 211
von Neumann, John, 72, 74, 82, 192, 193
von Opel, Fritz, 24, 25
Voodoo (See F-101B)
Vought-Sikorsky aircraft, 15, 39, 41
VTOL, 62, 66, 75, 80, 113

Wac rocket, 51, 53, 56, 64
  (See Corporal, Bumper-Wac)
WADC, 74
Wade, L., 15
Wagner, Ray, 212
Wakeford, R. C., 210
Walcott, Charles D., 11, 22, 202, 205
Walker, Joseph A., 75, 78, 80, 121, 123, 125, 188
Walker, P. B., 197
Wallops Island (Va.), 50, 56, 57, 61, 62, 98, 99, 102
  rocket facility established, 50
  first launching, 50
Wallops Island rocket range (NACA/NASA), 67, 68, 69, 71, 72, 76, 79, 82, 108, 111, 112, 115, 122, 125, 128
Walther, H., 42, 45, 48

War-surplus aircraft, 20
Ward (USS Aaron), 43
Ward, Vernon G., 56
Warheads, 48
Warner, Edward P., 21, 37, 175, 177, 196, 198, 205
Warsitz, Erich, 35, 38
Washington, Univ. of, 22
Washington Navy Yard, 14, 16, 19
*Washington,* USS *George,* 110, 116
Wasp rocket, 84
Wasp waist shape, 68
Wasserfall rocket, 47
Wasserkuppe, 24
Water landings, 11
Water performance of seaplanes, 28
Waterman, Alan T., 101
Watkins, George C., 97
Watkins, George E., 160
Watson, Thomas A., 3
Wattendorf, Frank L., 50
WDD, 75, 77
Weather
    analysis of, 14, 20, 74, 99, 134, 186, 192, 212
    forecasting of, 33, 186
    maps, 29
    and meteoric dust, 73
    (See *Vanguard II, Tiros I*)
Weather Bureau (U.S.), 11, 14, 19, 24, 54, 74, 128, 129, 134, 183, 203–205
    history of, 212
    and aerial navigation, 6
    and NACA, 5, 6
Webster, William, 205
Wedell, J. R., 158
Weems, P. V. D., 195
Weick, Fred E., 24, 31, 191
Weich airplane, 31
Weightlessness, 40, 64, 67, 68, 70, 78, 79, 101, 119
Welch, George, 192
Wells, Edward C., 194
Wendel, Fritz, 37, 158
Wesleyan, Univ., 180
West, Jake C., 51
West Germany, 132
Wester, Billy I., 76
Western Electric Co., 49, 111
Westinghouse Electric, 41, 44, 45, 52, 63, 66, 134, 195
Westover, Oscar, 92, 205
Westover AFB, 92
Wetmore, Alexander, 205
Wexler, Harry, 48, 74, 183
Weyerbacher, Ralph D., 205
Wheeler, J. V., 65
Wheelon, O. A., 197
Whitcomb, Richard T., 68, 174, 187, 199
White, Alice, 97
White, C. S., 212
White, Robert M., 123, 126
White, Stanley C., 172
White, Thomas D., 101, 205
White Sands Proving Ground (WSPG), 49, 50, 51, 54, 55, 58, 60, 64, 65, 67, 69, 70–72, 76, 82, 105, 130, 132
Whiting, Kenneth, 16
Whitman, Walter G., 205
Whitnah, Donald R., 212
Whitney, J. L., 8
Whittle, Frank, 27, 35, 41, 44, 55, 71, 172, 177
Wigner, Eugene, 38
Wiley, Herbert V., 179
Wilhelmshaven, 44
Wilkinson, Paul H., 212
Willett, Harold G., 190, 192

Williams, Alford J., 17, 33, 157
Williams, B., 212
Williams, K. G., 212
Williams, Walter, 55
Willis, Bailey, 13
Willow Grove (Pa.), 27
Wilmer, Wm. H., 7
Wilson, Charles E., 72, 73, 78, 80, 83,
Wilson, E. B., 4
Wilson, J. H., 12
Wilson, President Woodrow, 3, 5
"Window," 45
Wind-shear, 115
Wind tunnels, 1, 8, 12, 13, 23, 34, 35, 36, 39, 41, 44, 49, 52, 55, 56, 63
    first NACA, 8
    first full-scale, 36
    first hypersonic flow, 58
    first full-scale supersonic, 49
    techniques of, 39, 47, 48, 55, 56, 63
Wing
    cantilever, 4, 11
    delta, 72, 128
    flaps, 30
    flow-method, 47, 48
    slots, 26
    swept back, 48, 49, 50
    swept forward, 52
    variable sweep, 67
Winkler, Johannes, 27
*Winnie Mae,* 28, 30
Winslow, R. W., 82
Winzen Research, Inc., 77, 114, 123, 163
Wisconsin, Univ. of, 128
WNBT (N.Y.), 42
Wobus, H. B., 183
Wood, Donald H., 27
Wood, Floyd B., 48
Wood, Homer J., 197
Woolams, Jack, 53
Woolrich Arsenal (Br.), 32
Woomera Rocket Range (Austr.), 98, 101, 107, 112, 128, 131, 134
"World Cruiser," 18
World flight records, 153–160
World War I, x, 1, 3; 9, 13, 15, 209
    (See aircraft types by name)
World War II, 27, 34, 35–38, 44, 50, 51
    (See 1939–45 entries)
Wright, Orville, iv, 26, 34, 36, 59, 93, 173, 176, 181, 205, 210
Wright, Ray H., 56
Wright, Theodore P., 177, 196, 205, 212
Wright, Wilbur, iv, 26, 36, 93, 155, 181
Wright Air Development Center, 101
Wright Apache airplane, 23–26
Wright Brothers Lecturers, 37, 197–198
Wright Brothers Trophy, 198
Wright engines, 15, 27, 30 (See Curtiss-Wright)
Wright Field (Ohio), 17, 20, 24, 31, 33, 35, 37, 41, 43, 44
Wright-Martin Aircraft Corp., 5
Wright-Patterson AFB, 62, 64, 70, 182
Wyld, James, 36, 42
Wylie, F. J., 195

X-radiation, 93
X-1 rocket airplane, 47, 49, 62, 55, 51–59, 61, 68, 188, 207, 211
    first flight, 55
    first supersonic flight, 58
X-1A, 71, 75, 79
X-2, 52, 69, 72, 75, 82, 83, 178
X-3, 70, 75
X-4, 61
X-5, 67

X-10, 73, 76
X-13, 80, 83, 85
X-15, 75, 76, 79, 83, 85, 86, 93, 100, 103, 107, 110, 114, 117, 119, 120, 121, 123, 126, 127, 131, 133, 207, 211
X-17, 82, 86, 107
X-19A engine, 45
XB-15, first flight, 35
XB-29, 44
XB-43, 54
XB-52, 69
XC-35, 35
XC-99, 50
XC-123D, 80
XCAL-200 engine, 46
XCO5 engine, 24
XF2A-1 airplane, 36
XF2Y-1, 67, 72, 75
XF4-D, 73, 80
XF8U-1, 77
XF-85, 65
XF-91, 56, 70
XF-104, 72, 78
XF-105, 80
XFY-1, 66, 178
XFD-1, 45, 54
XFF-1, 27
XH-12, 61
XH-17, 70
XLR-11, 131
XLR-99, 119, 124, 131
XOP-1, 27
XP-40, 36, 39, 43
XP-47, 41
XP-59, 42, 43
XP-59A, 44, 52
XPB2Y-1, 34
XPBM-1, 35

XPY-1, 23
XS-2 (See X-2)
XSAM (Talos), 47, 53

YF-105A, 62
YZP6-Z, 74
*Yankee Clipper,* 38
Yates, Donald N., 121
Yeager, Charles E., 58, 59, 61, 68, 175, 178, 184
York, Herbert F., 96, 105, 113, 127, 128
Young, David A., 176
Young, Leo C., 15
Young, Raymond W., 185
Younger, John E., 195
Yucca Flat, 87

ZPN-1, 69
ZPN-2, 85
ZR-1 (Shenandoah), 17
ZR-3 (Los Angeles), 15, 18, 29
Zahm, A. F., 16
Zand, Stephen J., 196
Zeppelin, Count von, 1
Zeppelin (dirigibles)
  in atomic age, 210
  in World War I, 3-5
  history of 210
  -type airships, 17
  (See *Akron, Graf Zeppelin, Los Angeles, Macon*)
Zeppelin Co., 15
Zero-length launcher, 70
Ziegler, Jean, 69, 71, 72
Zimmerman, Charles H., 39, 181, 197
Ziolkovsky, Konstantin E., 33, 210
Ziolkovsky Gold Medal, 75
Zucrow, M. J., 189

www.ingramcontent.com/pod-product-compliance
Lightning Source LLC
Chambersburg PA
CBHW081721170526
45167CB00009B/3655